Introduction to Inverse Problems for Differential Equations

Alemdar Hasanov Hasanoğlu
Vladimir G. Romanov

Introduction to Inverse Problems for Differential Equations

Springer

Alemdar Hasanov Hasanoğlu
Department of Mathematics
University of Kocaeli
Izmit, Kocaeli
Turkey

Vladimir G. Romanov
Sobolev Institute of Mathematics
Novosibirsk
Russia

ISBN 978-3-319-62796-0   ISBN 978-3-319-62797-7 (eBook)
DOI 10.1007/978-3-319-62797-7

Library of Congress Control Number: 2017947470

Mathematics Subject Classification (2010): 65N21, 65F22, 35R25, 47J06, 47A52, 65R32, 65M32, 80A23

© Springer International Publishing AG 2017
This work is subject to copyright. All rights are reserved by the Publisher, whether the whole or part of the material is concerned, specifically the rights of translation, reprinting, reuse of illustrations, recitation, broadcasting, reproduction on microfilms or in any other physical way, and transmission or information storage and retrieval, electronic adaptation, computer software, or by similar or dissimilar methodology now known or hereafter developed.
The use of general descriptive names, registered names, trademarks, service marks, etc. in this publication does not imply, even in the absence of a specific statement, that such names are exempt from the relevant protective laws and regulations and therefore free for general use.
The publisher, the authors and the editors are safe to assume that the advice and information in this book are believed to be true and accurate at the date of publication. Neither the publisher nor the authors or the editors give a warranty, express or implied, with respect to the material contained herein or for any errors or omissions that may have been made. The publisher remains neutral with regard to jurisdictional claims in published maps and institutional affiliations.

Printed on acid-free paper

This Springer imprint is published by Springer Nature
The registered company is Springer International Publishing AG
The registered company address is: Gewerbestrasse 11, 6330 Cham, Switzerland

This book is dedicated to our wives, Şafak Hasanoğlu and Nina D. Romanova, who have been patient during all our lives together.

Preface

Mathematical models of the most physical phenomena are governed by initial and boundary value problems for partial differential equations (PDEs). Inverse problems governed by these equations arise naturally in almost all branches of science and engineering. The main objective of this textbook is to introduce students and researchers to inverse problems arising in PDEs. This book presents a systematic exposition of the main ideas and methods in treating inverse problems for PDEs arising in basic mathematical models, though we make no claim to cover all of the topics. More detailed additional information related to each chapter can be found in the following books/lecture notes/monographs of Aster, Borchers, and Thurber [2], Bal [6], Baunmeister [8], Beilina and Klibanov [9], Belishev and Blagovestchenskii [12], Chadan and Sabatier [17], Colton and Kress [19], Engl, Hanke, and Neubauer [23], Groetsch [31], Háo [36], Hofmann [43], Isakov [45, 46], Itou and Jin [48], Kabanikhin [50], Kaipio and E. Somersalo [51], Kirsch [54], Lavrentiev [58], Lavrentiev, Romanov, and Shishatski [59], Lavrentiev, Romanov, and Vasiliev [60], Lebedev, Vorovich, and Gladwell [61], Louis [63], Morozov [68], Nakamura and Potthast [72], Natterer [74], Ramm [84], Romanov [85, 86], Romanov and Kabanikhin [87], Schuster, Kaltenbacher, Hofmann, and Kazimierski [90], Tarantola [92], Tikhonov and Arsenin [97], Tikhonov, Concharsky, Stepanov, and Yagola [98], Vogel [102].

In Introduction, we discuss the nature of ill-posedness in differential and integral equations based on well-known mathematical models. Further, we pursue an in-depth analysis of a reason of ill-posedness of an inverse problem governed by integral operator. We tried to answer the question "why this problem is ill-posed?", by arriving to the physical meaning of the mathematical model, on one hand, and then explaining this in terms of compact operators, on the other hand. Main notions and tools, including best approximation, Moore–Penrose (generalized) inverse, singular value decomposition, regularization strategy, Tikhonov regularization for linear inverse problems, and Morozov's discrepancy principle, are given in Chap. 1. In Chap. 2, we tried to illustrate an implementation of all these notions and tools to inverse source problems with final overdetermination for evolution equations and to the backward problem for the parabolic equation, including some numerical reconstructions. The reason for the choice of this problem stems from the fact that historically, one of the first successful applications of inverse problems was Tikhonov's work [93] on inverse problem with final overdetermination for heat equation.

The second part of the book consists of almost independent six chapters. The choice of these chapters is motivated by the fact that the inverse coefficient and source problems considered here are based on the basic and commonly mathematical models governed by PDEs. These chapters describe not only these inverse problems, but also main inversion methods and techniques. Since the most distinctive features of any inverse problem related to PDEs are hidden in the properties of corresponding direct problems solutions, special attention is paid to the investigation of these properties.

Chapter 3 deals with some inverse problems related to the second-order hyperbolic equations. Starting with the simplest inverse source problems for the wave equation with the separated right-hand side containing spatial or time-dependent unknown source functions, we use the reflection method to demonstrate a method for finding the unknown functions based on integral equations. The next and more complex problem here is the problem of recovering the potential in the string equation. The direct problem is stated for semi-infinite string with homogeneous initial data and non-homogeneous Dirichlet data at the both ends. The Neumann output is used as an additional data for recovering an unknown potential. Using the method the successive approximations for obtained system of integral equations, we prove a local solvability for the considered inverse problem. Note that the typical situation for nonlinear inverse problems is that the only local solvability can be proved (see [88]). Nevertheless, for the considered inverse problem, the uniqueness and global stability estimates of solutions are derived. As an application, inverse coefficient problems for layered media are studied in the final part of Chap. 3.

Chapter 4 deals with inverse problems for the electrodynamic equations. These problems are motivated by geophysical applications. In the considered physical model, we assume that the space $\mathbb{R}^3$ is divided in the two half-spaces $\mathbb{R}^3_- =: \{x \in \mathbb{R}^3 | x_3 < 0\}$ and $\mathbb{R}^3_+ =: \{x \in \mathbb{R}^3 | x_3 > 0\}$. The domain $\mathbb{R}^3_-$ is filled by homogeneous non-conductive medium (air), while the domain $\mathbb{R}^3_+$ contains a ground which is a non-homogeneous medium with variable permittivity, permeability, and conductivity depending on the variable x_3 only. The tangential components of the electromagnetic field are assumed to be continuous across the interface $x_3 = 0$. The direct problem for electrodynamic equations with zero initial data and a dipole current applied at the interface is stated. The output data here is a tangential component of the electrical field given on the interface as a function of time $t > 0$. A method for reconstruction of one of the unknown coefficients (permittivity, permeability, or conductivity) is proposed when two others are given. This method is based on the Riemannian invariants and integral equations which lead to a well-convergent successive approximations. For the solutions of these inverse problems, stability estimates are stated.

Preface

Coefficient inverse problems for parabolic equations are studied in Chap. 5. First of all, the relationship between the solutions of direct problems corresponding to the parabolic and hyperbolic equations is derived. This relationship allows to show the similarity between the outputs corresponding inverse problems for parabolic and hyperbolic equations. Since this relationship is a special Laplace transform, it is invertible. Then it is shown that inverse problems for parabolic equations can be reduced to corresponding problems for hyperbolic equations, studied in Chap. 3. This, in particular, allows to use uniqueness theorems obtained for hyperbolic inverse problem. Further it is shown that the inverse problem of recovering the potential is closely related to the well-known inverse spectral problem for the Sturm–Liouville operator.

Chapter 6 deals with inverse problems for the elliptic equations. Here the inverse scattering problem for stationary Schrodinger equation is considered. For the sake of simplicity, we study this problem in Born approximation using the scattering amplitude for recovering a potential, following the approach proposed by R. Novikov [66]. Moreover, we study also the inverse problem which output is a measured value on a closed surface S of a trace of the solution of the problem for the Schrodinger equation with point sources located at the same surface. The later problem is reduced to the X-ray tomography problem. In the final part of this chapter, we define the Dirichlet-to-Neumann operator which originally has been introduced by J. Sylvester and G. Uhlmann [86]. Using this operator, we study the inverse problem of recovering the potential for an elliptic equation in the Born approximation and reduce it to the moment problem which has a unique solution.

In Chap. 7, inverse problems related to the transport equation without scattering are studied. We derive here the stability estimate for the solution of X-ray tomography problem and then inversion formula.

The inverse kinematic problem is studied in the final Chap. 8. On one hand, this is a problem of reconstructing the wave speed inside a domain from given travel times between arbitrary boundary points. On the other hand, this is also the problem of recovering conformal Riemannian metric via Riemannian distances between boundary points, which plays very important role in geophysics. It suffices to recall that much of our knowledge about the internal structure of the Earth are based on solutions of this problem. We derive first the equations for finding rays and fronts going from a point source. Then we consider one-dimensional and two-dimensional inverse problems and derive a stability estimates for the solutions of these problems.

For the convenience of the reader, a short review related to invertibility of linear, in particular compact, operators is given in Appendix A. Some necessary energy estimates for a weak and regular weak solutions of a one-dimensional parabolic equation are given in Appendix B.

The presentation of the material, especially Introduction and Part I, is self-contained and is intended to be accessible also to undergraduate and beginning graduate students, whose mathematical background includes only basic courses in advanced calculus, PDEs, and functional analysis. The book can be used as a backbone for a lecture on inverse and ill-posed problems for partial differential equations.

This book project took us about two and a half years to complete, starting from May, 2014. We would not be able to finish this book without support of our family members, students, and colleagues. We are deeply indebted to a number of colleagues and students who have read this material and given us many valuable suggestions for improving the presentation. We would like to thank our colleagues Onur Baysal, Andreas Neubauer, Roman Novikov, Burhan Pektaş, Cristiana Sebu, and Marian Slodička who have read various parts of the manuscript and made numerous suggestions for improvement. Our special thanks to anonymous reviewers who read the book proposal and the first version of the manuscript, as well as to the staff members of Springer, Dr. Martin Peters and Mrs. Ruth Allewelt, for their support and assistance with this project. We also are grateful to Mehriban Hasanoğlu for number of corrections.

We would be most grateful if you could send your suggestions and list of mistakes to the email addresses: alemdar.hasanoglu@gmail.com; romanov0511@gmail.com.

Izmit, Turkey
Novosibirsk, Russia
November 2016

Alemdar Hasanov Hasanoğlu
Vladimir G. Romanov

Contents

1 **Introduction Ill-Posedness of Inverse Problems for Differential and Integral Equations** 1
  1.1 Some Basic Definitions and Examples 1
  1.2 Continuity with Respect to Coefficients and Source: Sturm-Liouville Equation 9
  1.3 Why a Fredholm Integral Equation of the First Kind Is an Ill-Posed Problem? 13

Part I Introduction to Inverse Problems

2 **Functional Analysis Background of Ill-Posed Problems** 23
  2.1 Best Approximation and Orthogonal Projection 24
  2.2 Range and Null-Space of Adjoint Operators 31
  2.3 Moore-Penrose Generalized Inverse 33
  2.4 Singular Value Decomposition 38
  2.5 Regularization Strategy. Tikhonov Regularization 45
  2.6 Morozov's Discrepancy Principle 58

3 **Inverse Source Problems with Final Overdetermination** 63
  3.1 Inverse Source Problem for Heat Equation 64
    3.1.1 Compactness of Input-Output Operator and Fréchet Gradient 67
    3.1.2 Singular Value Decomposition of Input-Output Operator 72
    3.1.3 Picard Criterion and Regularity of Input/Output Data ... 79
    3.1.4 The Regularization Strategy by SVD. Truncated SVD ... 84
  3.2 Inverse Source Problems for Wave Equation 90
    3.2.1 Non-uniqueness of a Solution 93
  3.3 Backward Parabolic Problem 96

xi

3.4		Computational Issues in Inverse Source Problems	104
	3.4.1	Galerkin FEM for Numerical Solution of Forward Problems	105
	3.4.2	The Conjugate Gradient Algorithm	107
	3.4.3	Convergence of Gradient Algorithms for Functionals with Lipschitz Continuous Fréchet Gradient	112
	3.4.4	Numerical Examples	116

Part II Inverse Problems for Differential Equations

4 Inverse Problems for Hyperbolic Equations 123
4.1	Inverse Source Problems	123
	4.1.1 Recovering a Time Dependent Function	124
	4.1.2 Recovering a Spacewise Dependent Function	126
4.2	Problem of Recovering the Potential for the String Equation	128
	4.2.1 Some Properties of the Direct Problem	129
	4.2.2 Existence of the Local Solution to the Inverse Problem	133
	4.2.3 Global Stability and Uniqueness	138
4.3	Inverse Coefficient Problems for Layered Media	141

5 One-Dimensional Inverse Problems for Electrodynamic Equations 145
5.1	Formulation of Inverse Electrodynamic Problems	145
5.2	The Direct Problem: Existence and Uniqueness of a Solution	146
5.3	One-Dimensional Inverse Problems	155
	5.3.1 Problem of Finding a Permittivity Coefficient	155
	5.3.2 Problem of Finding a Conductivity Coefficient	160

6 Inverse Problems for Parabolic Equations 163
6.1	Relationships Between Solutions of Direct Problems for Parabolic and Hyperbolic Equations	163
6.2	Problem of Recovering the Potential for Heat Equation	166
6.3	Uniqueness Theorems for Inverse Problems Related to Parabolic Equations	168
6.4	Relationship Between the Inverse Problem and Inverse Spectral Problems for Sturm-Liouville Operator	171
6.5	Identification of a Leading Coefficient in Heat Equation: Dirichlet Type Measured Output	174
	6.5.1 Some Properties of the Direct Problem Solution	175
	6.5.2 Compactness and Lipschitz Continuity of the Input-Output Operator. Regularization	177
	6.5.3 Integral Relationship and Gradient Formula	183
	6.5.4 Reconstruction of an Unknown Coefficient	186

	6.6	Identification of a Leading Coefficient in Heat Equation: Neumann Type Measured Output...	191
		6.6.1 Compactness of the Input-Output Operator............	193
		6.6.2 Lipschitz Continuity of the Input-Output Operator and Solvability of the Inverse Problem..	197
		6.6.3 Integral Relationship and Gradient Formula...........	200
7	**Inverse Problems for Elliptic Equations**		205
	7.1	The Inverse Scattering Problem at a Fixed Energy.............	205
	7.2	The Inverse Scattering Problem with Point Sources	208
	7.3	Dirichlet to Neumann Map......................................	213
8	**Inverse Problems for the Stationary Transport Equations**.........		219
	8.1	The Transport Equation Without Scattering	219
	8.2	Uniqueness and a Stability Estimate in the Tomography Problem....................................	222
	8.3	Inversion Formula..	223
9	**The Inverse Kinematic Problem**......................................		227
	9.1	The Problem Formulation....................................	227
	9.2	Rays and Fronts	228
	9.3	The One-Dimensional Problem...............................	231
	9.4	The Two-Dimensional Problem	234

Appendix A: Invertibility of Linear Operators..................... 239

Appendix B: Some Estimates For One-Dimensional Parabolic Equation.................................... 247

References 253

Index 259

Chapter 1
Introduction Ill-Posedness of Inverse Problems for Differential and Integral Equations

1.1 Some Basic Definitions and Examples

Inverse problems arise in almost all areas of science and technology, in modeling of problems motivated by various physical and social processes. Most of these models are governed by differential and integral equations. If all the necessary inputs in these models are known, then the solution can be computed and behavior of the physical system under various conditions can be predicted. In terms of differential problems, the necessary inputs include such information as initial or boundary data, coefficients and force term, also shape and size of the domain. If all these data are enough to describe the system adequately, then it is possible to use the mathematical model for studying the physical system. For example, consider the case of the simplest one-dimensional model of steady-state heat transfer

$$\begin{cases} (Lu)(x) := -\big(k(x)u'(x)\big)' + q(x)u(x) = F(x), & x \in (0,\ell), \\ u(0) = 0, \quad (k(x)u'(x))_{x=\ell} = \varphi, \quad \varphi \in \mathbb{R}, \end{cases} \quad (1.1.1)$$

where the heat coefficients $(k(x), q(x))$, external heat source $F(x)$ and the heat flux φ at the end $x = \ell$ of the rod with length $\ell > 0$ are the given *input data*. If we assume that $k \in C^1(0, \ell) \cap C^0[0, \ell]$, $q, F \in C^0(0, \ell)$, then the solution $u(x)$ of the two-point boundary value problem (1.1.1) is evidently in $C^2(0, \ell) \cap C^1[0, \ell]$. This *solution exists, is unique and continuously depends on the above given input data*, in norms of appropriate spaces. If the rod is non-homogeneous and external heat source has a discontinuity, then the solution $u(x)$ of the boundary value problem (1.1.1) needs to be considered in larger class of functions. Specifically, if

$$\begin{aligned} k \in \mathcal{K} &:= \{k \in L^2(0, \ell) : 0 < c_0 \leq k(x) \leq c_1\}, \\ q \in \mathcal{Q} &:= \{q \in L^2(0, \ell) : 0 \leq q(x) \leq c_2\}, \\ F &\in L^2(0, \ell), \end{aligned} \quad (1.1.2)$$

© Springer International Publishing AG 2017
A. Hasanov Hasanoğlu and V.G. Romanov, *Introduction to Inverse Problems for Differential Equations*, DOI 10.1007/978-3-319-62797-7_1

then there exists a unique solution $u(x)$ of the boundary value problem (1.1.1) in $\mathcal{V}^1(0,\ell) := \{u \in H^1(0,\ell) : u(0) = 0\}$. Here and below $H^1(0,\ell)$ is the Sobolev space [1]. Moreover, this solution *continuously depends* on the above given *input data*, as we will show below. This means that the *parameter-to-solution maps*

$$\begin{aligned} A_k : k \in \mathcal{K} \subset L^2(0,\ell) \mapsto u \in \mathcal{V}^1(0,\ell), \\ A_q : q \in \mathcal{Q} \subset L^2(0,\ell) \mapsto u \in \mathcal{V}^1(0,\ell), \\ A_F : F \in L^2(0,\ell) \mapsto u \in \mathcal{V}^1(0,\ell) \end{aligned} \qquad (1.1.3)$$

are continuous. The sets $\mathcal{K}$ and $\mathcal{Q}$ are called the *sets of input data*.

The above mentioned three conditions (existence, uniqueness and continuous dependence) related to solvability of various initial boundary value problems has later been identified with the notion of a "*well-posed problem*". Jacques Hadamard played a pioneering role recognizing first and developing then the idea of well-posedness at the beginning of the twentieth century. But initially, in [32], Hadamard included only existence and uniqueness in the definition of well-posedness, insisting on the fact that continuous dependence on the initial data is important only for the Cauchy problem. This latter feature of a "well-posed problem" was added to Hadamard's definition by Hilbert and Courant in [21]. Many years later, Hadamard used this corrected definition in his book [33], referring to Hilbert and Courant.

We formulate the definition of well-posedness for a linear operator equation $Au = F$, $A \in \mathcal{L}(H, \tilde{H})$, where $\mathcal{L}(H, \tilde{H})$ denotes the set of all bounded linear operators and $H, \tilde{H}$ are Hilbert spaces.

Definition 1.1.1 (*well-posedness*) The linear equation/problem $Au = F$, $A \in \mathcal{L}(H, \tilde{H})$ is defined to be well-posed in Hadamard's sense, if
(i1) a solution of this equation exists, for all admissible data $F \in \tilde{H}$;
(i2) this solution is unique;
(i3) it depends continuously on the data.
Otherwise, the equation $Au = F$ is defined to be ill-posed in Hadamard's sense.

According to this definition the boundary value problem (1.1.1) is well-posed.

However, in practice it is not possible to measure experimentally all these inputs, since in real physical systems the inputs can be defined as *measurable* and *unmeasurable* ones. Instead, it is possible to measure experimentally certain (additional) outputs of the system and use this information together with other inputs to recover the missing, *unmeasurable input*. Hence, in a broad sense, *an inverse problem can be defined as the problem of determining unmeasurable parameters of a system from measurable parameters, by using a mathematical/physical model.*

Let us consider the case where $q \in \mathcal{Q}$, $F \in L^2(0,\ell)$, $\varphi \in \mathbb{R}$ are the given inputs, but the coefficient $k(x)$ in (1.1.1) is unknown and needs to be determined from some additional (physically reasonable) condition(s).

Example 1.1.1 Parameter identification (or inverse coefficient) problem for the Sturm-Liouville operator $Lu := -(ku')' + qu$

1.1 Some Basic Definitions and Examples

Let $q(x) \equiv 0$ in (1.1.1) and the function $F(x)$ be given as follows:

$$F(x) = \begin{cases} 0, & x \in [0, \xi], \\ 1, & x \in (\xi, \ell], \end{cases}$$

where $\xi \in [0, \ell]$ is an arbitrary parameter. Then solution to problem (1.1.1) depends on the parameter $\xi \in [0, \ell]$, that is, $u = u(x, \xi)$. We assume that the *measured value* of $u(x, \xi)$ at $x = \ell$ is given as an additional information, i.e. as a *measured output data f*, that is,

$$f(\xi) := u(\ell, \xi), \quad \xi \in [0, \ell]. \tag{1.1.4}$$

Based on this information we are going to identify the unknown coefficient $k \in \mathcal{K}$, i.e. the *input data*. We define the *problem of identifying the unknown coefficient* $k(x)$ in (1.1.1) from the output data $f \in \mathcal{F} \subset L^2(0, \ell)$ given by (1.1.4) as a *parameter identification (or inverse coefficient) problem* for the Sturm-Liouville operator. In this context, for the given inputs $k(x)$, $q(x)$, $F(x)$ and φ, the boundary value problem (1.1.1) is defined as a *direct problem*.

We formulate the inverse problem of identifying the unknown coefficient $k(x)$ in terms of an operator equation. Let $k \in \mathcal{K}$ be a given admissible coefficient. Since there exists a unique solution $u := u(x, \xi; k)$ of the boundary value problem (1.1.1) in $\mathcal{V}^1(0, \ell)$, for each $k \in \mathcal{K}$ the trace $u(x, \xi; k)_{x=l}$ is uniquely determined. We define this mapping as *an input-output operator* $\Phi : k \mapsto f$ as follows:

$$\Phi[k](\xi) := u(x, \xi; k)_{x=l}, \ u \in \mathcal{V}^1(0, \ell), \ k \in \mathcal{K}, \tag{1.1.5}$$

and reformulate this inverse problem as the problem of solving the following operator equation

$$\Phi[k] = f, \ k \in \mathcal{K}, \ f \in \mathcal{F}. \tag{1.1.6}$$

Let us analyze now the inverse coefficient problem. Taking into account conditions (1.1.3), we integrate Eq. (1.1.1) on (x, ℓ) and use the boundary condition $(k(x)u'(x))_{x=\ell} = \varphi$. Then we obtain:

$$k(x)u'(x, \xi) = \varphi + \psi(x, \xi), \tag{1.1.7}$$

where

$$\psi(x, \xi) = \begin{cases} \ell - \xi, & x \in [0, \xi], \\ \ell - x, & x \in (\xi, \ell], \ \xi \in [0, \ell]. \end{cases} \tag{1.1.8}$$

Integrating (1.1.7) and using the boundary condition $u(0) = 0$ we get:

$$u(x, \xi) = \int_0^x \frac{\varphi + \psi(\tau, \xi)}{k(\tau)} d\tau, \quad x \in [0, \ell]. \tag{1.1.9}$$

Hence, the formula for finding $k(x)$ is

$$\begin{aligned} f(\xi) &= \int_0^\ell \frac{\varphi + \psi(\tau, \xi)}{k(\tau)} d\tau \\ &= \int_0^\xi \frac{\varphi + \ell - \xi}{k(\tau)} d\tau + \int_\xi^\ell \frac{\varphi + \ell - \tau}{k(\tau)} d\tau, \quad \xi \in [0, \ell]. \end{aligned} \tag{1.1.10}$$

Taking the first and then the second derivative of both sides we find:

$$\begin{aligned} f'(\xi) &= -\int_0^\xi \frac{d\tau}{k(\tau)}, \\ f''(\xi) &= -\frac{1}{k(\xi)}, \quad \xi \in [0, \ell]. \end{aligned} \tag{1.1.11}$$

It follows from (1.1.11) that $f'(0) = 0$ and $f \in H^2(0, \ell)$. Furthermore, $k \in \mathcal{K}$ implies: $0 < 1/c_1 \leq -f''(\xi) \leq 1/c_0$. Then, introducing the subset $\mathcal{V}^2(0, \ell) := \{f \in H^2(0, \ell) : f'(0) = 0\}$, where $H^2(0, \ell)$ is the Sobolev space, we conclude that if $k \in \mathcal{K} := \{k \in L^2(0, \ell) : 0 < c_0 \leq k(x) \leq c_1\}$, the *set of admissible outputs* is

$$\mathcal{F} := \{f \in \mathcal{V}^2(0, \ell) : 0 < 1/c_1 \leq -f''(\xi) \leq 1/c_0\}. \tag{1.1.12}$$

Moreover, the unknown coefficient is defined by the *inversion formula*:

$$k(\xi) = -\frac{1}{f''(\xi)}, \quad \xi \in [0, \ell]. \tag{1.1.13}$$

We can deduce from formula (1.1.13) the following estimate: if $k_i \in \mathcal{K}$ and $f_i \in \mathcal{F}, i = 1, 2$, then

$$\|k_1 - k_2\|_{L^2(0,\ell)} \leq \frac{1}{c_1^2} \|f_1 - f_2\|_{H^2[0,\ell]}, \quad c_1 > 0. \tag{1.1.14}$$

This estimate is called the *stability estimate* of the inverse problem.

As noted above, under conditions (1.1.2) there exists a unique solution $u(x)$ of the boundary value problem (1.1.1) in $H^1(0, \ell)$ for each $k \in \mathcal{K} \subset L^2(0, \ell)$. Moreover, as it is shown in the next section, this solution continuously depends on the input data $k(x)$, $F(x)$ and φ. Therefore, the *direct problem* is *well-posed* from $L^2(0, \ell)$ to $H^1(0, \ell)$, by Definition 1.1.1.

Consider now the inverse problem (1.1.6), i.e. the problem of determining the unknown $k \in \mathcal{K} \subset L^2(0, \ell)$ from the given output $f \in L^2(0, \ell)$. Assume first the input-output operator (1.1.5) acts from $L^2(0, \ell)$ to $L^2(0, \ell)$, that is, the range of the operator Φ is the entire space: $\mathcal{R}(\Phi) = L^2(0, \ell)$. We may use formula (1.1.10) to get the explicit form of this operator:

1.1 Some Basic Definitions and Examples

$$\Phi[k](\xi) := \int_0^\ell \frac{\varphi + \psi(\tau, \xi)}{k(\tau)} d\tau, \quad x \in [0, \ell].$$

Using formula (1.1.8) in the above integral we may derive, after elementary transformations, the following explicit form of the input-output operator:

$$\Phi[k](\xi) := \varphi \int_0^\ell \frac{d\eta}{k(\eta)} + \int_\xi^\ell \int_0^\eta \frac{1}{k(\tau)} d\tau \, d\eta, \quad x \in [0, \ell]. \qquad (1.1.15)$$

It follows from the theory of integral operators [103] that in this case the input-output operator (1.1.15) is compact. As we will prove in Lemma 1.3.1, if the operator $\Phi : L^2(0, \ell) \mapsto L^2(0, \ell)$ is compact, then the problem $\Phi[k] = f$ is ill-posed. Thus, *if the input-output operator is defined from $L^2(0, \ell)$ to $L^2(0, \ell)$, then the inverse problem (1.1.6) is ill-posed.*

Assume now that the input-output operator is defined from $L^2(0, \ell)$ not to $L^2(0, \ell)$, but to $H^1(0, \ell) \subset L^2(0, \ell)$, that is, $\mathcal{R}(\Phi) = H^1(0, \ell)$. The right hand side of the first formula of (1.1.11) implies that the *range restriction* $\Phi : L^2(0, \ell) \mapsto H^1(0, \ell) \subset L^2(0, \ell)$ is well-defined, since $k \in \mathcal{K} \subset L^2(0, l)$. Again, it follows from the theory of integral operators that in this case the input-output operator (1.1.15) is also compact. As a result, we conclude that *if the input-output operator (1.1.15) is defined from $L^2(0, \ell)$ to $H^1(0, \ell)$, then the inverse problem (1.1.6) is still ill-posed.*

We continue the range restriction, assuming finally that the input-output operator is defined from $L^2(0, \ell)$ to $\mathcal{F} \subset H^2[0, \ell]$, where $\mathcal{F}$ is the set of outputs defined by (1.1.12), that is, $\Phi : L^2(0, \ell) \mapsto \mathcal{F} \subset H^2[0, \ell]$. This mapping is well-defined, due to the second formula of (1.1.11), which also allows to derive the explicit form of the inverse operator $\Phi^{-1} : \mathcal{R}(\Phi) = \mathcal{F} \subset H^2[0, \ell] \mapsto L^2(0, \ell)$:

$$k(x) := (\Phi^{-1} f)(x) := -\frac{1}{f''(x)}, \quad x \in [0, \ell].$$

Thus, *if the input-output operator is defined as $\Phi : L^2(0, \ell) \mapsto \mathcal{F} \subset H^2[0, \ell]$, i.e. if $\mathcal{R}(\Phi) = \mathcal{F}$, then the inverse problem defined by (1.1.6) or by (1.1.1) and (1.1.4) is well-posed.* □

The above analysis of the simplest inverse coefficient problem shows some distinguished features of inverse problems. The first feature is that if even the direct problem (1.1.1) is *linear*, the *inverse coefficient problem is nonlinear*. Remark that, inverse source problems corresponding to linear direct problems, in particular, considered in Chap. 3, are linear. The second feature is that besides the *linear ill-posedness* arising from the differentiation operation in (1.1.13), there is a *nonlinear ill-posedness* due to the presence of the quotient. Specifically, in practice, the output $f \in L^2(0, \ell)$ is obtained from measurements and always contains a random noise. As a result, errors at small values of $f''(x)$ in (1.1.13) are amplified much stronger which can lead to instability. These two features show that inverse coefficient problems are nonlinear and most sensitive, while so-called inverse source problems are less sensitive and linear.

Another observation from above analysis is that, *depending on the choice of functional spaces*, an inverse problem can be ill-posed or well-posed. To explain the cause of this change in the framework of the linear theory of ill-posed problems, given in Chap. 2, we define the new input-output operator as follows: $\hat{\Phi}[k] = \Phi[1/k]$, where Φ is defined by (1.1.5). Then the inverse problem becomes a linear one:

$$\hat{\Phi}[k](\xi) := u(x, \xi; 1/k)_{x=l}, \ u \in \mathcal{V}^1(0, \ell), \ k \in \mathcal{K}.$$

The first statement here is that the restriction of $\mathcal{R}(\hat{\Phi})$ results in the restriction of the null-space $\mathcal{N}(\hat{\Phi}^*)^\perp$ of the adjoint operator $\hat{\Phi}^*$, since $\mathcal{R}(\hat{\Phi}) = \mathcal{N}(\hat{\Phi}^*)^\perp$, as proved in Sect. 2.2. But what does it mean in terms of differential operators? Picard's Theorem in Sect. 2.4 says that the condition $f \in \mathcal{N}(\hat{\Phi}^*)^\perp$ is one of the necessary and sufficient conditions for solvability of the operator equation $\hat{\Phi}[k] = f$. Hence the range restriction is the restriction of the class $\{f\}$ of the right hand side functions, in order to get better functions to ensure a convergent SVD expansion. In terms of differential problems this is equivalent to increasing the smoothness of the function $f(x)$, i.e. restricting the class of outputs, as stated in the Sect. 3.1.3. The second statement is that the range restriction, step-by-step from $L^2(0, \ell)$ to $H^2[0, \ell]$, led us to the fact that the range $\mathcal{R}(\hat{\Phi}) = \mathcal{F}$ of the input-output operator became *compact* in $L^2(0, \ell)$. Indeed, Rellich's lemma for Sobolev spaces asserts that if $\Omega \subset \mathbb{R}^n$ is a bounded domain, then the embedding $H^1(\Omega) \hookrightarrow L^2(\Omega)$ is a compact operator. This lemma implies, in particular, that every bounded set in $H^1(\Omega)$ is compact in $L^2(\Omega)$. The range restriction of the input-output operator $\hat{\Phi} : L^2(0, \ell) \mapsto L^2(0, \ell)$ from $L^2(0, \ell)$ to the set $\mathcal{F} \subset H^2[0, \ell] \subset L^2(0, \ell)$ means that we *defined the domain $\mathcal{D}(\hat{\Phi}^{-1}) := \mathcal{F}$ of the inverse operator to be compact in $L^2(0, \ell)$* (moreover, in $H^1(0, \ell)$). In this compact set $\mathcal{F}$, the stability estimate (1.1.14) for the operator $\hat{\Phi}$ can be treated also as the Lipschitz continuity of the inverse operator $\hat{\Phi}^{-1} : f \mapsto k$:

$$\left\| \hat{\Phi}^{-1} f_1 - \hat{\Phi}^{-1} f_2 \right\|_{L^2(0,\ell)} \leq \| f_1 - f_2 \|_{H^2[0,\ell]}, \text{ for all } f_1, f_2 \in \mathcal{F}.$$

Thus, a common situation in inverse and ill-posed problems rather is that *compactness plays a dual role in the ill-posedness of inverse problems*. Whilst compactness, as a property of the operator, plays a negative role, making actually the problem worse, i.e. ill-posed, as a property of the domain $\mathcal{D}(\hat{\Phi}^{-1}) := \mathcal{F}$ of the inverse operator plays an essential positive role. This positive role has first been discovered by Tikhonov [94, 97]. The fundamental Tikhonov's lemma on the continuity of the inverse of an operator, which is injective, continuous and defined on a compact set, clearly illustrates this positive role and until now is used as an important tool in regularization of ill-posed problems.

Lemma 1.1.1 (Tikhonov) *Let H and $\tilde{H}$ be metric spaces and $A : U \subset H \mapsto \tilde{H}$ be a one-to-one continuous operator with $A(U) = V$. If $U \in H$ is a compact set, then the inverse operator $A^{-1} : V \subset \tilde{H} \mapsto U \subset H$ is also continuous.*

1.1 Some Basic Definitions and Examples

Proof Let $v \in V$ be any element. Since the operator $A : U \subset H \mapsto V \subset \tilde{H}$ is one-to-one, there exists a unique element $u \in U$ such that $u = A^{-1}v$. Now suppose, contrary to the assertion, that the inverse operator $A^{-1} : V \subset \tilde{H} \mapsto U \subset H$ is not continuous. This implies that there exists a positive number $\epsilon > 0$ and an element $v_\delta \in V$ such that for all $\delta > 0$ the following conditions hold:

$$\rho_{\tilde{H}}(v, v_\delta) < \delta, \quad \text{but} \quad \rho_H(A^{-1}v, A^{-1}v_\delta) \geq \epsilon. \tag{1.1.16}$$

Due to the arbitrariness of $\delta > 0$, there exists a sequence of positive numbers $\{\delta_n\}_{n=1}^\infty$ such that $\delta_n \to 0^+$, as $n \to \infty$. With (1.1.16) this implies that the corresponding sequence of elements $\{v_{\delta_n}\} \subset V$ satisfy the conditions:

$$\rho_{\tilde{H}}(v, v_{\delta_n}) < \delta_n, \quad \text{but} \quad \rho_H(A^{-1}v, A^{-1}v_{\delta_n}) \geq \epsilon. \tag{1.1.17}$$

Taking the limit here, as $n \to \infty$ we conclude

$$\lim_{n \to \infty} \rho_{\tilde{H}}(v, v_{\delta_n}) = 0. \tag{1.1.18}$$

Being a subset of the compact U, the sequence $\{u_{\delta_n}\}_{n=1}^{n=\infty} \subset U$, $u_{\delta_n} := A^{-1}v_{\delta_n}$, has a convergent subsequence $\{u_{\delta_m}\}_{m=1}^\infty \subset \{u_{\delta_n}\}_{n=1}^\infty$. Then there exists an element $\tilde{u} \in U$ such that $\lim_{m \to \infty} \rho_H(\tilde{u}, u_{\delta_m}) = 0$. By the continuity of the operator $A : U \subset H \mapsto V \subset \tilde{H}$ this implies:

$$\lim_{m \to \infty} \rho_{\tilde{H}}(A\tilde{u}, Au_{\delta_m}) = 0. \tag{1.1.19}$$

By (1.1.18), $\lim_{m \to \infty} \rho_{\tilde{H}}(v, v_{\delta_m}) = 0$, where $v_{\delta_m} := Au_{\delta_m}$. With (1.1.19) this implies that $A\tilde{u} = v = Au$. Taking now into account the fact that the operator $A : U \subset H \mapsto V \subset \tilde{H}$ is one-to-one, we find $\tilde{u} = u$. On the other hand, $u = A^{-1}v$, $u_{\delta_m} := A^{-1}v_{\delta_m}$, by definition, and we deduce from the second assertion of (1.1.17) that $\rho_H(u, u_{\delta_m}) \geq \epsilon$. Taking a limit here as $m \to \infty$ we get: $\rho_H(u, \tilde{u}) \geq \epsilon$. This contradicts the previous assertion $\tilde{u} = u$ and completes the proof. $\square$

Let us return now to Definition 1.1.1. The first condition (existence) means that the operator $A \in \mathcal{L}(H, \tilde{H})$ is surjective, the second condition (uniqueness) means that this operator is injective, and the last condition means that the inverse operator $A^{-1} : \tilde{H} \mapsto H$ is continuous. In terms of operator theory the first condition (existence) is equivalent to the condition $\mathcal{R}(A) = \tilde{H}$, where $\mathcal{R}(A)$ is the range of the operator A. The second condition (uniqueness) is equivalent to the condition $\mathcal{N}(A) = \{0\}$, where $\mathcal{N}(A) := \{u \in \mathcal{D}(A) : Au = 0\}$ is the kernel of the operator A. It is important, from applications point of view, to note that in finite dimensional spaces the conditions **(p1)** and **(p2)** are equivalent, as the Halmos's theorem below shows.

Theorem 1.1.1 *Let $A : H \mapsto H$ be a linear operator defined on the finite dimensional space H. Then the following assertions are equivalent:*

(**H1**) $\mathcal{N}(A) = \{0\}$, i.e. A is injective;
(**H2**) A is surjective.

Hence, in a finite dimensional space the corresponding inverse problem $Au = F$ has a unique solution $u = A^{-1}F$, for each $F \in H$, i.e. the inverse operator $A^{-1} : H \mapsto H$ exists (but may not be continuous!), if only one of the properties (**H1**)–(**H2**) holds. However, in infinite-dimensional spaces these properties are not equivalent, as the following example shows.

Example 1.1.2 The case when injectivity does not imply surjectivity

Let $A : C([0, 1]) \mapsto C([0, 1])$ be a linear operator defined by

$$(Au)(x) := \int_0^x u(\xi) d\xi, \quad u \in C([0, 1]). \tag{1.1.20}$$

Evidently, $(Au)(0) = 0$, for all $u \in C([0, 1])$, so the range $\mathcal{R}(A)$ of the operator A is a proper subset of $C([0, 1])$, i.e. $C([0, 1]) \setminus \mathcal{R}(A) \neq \varnothing$. This means operator A is not surjective. Obviously, this operator is injective, since differentiating the equation $Au = 0$, where A is defined by (1.1.20), we get $u(x) = 0$. So, $Au = 0$ implies $u(x) = 0$ in $C([0, 1])$. □

In view of Hadamard's definition, we can distinguish the following three types of *Hadamard's ill-posedness*:

(**p1**) *Non-existence (A is not surjective)*;
(**p2**) *Non-uniqueness (A is not injective)*;
(**p3**) *Instability (the inverse operator A^{-1} is not continuous)*.

For an adequate mathematical model of a physical process it is reasonable to require an *existence of a solution* in an appropriate class of functions, at least for an exact data. With regard to the uniqueness of a solution, this is the most important issue in inverse problems theory and is often not easy to prove. In the case when the uniqueness can not be guaranteed by given data, one needs either to impose an additional data or to restrict the set of admissible solutions using a-priori information on the solution. In many applied problems the non-uniqueness can be used as an advantage to obtain a desired solution among several ones. Nevertheless, *the main issue in inverse and ill-posed problems is usually stability, i.e. continuous dependence of the solution on measured output*. Moreover, this datum always contain random noise, hence the equation $Au = F$ cannot be satisfied exactly in general. Even if the noise level in a measured output data is small, many algorithms developed for well-posed problems do not work in case of a violation of the third condition (**p3**) due to round-off errors, if they do not address the instability. An algorithm using differentiation may serve as an example to this, since differentiation has the properties of an ill-posed problem. This is the reason why regularization methods play a central role in the theory and applications of inverse and ill-posed problems to overcome the above mentioned instabilities [23, 54].

1.2 Continuity with Respect to Coefficients and Source: Sturm-Liouville Equation

This introductory section familiarizes the reader with some aspects of continuous dependence of the solution of the boundary problem (1.1.1) on the coefficients $k \in \mathcal{K}$, $q \in \mathcal{Q}$ and the source function $F \in L^2(0, \ell)$. It is important to distinguish the character of continuity with respect to different coefficients and also the source function in inverse/identification problems, as well as in optimal control governed by PDEs. Let us define the weak solution of problem (1.1.1) as the solution $u \in \mathcal{V}^1(0, \ell)$ of the integral identity:

$$\int_0^\ell [k(x)u'v' + q(x)uv]dx = \int_0^\ell f(x)v dx + \varphi v(b), \quad \forall v \in \mathcal{V}^1(0, \ell). \quad (1.2.1)$$

Denote by $C^{0,\lambda}[0, \ell]$ the Hölder space of functions with exponent $\lambda \in (0, 1]$, that is, there exists a positive constant M such that

$$|u(x_1) - u(x_2)| \leq M|x_1 - x_2|^\lambda, \quad \text{for all } x_1, x_2 \in [0, \ell].$$

Lemma 1.2.1 *Let conditions (1.1.2) hold. Then there exists a unique solution $u \in \mathcal{V}^1(0, \ell)$ of the boundary value problem (1.1.1). This solution is Hölder continuous with exponent $\lambda = 1/2$, i.e. for all $x_1, x_2 \in [0, \ell]$,*

$$|u(x_1) - u(x_2)| \leq \frac{\ell}{\sqrt{2}\, c_0} \left[\|F\|_{L^2(0,\ell)} + \sqrt{2/\ell}\, |\varphi| \right] |x_1 - x_2|^{1/2}. \quad (1.2.2)$$

Moreover, the following estimate holds:

$$\|u\|_{\mathcal{V}^1(0,\ell)} \leq \frac{\ell^2 + 2}{\sqrt{2}\, c_0} \left[\|F\|_{L^2(0,\ell)} + \sqrt{2/\ell}\, |\varphi| \right] \quad (1.2.3)$$

Proof We use the integral identity (1.2.1) to introduce the symmetric bilinear $a : \mathcal{V}^1(0, \ell) \times \mathcal{V}^1(0, \ell) \mapsto \mathbb{R}$ and linear $b : \mathcal{V}^1(0, \ell) \mapsto \mathbb{R}$ functionals:

$$a(u, v) := \int_0^\ell [k(x)u'v' + q(x)uv]dx,$$
$$b(u) := \int_0^\ell F(x)v dx + \varphi v(b), \quad u, v \in \mathcal{V}^1(0, \ell).$$

By conditions (1.1.2), $a(u, v)$ is a strongly positive bounded bilinear functional and $b(v)$ is a bounded linear functional. Then, by Variational Lemma (Theorem 1.1.1, Sect. 1.1) there exists a unique solution $u \in \mathcal{V}^1(0, \ell)$ of the variational problem

$$a(u, v) = b(v), \quad \forall v \in \mathcal{V}^1(0, \ell).$$

Let us prove now the Hölder continuity of the solution. For any $x_1, x_2 \in [0, \ell]$ we have:

$$|u(x_1) - u(x_2)| = \left|\int_{x_1}^{x_2} u'(\xi)d\xi\right|$$
$$\leq \left(\int_{x_1}^{x_2}(u'(\xi))^2 d\xi\right)^{1/2} |x_1 - x_2|^{1/2} \leq \|u'\|_{L^2(0,\ell)} |x_1 - x_2|^{1/2}. \quad (1.2.4)$$

To estimate the norm $\|u'\|_{L^2(0,\ell)}$ we use the energy identity

$$\int_0^\ell [k(x)(u'(x))^2 + q(x)(u(x))^2]dx = \int_0^\ell F(x)u\,dx + \varphi u(b), \; u \in \mathcal{V}^1(0, \ell)$$

and the Poincaré inequality $\|u\|_{L^2(0,\ell)} \leq \left(\ell/\sqrt{2}\right) \|u'\|_{L^2(0,\ell)}$. We have:

$$\|u'\|_{L^2(0,\ell)}^2 \leq \frac{1}{c_0} \int_0^\ell [k(x)(u'(x))^2 + q(x)(u(x))^2]dx$$
$$= \frac{1}{c_0}\left[\int_0^\ell F(x)u\,dx + \varphi \int_0^\ell u'(x)dx\right]$$
$$\leq \frac{1}{c_0}\left[\|F\|_{L^2(0,\ell)}\|u\|_{L^2(0,\ell)} + \sqrt{\ell}\,|\varphi|\|u'\|_{L^2(0,\ell)}\right]$$
$$\leq \frac{1}{c_0}\left[\frac{\ell}{\sqrt{2}}\|F\|_{L^2(0,\ell)} + \sqrt{\ell}\,|\varphi|\right]\|u'\|_{L^2(0,\ell)}.$$

After dividing both sides by $\|u'\|_{L^2(0,\ell)} \neq 0$, we obtain

$$\|u'\|_{L^2(0,\ell)} \leq \frac{1}{c_0}\left[\frac{\ell}{\sqrt{2}}\|F\|_{L^2(0,\ell)} + \sqrt{\ell}\,\varphi\right], \quad (1.2.5)$$

which with (1.2.4) leads the desired estimate (1.2.2).

To prove estimate (1.2.3) one needs to use in (1.2.5) the inequality $\|u'\|_{L^2(0,\ell)} \geq \beta_0 \|u\|_{\mathcal{V}^1(0,\ell)}$ with $\beta_0 = \ell/\sqrt{\ell^2 + 2}$, which follows from the Poincaré inequality. □

It follows from estimate (1.2.2) that the weak solution $u \in \mathcal{V}^1(0, \ell)$ of the boundary value problem (1.1.1) belongs to the Hölder space $C^{0,\lambda}[0, \ell]$, with exponent $\lambda \in (0, 1/2]$. Furthermore, estimate (1.2.3) means continuity of this weak solution with respect to the source term $F \in L^2(0, \ell)$ and Neumann boundary data $\varphi \in \mathbb{R}$.

The following theorem shows that the nature of continuity of the weak solution $u \in \mathcal{V}^1(0, \ell)$ of the boundary value problem (1.1.1), with respect to the coefficients $k(x)$ and $q(x)$ is different.

Theorem 1.2.1 *Let conditions (1.1.2) hold. Assume that $\{k_n\} \subset \mathcal{K}$, $\{q_n\} \subset \mathcal{Q}$ and $\{F_n\} \subset L^2(0, \ell)$ are the sequences of coefficients and source functions. Denote by $\{u_n\} \subset \mathcal{V}^1(0, \ell)$ the sequence of corresponding weak solutions, that is, for each*

1.2 Continuity with Respect to Coefficients and Source: Sturm-Liouville Equation

$n = 1, 2, 3, \ldots$, the function $u_n(x) := u[x; k_n, q_n, F_n]$ is the weak solution of the boundary value problem

$$\begin{cases} -(k_n(x)u_n'(x))' + q_n(x)u_n(x) = F_n(x), & x \in (0, \ell), \\ u_n(0) = 0, \quad (k_n(x)u_n'(x))_{x=\ell} = \varphi, \quad \varphi \in \mathbb{R}. \end{cases} \quad (1.2.6)$$

If

$$\begin{cases} \frac{1}{k_n(x)} \rightharpoonup \frac{1}{k(x)} & \text{weakly in } L^2(0, \ell), \\ q_n(x) \rightharpoonup q(x) & \text{weakly in } L^2(0, \ell), \\ F_n(x) \rightharpoonup f(x) & \text{weakly in } L^2(0, \ell), \quad \text{as } n \to \infty, \end{cases} \quad (1.2.7)$$

then the sequence of solutions of problem (1.2.6) converges to the solution $u \in \mathcal{V}^1(0, \ell)$, $u(x) := u[x; k, q, f]$, of the boundary value problem (1.1.1), in the norm of $C^{0,\lambda}$, with $0 < \lambda < 1/2$, as $n \to \infty$.

Proof First we derive an integral representation for the solution of the boundary value problem (1.1.1). To this aim we integrate equation (1.1.1) on $[x, b]$ and use the Neumann boundary condition $(k(x)u'(x))_{x=b} = \varphi$, then integrate again both sides on $[a, x]$ and use the Dirichlet boundary condition $u(a) = 0$. This yields:

$$\begin{aligned} u(x) &= \varphi \int_0^x \frac{dt}{k(t)} + \int_0^x \frac{P(t)}{k(t)} dt, \\ P(t) &:= \int_t^\ell [F(\xi) - q(\xi)u(\xi)] d\xi. \end{aligned} \quad (1.2.8)$$

Now we use estimate (1.2.3) for the solution u_n of problem (1.2.6):

$$\|u_n\|_{\mathcal{V}^1(0,\ell)} \leq \frac{\ell^2 + 2}{\sqrt{2c_0}} \left[\|F_n\|_{L^2(0,\ell)} + \sqrt{2/\ell} \, |\varphi| \right]. \quad (1.2.9)$$

Since $F_n \rightharpoonup F$ weakly in $L^2(0, \ell)$, as $n \to \infty$, due to the weak convergence criteria [64] we have:

(a) $\{\|F_n\|_0\}$, is uniformly bounded;
(b) $\int_0^x F_n(\tau) d\tau \to \int_0^x F(\tau) d\tau$, for all $x \in (0, \ell]$.

The uniform boundedness of the sequence $\|F_n\|_0$ with estimate (1.2.9) implies the uniform boundedness of the sequence $\{u_n\} \subset \mathcal{V}^1(0, \ell)$ in H^1-norm. Since every bounded sequence in $H^1(0, \ell)$ is a compact in $C^{0,\lambda}[0, \ell]$ for $0 < \lambda < 1/2$, the sequence $\{u_n\}$ is a compact in $C^{0,\lambda}[0, \ell]$. Hence we can extract a subsequence $\{u_m\} \subset \mathcal{V}^1(0, \ell)$ that converges to a function $u \in C^{0,\lambda}[0, \ell]$ in $C^{0,\lambda}$-norm. Taking into account that the convergence in $C^{0,\lambda}[0, \ell]$ is a uniform convergence, we have $u_m(x) \to u(x)$ uniformly, for all $x \in [0, \ell]$, as $n \to \infty$.

We prove finally that the limit function $u(x)$ is the solution of the boundary value problem (1.1.1) corresponding to the limit functions $k(x), q(x)$ and $F(x)$ in (1.2.7). The integral representation (1.2.8) for $u_m \in \mathcal{V}^1(0, \ell)$, $m = 1, 2, 3, \ldots$, yields:

$$u_m(x) = \varphi \int_0^x \frac{dt}{k_m(t)} + \int_0^x \frac{P_m(t)}{k_m(t)} dt,$$
$$P_m(t) := \int_t^\ell [F_m - q_m u_m] d\xi. \tag{1.2.10}$$

According to the weak convergence $F_m(t) \rightharpoonup F(t)$ in $L^2(0, \ell)$ we have:

$$\int_t^b F_m(t) dt \to \int_t^b F(t) dt, \text{ for all } t \in [0, \ell], \text{ as } m \to \infty. \tag{1.2.11}$$

Due to the fact that $q_m(\xi) \rightharpoonup q(\xi)$ weakly in $L^2(0, \ell)$ and $u_m(\xi) \to u(\xi)$, in $C^{0,\lambda}$-norm, we conclude $q_m(\xi) u_m(\xi) \rightharpoonup q(\xi) u(\xi)$ weakly in $L^2(0, \ell)$. Hence

$$\int_t^b q_m(\xi) u_m(\xi) dt \to \int_t^b q(\xi) u(\xi) dt, \quad \forall t \in [0, \ell], \text{ as } n \to \infty,$$

which, with (1.2.11), implies $P_m(t) \to P(t)$, as $n \to \infty$. Now we prove that

$$\int_0^x \frac{P_m(t)}{k_m(t)} dt \to \int_0^x \frac{P(t)}{k(t)} dt, \quad \forall x \in [0, \ell], \text{ as } n \to \infty. \tag{1.2.12}$$

Indeed,

$$\left| \int_0^x \frac{P_m(t)}{k_m(t)} dt - \int_0^x \frac{P(t)}{k(t)} dt \right|$$
$$\leq \left| \int_0^x \frac{P_m(t) - P(t)}{k_m(t)} dt \right| + \left| \int_0^x \left[\frac{1}{k_m(t)} - \frac{1}{k(t)} \right] P(t) dt \right|$$
$$\leq \frac{1}{c_0} \int_0^x |P_m(t) - P(t)| dt + \left| \int_0^x \left[\frac{1}{k_m(t)} - \frac{1}{k(t)} \right] P(t) dt \right|, \ c_0 > 0.$$

The first right hand side term tends to zero due to the above convergence $P_m(t) \to P(t)$ in (1.2.12) as $m \to \infty$, and the second term also tends to zero due to the weak convergence criteria. Therefore,

$$\int_0^x \frac{dx}{k_m(x)} \to \int_0^x \frac{dx}{k(x)},$$
$$\int_0^x \frac{P_m(t)}{k_m(t)} dt \to \int_0^x \frac{P(t)}{k(t)} dt, \text{ as } n \to \infty,$$

for all $x \in [0, \ell]$, and the function $u_m(x)$, represented by (1.2.10) converges to the solution $u(x)$ represented by the integral representation (1.2.8), for all $x \in [0, \ell]$, as $m \to \infty$. Since the problem (1.1.1) has a unique solution, the sequence $\{u_n\} \subset \mathcal{V}^1(0, \ell)$ converges to the solution $u(x)$ of problem (1.1.1), for all $x \in [0, \ell]$, as $n \to \infty$, and we have the proof. □

Theorem 1.2.1 clearly illustrates that the nature of convergence of the leading coefficient $k(x)$ differs from the nature of convergence of the sink term $q(x)$. A necessary condition for the convergence of the sequence of solutions $\{u_n\}$ is the convergence of the sequence $1/k_n(x) \rightharpoonup 1/k(x)$, but not the convergence of the sequence of coefficients $k_n(x) \rightharpoonup k(x)$, while in case of the sink term $q(x)$, the necessary condition is the convergence of the sequence of coefficients $q_n(x) \rightharpoonup q(x)$. Remark that we met the term $1/k(x)$ in the Example 1.1.1.

1.3 Why a Fredholm Integral Equation of the First Kind Is an Ill-Posed Problem?

Consider the problem of solving the Fredholm Ill-Posed integral equation of the first kind

$$\int_0^1 K(x,y)u(y)dy = F(x), \quad x \in [0,1], \tag{1.3.1}$$

where $u \in C[0,1]$ is the unknown function, the kernel $K : [0,1] \times [0,1] \to \mathbb{R}$ is a continuous function and $F(x)$ is a given function in $C[0,1]$. It is known that if $\mathcal{D} := \{v \in C[0,1] : \|v\|_{C[0,1]} \leq M, M > 0\}$, then the linear continuous operator $A : \mathcal{D} \subset C[0,1] \mapsto C[0,1]$, defined as the integral operator

$$(Au)(x) := \int_0^1 K(x,y)u(y)dy, \tag{1.3.2}$$

is a linear compact operator from $C[0,1]$ to $C[0,1]$, i.e. transforms each bounded set in $C[0,1]$ to a relatively compact set in $C[0,1]$ (see, for instance, the reference [103]). Let us show that the problem (1.3.1) is ill-posed in sense of the third Hadamard's condition (**p3**).

Let $u := u(x; F)$ be a solution of (1.3.1) for the given $F \in C[0,1]$. To show that the dependence $u(\cdot\,; F)$ is not continuous in $C[0,1]$, we define the sequence of continuous functions

$$\varepsilon_n(x) = \int_0^1 K(x,y)\sin(n\pi y)dy, \quad n = 1, 2, 3, \ldots . \tag{1.3.3}$$

By the continuity of the kernel $K(x,y)$, $\|\varepsilon_n\|_{C[0,1]} \to 0$, as $n \to \infty$. Now we define the "perturbed" source functions $F_n(x) = F(x) + \varepsilon_n(x)$, $n = 1, 2, 3, \ldots$. Then the function $u_n(x) = u(x) + \sin(n\pi y)$ will be a solution of the "perturbed" problem

$$\int_0^1 K(x,y)u_n(y)dy = F_n(x), \quad x \in [0,1],$$

for each $n = 1, 2, 3, \ldots$. Evidently, the norm $\|F - F_n\|_{C[0,1]} = \|\varepsilon_n\|_{C[0,1]}$ tends to zero, as $n \to \infty$, although $\|u_n - u\|_{C[0,1]} = \|\sin(n\pi y)\|_{C[0,1]} = 1$, for all $n = 1, 2, 3, \ldots$. This shows that if the Fredholm operator (1.3.2) is defined as $A : C[0,1] \mapsto C[0,1]$, then *problem (1.3.1) is ill-posed*.

Note that the same conclusion is still hold, if $K \in L^2((0, 1) \times (0, 1))$ and the Fredholm operator (1.3.2) is defined as $A : L^2(0, 1) \mapsto L^2(0, 1)$. In this case the integral (1.3.3) tends to zero as $n \to \infty$, by the Riemann-Lebesgue Lemma. □

To answer the question "*why the problem* (1.3.1) *is ill-posed?*" we need to study this problem deeper, by arriving to the physical meaning of the mathematical model.

Example 1.3.1 Relationship between the Fredholm integral equation and the differential problem.

Let us analyze again problem (1.3.1) assuming that the kernel is given by the formula

$$\mathring{K}(x, y) = \begin{cases} (1-x)y, & 0 \leq y \leq x, \\ x(1-y), & x \leq y \leq 1. \end{cases} \tag{1.3.4}$$

It is easy to verify that

$$\int_0^1 \mathring{K}(x, y) \sin(n\pi y)\, dy = \frac{1}{(n\pi)^2} \sin(n\pi x), \quad n = 1, 2, \ldots. \tag{1.3.5}$$

Equation (1.3.5) means that the numbers $\{(n\pi)^{-2}\}_{n=1}^\infty$ are eigenvalues of the integral operator

$$(\mathring{A}u)(x) := \int_0^1 \mathring{K}(x, y) u(y) dy, \tag{1.3.6}$$

and $\{\sqrt{2} \sin(n\pi x)\}_{n=1}^\infty$ are the orthonormal eigenfunctions. Using (1.3.4) and (1.3.5) we can define the Fourier Sine series representation for the kernel $\mathring{K}(x, y)$:

$$\mathring{K}(x, y) = 2 \sum_{n=1}^\infty \frac{\sin(n\pi x) \sin(n\pi y)}{(n\pi)^2}. \tag{1.3.7}$$

To solve now the Fredholm integral equation

$$(\mathring{A}u)(x) := \int_0^1 \mathring{K}(x, y) u(y) dy = F(x), \quad x \in [0, 1], \tag{1.3.8}$$

we use the Fourier Sine series for the functions $F(x)$ and $u(x)$:

$$F(x) = \sqrt{2} \sum_{n=1}^\infty F_n \sin(n\pi x), \quad u(x) = \sqrt{2} \sum_{n=1}^\infty u_n \sin(n\pi x),$$

1.3 Why a Fredholm Integral Equation of the First Kind Is an Ill-Posed Problem?

where

$$F_n = \sqrt{2}\int_0^1 F(\xi)\sin(n\pi\xi)d\xi, \quad u_n = \sqrt{2}\int_0^1 u(\xi)\sin(n\pi\xi)d\xi$$

are the Fourier coefficients. Substituting these in (1.3.8) and using (1.3.5) we deduce:

$$\sum_{n=1}^{\infty} \frac{u_n}{(n\pi)^2}\sin(n\pi x) = \sum_{n=1}^{\infty} F_n \sin(n\pi x).$$

This implies:

$$u_n = (n\pi)^2 F_n, \quad n = 1, 2, \ldots. \tag{1.3.9}$$

The relationship (1.3.9) can be treated as an *input-output relationship* for problem (1.3.8).

Thus, the Fourier series solution of the Fredholm integral equation (1.3.8) with the kernel given by (1.3.4) is the function

$$u(x) = \sqrt{2}\sum_{n=1}^{\infty}(n\pi)^2 F_n \sin(n\pi x), \quad x \in [0, 1], \tag{1.3.10}$$

if the series converges in the considered solution set $C[0, 1]$. However, there are very simple cases where this fails to happen, even in $L^2[0, 1]$. Indeed, let $F(x) \equiv 1$, $x \in [0, 1]$. Calculating the Fourier coefficients F_n we get:

$$F_n = \sqrt{2}\,\frac{[1-(-1)^n]}{n\pi}, \quad n = 1, 2, 3, \ldots.$$

This means the series (1.3.10) fails to converge. □

The above example tells us that problem (1.3.1) or (1.3.8) *may not have a solution for each function* $F(x)$ *from* $C[0, 1]$.

To understand the reason of this phenomenon, we interchange roles of $u(x)$ and $F(x)$ in problem (1.3.8), *assuming now that $F(x)$ is the unknown function and $u(x)$ is the given one.* Differentiating the left hand side of (1.3.8) and taking into account (1.3.4), we get

$$\frac{d^2}{dx^2}\int_0^1 \mathring{K}(x, y)u(y)dy = \frac{d}{dx}\left(-\int_0^x yu(y)dy + \int_x^1 (1-y)u(y)dy\right)$$
$$= -u(x).$$

Then we obtain the following formal equation $-F''(x) = u(x)$, $x \in (0, 1)$ with respect to the unknown function $F(x)$. Note that in terms of problem (1.3.8) this equation, in particular, implies a necessary condition for the existence of a solution in $C[0, 1]$: *the function $F(x)$ should belong to the space $C^2[0, 1]$.*

Clearly, function $F(x)$ should also satisfy the boundary conditions $F(0) = F(1) = 0$, as the integral Eq. (1.3.8) with the kernel given by formula (1.3.7) shows. Thus, if we assume in (1.3.8) that $F(x)$ is unknown function and $u(x)$ is the given function, then we conclude that $F(x)$ is the solution to the boundary value problem

$$\begin{cases} -F''(x) = u(x), \; x \in (0, 1), \\ F(0) = F(1) = 0. \end{cases} \quad (1.3.11)$$

Evidently, $\overset{\circ}{K}(x, y)$ is the Green's function for the operator $-d^2/dx^2$ under the boundary conditions (1.3.11).

Let us compare now problems (1.3.8) and (1.3.11), taking into account the swapping of the functions $u(x)$ and $F(x)$. It follows from the above considerations that, problems (1.3.8) and (1.3.11) can be defined as *inverse to each other*, as stated in [53]. Then, it is natural to ask the question: which problem is the direct (i.e. original) problem, and which problem is the inverse problem? To answer this question, we need to go back to the physical model of the problem. The boundary value problem (1.3.11) is the simplest mathematical model of deflection of a string, occupying the interval [0, 1]. The Dirichlet conditions in (1.3.11) mean that the string is clamped at its end points. In this model, the function $u(x)$, as a given right hand side of the differential Eq. (1.3.11), represents a given pressure, and the function $F(x)$, as a solution of the boundary value problem (1.3.11), represents the corresponding deflection. The unique (classical) solution of the two-point boundary value problem (1.3.11) is the function

$$F(x) = \int_0^1 \overset{\circ}{K}(x, y) u(y) dy, \; x \in [0, 1], \quad (1.3.12)$$

where the kernel $\overset{\circ}{K}(x, y)$ is the Green's function defined by (1.3.4). Hence *each pressure $u \in C[0, 1]$, defines uniquely the deflection function $F \in C^2[0, 1]$, $F(0) = F(1) = 0$.* In other words, the boundary value problem (1.3.11) is a *well-posed problem*, with the unique solution (1.3.12). On the other hand, as we have seen above, the integral equation (1.3.8) may not have a solution for each continuous function $F(x)$ (deflection). *The physical interpretation is clear: each (admissible) pressure generates a unique deflection, but an arbitrary function cannot be regarded as a deflection.* Applied to the integral equation (1.3.8) this means that in order to $F(x)$ be a possible deflection it needs, at least, to satisfy the rigid clamped boundary conditions $F(0) = F(1) = 0$ and to have continuous second derivative. This is a reason, in the language of the physical model, why a Fredholm integral equation of the first kind is an ill-posed problem.

To finish the above analysis, now we return to the integral equation (1.3.8) with the kernel (1.3.4) and ask: what type of functions $F(x)$ are admissible in order to get a convergent series (1.3.10)?

1.3 Why a Fredholm Integral Equation of the First Kind Is an Ill-Posed Problem?

Example 1.3.2 Existence and non-existence of a solution of the Fredholm integral equation.

The solution of the integral equation (1.3.8) with the kernel (1.3.4) is the function $u(x)$ given by series (1.3.10). First, we assume that $F(x)$ in (1.3.8) is given by formula

$$F(x) = \begin{cases} x/2, & 0 \le x \le 1/2, \\ (1-x)/2, & 1/2 < x \le 1. \end{cases} \tag{1.3.13}$$

Note that this function is continuous, but not continuously differentiable. Calculating the Fourier sine coefficients we get:

$$F_n = \frac{\sqrt{2}}{(n\pi)^2} \sin\left(\frac{n\pi}{2}\right) = \frac{\sqrt{2}}{(n\pi)^2} \begin{cases} 0, & n = 2k, \\ (-1)^{n-1}, & n = 2k-1, \end{cases} k = 1, 2, 3, \ldots.$$

Substituting this in (1.3.10) we obtain the series solution of problem (1.3.8):

$$u(x) = 2 \sum_{n=1}^{\infty} (-1)^{n-1} \sin[(2n-1)\pi x], \quad x \in [0, 1]. \tag{1.3.14}$$

This is exactly the Fourier sine series expansion of Dirac delta function $\delta(x - 1/2)$, which not only does not satisfy the above differentiability conditions, but also is not even a regular generalized function. Hence, the function (1.3.14) corresponding to $F(x)$, given by formula (1.3.13), is not regarded as a solution of the Fredholm integral equation (1.3.8) in the above mentioned sense.

Let us use in (1.3.8) more smooth input data:

$$F(x) = x(1-x)/2, \quad x \in [0, 1], \tag{1.3.15}$$

restricting the class of functions $\{F(x)\}$. Evidently, this function has continuous derivatives up to order 2 and satisfies the conditions $F(0) = F(1) = 0$. The Fourier sine coefficients of function (1.3.15) are

$$F_n = \sqrt{2} \, \frac{2[1 - (-1)^n]}{(n\pi)^3}, \quad n = 1, 2, 3, \ldots.$$

Substituting this in (1.3.10) we obtain the series solution:

$$u(x) = \frac{2}{\pi} \sum_{n=1}^{\infty} \frac{[1 - (-1)^n]}{n} \sin(n\pi x) = \frac{4}{\pi} \sum_{m=1}^{\infty} \frac{\sin[(2m-1)\pi x]}{2m-1}, \quad x \in [0, 1].$$

This is the Fourier sine series expansion of the function $u(x) \equiv 1$, $x \in [0, 1]$. Hence, for the smooth enough function $F(x) = x(1-x)/2$, the solution of the Fredholm integral equation of the first kind with the kernel (1.3.4), is the function $u(x) \equiv 1$, $x \in [0, 1]$. □

Let us return now to the last two examples and try to reformulate in terms of compact operators the above conclusion "*every (admissible) pressure generates a deflection, but an arbitrary function cannot be regarded as a deflection*", obtained in terms of a physical model.

Definition 1.3.1 Let B and $\tilde{B}$ be normed spaces, and $A : B \mapsto \tilde{B}$ be a linear operator. A is called a compact operator if the set $\{Au : \|u\|_B \le 1\} \subset \tilde{B}$ is a pre-compact, i.e. has compact closure in $\tilde{B}$.

This definition is equivalent to the following one: for every bounded sequence $\{u_n\} \subset B$, the sequence of images $\{Au_n\} \subset \tilde{B}$ has a subsequence which converges to some element of $\tilde{B}$. Compact operators are also called completely continuous operators.

The lemma below explains the above conclusion, in terms of compact operators. It simply asserts that *any neighborhood of an arbitrary element $F \in \mathcal{R}(A)$ from the range of a linear compact operator A might not have a preimage in H*. Remember that if $K(x, y)$ in (1.3.1) is a bounded Hilbert-Schmidt kernel, i.e. if $K \in L^2((0, 1) \times (0, 1))$, then the integral operator $A : L^2(0, 1) \mapsto L^2(0, 1)$, defined by (1.3.2), is a compact operator.

Lemma 1.3.1 *Let $A : \mathcal{D}(A) \subset H \mapsto \tilde{H}$ be a compact operator between two infinite-dimensional Hilbert spaces H and $\tilde{H}$, with bounded domain $\mathcal{D}(A)$. Assume that $f \in \mathcal{R}(A)$ is an arbitrary element from the range $\mathcal{R}(A) \subset \tilde{H}$ of the operator A. Then for any $\varepsilon > 0$ there exists an element $f_0 \in V_\varepsilon(f) := \{g \in \tilde{H} : \|g - f\|_{\tilde{H}} \le \varepsilon\}$ such that $f_0 \notin \mathcal{R}(A)$.*

Proof Let $f \in \mathcal{R}(A)$ be an arbitrary element. To prove the lemma assume on the contrary that there exists $\varepsilon_0 > 0$ such that $V_{\varepsilon_0}(f) \subset \mathcal{R}(A)$. Let $\{f_i\}_{i=1}^\infty \subset \tilde{H}$ be an orthonormal basis. We define the sequence $\{g_i\}_{i=1}^\infty \subset \tilde{H}$, with $g_i := f + \varepsilon_0 f_i/2$. Evidently, $\{g_i\}_{i=1}^\infty \subset V_{\varepsilon_0}(f)$, since $\|g_i - f\|_{\tilde{H}} = (\varepsilon_0/2)\|f_i\|_{\tilde{H}} = \varepsilon_0/2 < \varepsilon_0$, which means the sequence $\{g_i\}_{i=1}^\infty$ is bounded in $\tilde{H}$. However, for all $i, j = \overline{1, \infty}$, $\|g_i - g_j\|_{\tilde{H}} = \varepsilon_0/\sqrt{2}$ and this sequence does not contain any Cauchy subsequence. Hence, the set $V_{\varepsilon_0}(f) \subset \mathcal{R}(A)$ is not a precompact. On the other hand, $\mathcal{D}(A) \subset H$ is a bounded set and A is a compact operator. Hence, as an image $V_{\varepsilon_0}(f) := A(\mathcal{D}_0)$ of a bounded set $\mathcal{D}_0 \subset \mathcal{D}(A)$, the set $V_{\varepsilon_0}(f) \subset \mathcal{R}(A)$ needs to be a precompact. This contradiction completes the proof of the lemma. □

Corollary 1.3.1 *Let conditions of Lemma 1.3.1 hold. Assume, in addition, that $A : \mathcal{D}(A) \subset H \mapsto \tilde{H}$ is a linear injective operator. Then the inverse operator $A^{-1} : \mathcal{R}(A) \mapsto \mathcal{D}(A)$ is discontinuous, that is, the problem $Au = f$, $u \in \mathcal{D}$, is ill-posed.*

Proof Assume on the contrary that the inverse operator $A^{-1} : \mathcal{R}(A) \mapsto \mathcal{D}(A)$ is continuous. Then, as an image $A^{-1}(\mathcal{R}(A)) := \mathcal{D}(A)$ (under the inverse operator A^{-1}) of the compact set $\mathcal{R}(A)$, the set $\mathcal{D}(A)$ needs to be compact, while it is a bounded set. □

1.3 Why a Fredholm Integral Equation of the First Kind Is an Ill-Posed Problem?

These results give us some insights into the structure and dimension of the range $\mathcal{R}(A)$ of a linear compact operator. In addition to above obtained result that the problem (1.3.1) is ill-posed in sense of the *third Hadamard's condition* (**p3**), we conclude from Corollary 1.3.1 that this problem is ill-posed also *in sense of the first Hadamard's condition* (**p1**). The reason is that, the range $\mathcal{R}(A)$ of a linear compact operator is not dense everywhere, by the assertion of Lemma 1.3.1. We will show in the Appendix A that the range $\mathcal{R}(A)$ of a linear compact operator A, defined on a Hilbert space H, is "almost finite-dimensional", i.e. can be approximated to any given accuracy by a finite dimensional subspace in $\mathcal{R}(A)$. Note that, if the range $\mathcal{R}(A)$ of a bounded linear operator is finite dimensional, then it is a compact operator. This follows from the Heine-Borel Theorem, since the closure $\overline{A(\mathcal{D})} \subset \tilde{H}$ of the image $A(\mathcal{D})$ of a bounded set $\mathcal{D} \subset H$ is closed and bounded in the finite dimensional subspace $\mathcal{R}(A) \subset \tilde{H}$, so, is compact.

Finally, note that Lemma 1.3.1 also asserts that if a compact operator has a bounded inverse, then H must be finite dimensional. Then it can be shown that, if H is infinite dimensional, then there is no injective compact operator from H onto $\tilde{H}$. Details of these assertions will be analyzed in the Appendix A.

Part I
Introduction to Inverse Problems

Chapter 2
Functional Analysis Background of Ill-Posed Problems

The main objective of this chapter is to present some necessary results of functional analysis, frequently used in study of inverse problems. For simplicity, we will derive these results in Hilbert spaces. Let H be a vector space over the field of real ($\mathbb{R}$) or complex ($\mathbb{C}$) numbers. Recall that the mapping $(\cdot,\cdot)_H$ defined on $H \times H$ is called an *inner product* (or *scalar product*) of two elements of H, if the following conditions are satisfied:

(i1) $(u_1 + u_2, v)_H = (u_1, v)_H + (u_2, v)_H$, $\forall u_1, u_2, v \in H$;
(i2) $(\alpha u, v)_H = \alpha (u, v)_H$, $\forall \alpha \in \mathbb{C}$, $\forall u, v \in H$;
(i3) $(u, v)_H = \overline{(v, u)_H}$, $\forall u, v \in H$;
(i4) $(u, u)_H \geq 0$, $\forall u \in H$ and $(u, u)_H = 0$ iff $u = 0$.

Through the following, we will omit the subscript H in the scalar (dot) product and the norm whenever it is clear from the text.

The vector space H together with the inner product is called an *inner product space* or a *pre-Hilbert space*. The norm in a pre-Hilbert space is defined by the above introduced scalar product: $\|u\|_H := (u, u)^{1/2}$, $u \in H$. Hence a pre-Hilbert space H is a normed space. If, in addition, a pre-Hilbert space is complete, it is called a Hilbert space. Thus, a Hilbert space is a Banach space, i.e. complete normed space, with the norm defined via the scalar product.

A Hilbert space is called infinite dimensional (finite dimensional) if the underlying vector space is infinite dimensional (finite dimensional). The basic representative of infinite dimensional space Hilbert in the weak solution theory of PDEs is the space of square integrable functions

$$L^2(a,b) := \{u : (a,b) \mapsto \mathbb{R} : \int_a^b u^2(x)dx < +\infty\},$$

with the scalar product

© Springer International Publishing AG 2017
A. Hasanov Hasanoğlu and V.G. Romanov, *Introduction to Inverse Problems for Differential Equations*, DOI 10.1007/978-3-319-62797-7_2

$$(u, v)_{L^2(a,b)} := \int_a^b u(x)v(x)dx, \ u, v \in L^2(a, b).$$

A finite dimensional analogue of this space of square-summable sequences in $\mathbb{R}^n$:

$$l^2 := \{x := (x_1, x_2, ..., x_n) \in \mathbb{R}^n : \sum_{k=1}^n x_k^2 < +\infty\},$$

with the scalar product

$$(x, y)_{l^2} := \sum_{k=1}^n x_k y_k, \ x, y \in \mathbb{R}^n.$$

2.1 Best Approximation and Orthogonal Projection

Let H be an inner product space. The elements $u, v \in H$ are called *orthogonal*, if $(u, v) = 0$. This property is denoted by $u \perp v$. Evidently, $0 \perp v$, for any $v \in H$, since $(0, v) = 0, \forall v \in H$. Let $U \subset H$ be a non-empty subset and $u \in H$ an arbitrary element. If $(u, v) = 0$ for all $v \in U$, then the element $u \in H$ is called orthogonal to the subset U and is denoted by $u \perp U$. The set of all elements $u \in H$ orthogonal to $U \subset H$ is called an *orthogonal complement* of U and is denoted by $U^\perp$:

$$U^\perp := \{u \in H : (u, v) = 0, \ \forall v \in U\}. \tag{2.1.1}$$

The subsets $U, V \subset H$ are called orthogonal subsets of H if $(u, v) = 0$ for all $u \in U$ and $v \in V$. This property is denoted by $U \perp V$. Evidently, if $U \perp V$, then $U \cap V = \{0\}$.

Theorem 2.1.1 *Let $U \subset H$ be a subset of an inner product space H. Then an orthogonal complement $U^\perp$ of U is a closed subspace of H. Moreover, the following properties are satisfied:*

(**p1**) $U \cap U^\perp \subset \{0\}$ *and* $U \cap U^\perp = \{0\}$ *iff U is a subspace;*
(**p2**) $U \subset \left(U^\perp\right)^\perp =: U^{\perp\perp}$, $\left(U^\perp\right)^\perp := \{u \in H : (u, v) = 0, \ \forall v \in U^\perp\}$;
(**p3**) *If $U_1 \subset U_2 \subset H$, then $U_2^\perp \subset U_1^\perp$.*

Proof Evidently, $\alpha u + \beta v \in U^\perp$, for all $\alpha, \beta \in \mathbb{C}$ and $u, v \in U$, by the above definition of the scalar product: $(\alpha u + \beta v, w) = \alpha(u, w) + \beta(v, w) = 0$, for all $w \in U$. This implies that $U^\perp$ is a subspace of H. It is easy to prove, by using continuity of the scalar product, that if $\{u_n\} \subset U^\perp$ and $u_n \to u$, as $n \to \infty$, then $u \in U^\perp$. So, $U^\perp$ is closed. To prove (**p1**), we assume that there exists an element $u \in U \cap U^\perp$. Then, by definition (2.1.1) of $U^\perp$, $(u, u) = 0$, which implies $u = 0$. Hence $U \cap U^\perp \subset \{0\}$.

2.1 Best Approximation and Orthogonal Projection

If, in addition, U is a subspace of H, then $0 \in U$ and it yields $U \cap U^\perp = \{0\}$. To prove (**p2**) we assume in contrary, that there exists an element $u \in U$ such that $u \notin \left(U^\perp\right)^\perp$. This means an existence of such an element $v \in U^\perp$ that $(u, v) \neq 0$. On the other hand, for all $u \in U$ and $v \in U^\perp$, we have $(u, v) = 0$, which is a contradiction. Hence $U \subset \left(U^\perp\right)^\perp$. Finally, to prove (**p3**), let $v \in U_2^\perp$ be an arbitrary element. Then $(u, v) = 0$, for all $u \in U_2$. Since $U_1 \subset U_2$, this holds for all $u \in U_1$ as well. Hence for any element $v \in U_2^\perp$, the condition $(u, v) = 0$ holds for all $u \in U_1$. This implies, by the definition, that $v \in U_1^\perp$, which completes the proof. $\square$

Definition 2.1.1 (*Best approximation*) Let $U \subset H$ be a subset of an inner product space H and $v \in H$ be a given element. If

$$\|v - u\|_H = \inf_{w \in U} \|v - w\|_H, \qquad (2.1.2)$$

then $u \in U$ is called the best approximation to the element $v \in H$ with respect to the set $U \subset H$.

The right hand side of (2.1.2) is a distance between a given element $v \in H$ and the set $U \subset H$. Hence, the best approximation is an element with the smallest distance to the set $U \subset H$.

This notion plays a crucial role in inverse problems theory and applications, since a measured output data can only be given with some measurement error. First we will prove that if $U \subset H$ is a closed linear subspace, then the best approximation is determined uniquely. For this aim we will use the main theorem on quadratic variational problems and its consequence, called perpendicular principle [103].

Theorem 2.1.2 *Let $a : H \times H \mapsto \mathbb{R}$ be a symmetric, bounded, strongly positive bilinear form on a real Hilbert space H, and $b : H \mapsto \mathbb{R}$ be a linear bounded functional on H. Then*
(i) The minimum problem

$$J(v) = \min_{w \in H} J(w), \quad J(w) := \frac{1}{2} a(w, w) - b(w) \qquad (2.1.3)$$

has a unique solution $v \in H$.
(ii) This minimum problem is equivalent to the followig variational problem: Find $v \in H$ such that

$$a(v, w) = b(w), \; \forall w \in H. \qquad (2.1.4)$$

We use this theorem to prove that in a closed linear subspace U of a real or complex Hilbert space H, the best approximation is uniquely determined.

Theorem 2.1.3 *Let U be a closed linear subspace of a Hilbert space H and $v \in H$ be a given element. Then the best approximation problem (2.1.2) has a unique solution $u \in U$. Moreover $v - u \in U^\perp$.*

Proof We rewrite the norm $\|v - u\|_H$ as follows:

$$\|v - u\|_H := (v, v) - (v, u) - (u, v) + (u, u)$$
$$= a(v, v) + 2\left[\frac{1}{2}a(u, u) - b(u)\right], \qquad (2.1.5)$$

where

$$a(u, w) := Re(u, w), \quad b(u) := \frac{1}{2}[(v, u) + (u, v)] = Re(v, u)$$

The right hand side of (2.1.5) shows that the best approximation problem (2.1.5) is equivalent to the variational problem (2.1.3) with the above defined bilinear and linear forms. Then it follows from Theorem 2.1.2 that if H is a real Hilbert space, then there exists a unique best approximation $v \in U$ to the element $u \in H$.

If H is a complex Hilbert space, then we can introduce the new scalar product $(v, w)_* := Re(v, w)$, $v, w \in H$ and again apply Theorem 2.1.2.

We prove now that if $u \in U$ is the best approximation to the element $v \in H$, then $v - u \perp U$. Indeed, it follows from (2.1.2) that

$$\|v - u\|_H^2 \leq \|v - (u + \lambda w)\|_H^2, \quad \forall \lambda \in \mathbb{C}, \ w \in U.$$

This implies,

$$(v - u, v - u) \leq (v - u, v - u) - \overline{\lambda}(v - u, w) - \lambda(w, v - u) + |\lambda|^2(w, w)$$

or

$$0 \leq -\overline{\lambda}(v - u, w) - \lambda(w, v - u) + |\lambda|^2(w, w).$$

Assume that $v - u \neq 0$, $w \neq 0$. Then taking $\lambda = (w, v - u)/\|w\|^2$ we obtain: $0 \leq -|(v - u, w)|^2$, which means that $(v - u, w) = 0$, $\forall w \in U$. Note that this remains true also if $v - u = 0$. □

Corollary 2.1.1 (Orthogonal decomposition) Let U be a closed linear subspace of a Hilbert space H. Then there exists a unique decomposition of a given arbitrary element $v \in H$ of the form

$$v = u + w, \ u \in U, \ w \in U^\perp. \qquad (2.1.6)$$

Existence of this orthogonal decomposition follows from Theorem 2.1.3. We prove the uniqueness. Assume, in contrary, that there exists another decomposition of $v \in H$ such that

$$v = u_1 + w_1, \ u_1 \in U, \ w_1 \in U^\perp.$$

2.1 Best Approximation and Orthogonal Projection

Since U is a linear subspace of H, we have $u - u_1 \in U$, $w - w_1 \in U^\perp$, and

$$(u - u_1) + (w - w_1) = 0.$$

Multiplying scalarly both sides by $u - u_1$ we get

$$(u - u_1, u - u_1) + (w - w_1, u - u_1) = 0,$$

which implies $u = u_1$, since the second term is zero due to $w - w_1 \in U^\perp$. In a similar way we conclude that $w = w_1$. This completes the proof. □

The orthogonal decomposition (2.1.6) can also be rewritten in terms of the subspaces U and $U^\perp$ as follows:

$$H = U \oplus U^\perp.$$

Example 2.1.1 Let $H := L^2(-1, 1)$, $U := \{u \in L^2(-1, 1) : u(-x) = u(x)$, a.e. in $(-1, 1)\}$ be the set of even functions and $U^\perp := \{u \in L^2(-1, 1) : u(-x) = -u(x)$, a.e. in $(-1, 1)\}$ be the set of odd functions. Then $L^2(-1, 1) = U \oplus U^\perp$. Remark that for any $v \in L^2(-1, 1)$, $v(x) = [v(x) + v(-x)]/2 + [v(x) - v(-x)]/2$.

Corollary 2.1.1 shows that there exists a mapping which uniquely transforms each element $v \in H$ to the element u of a closed linear subspace U of a Hilbert space H. This assertion is called *Orthogonal Projection Theorem*.

Definition 2.1.2 The operator $P : H \mapsto U$, with $Pv = u$, in the decomposition (2.1.6) which maps each element $v \in H$ to the element $u \in U$ is called the projection operator or orthogonal projection.

Using this definition, we may rewrite the best approximation problem (2.1.2) as follows:

$$\|v - Pv\|_H = \inf_{w \in U} \|v - w\|_H.$$

We denote by $\mathcal{N}(P) := \{v \in H : Pv = 0\}$ and $\mathcal{R}(P) := \{Pv : v \in H\}$ the nullspace and the range of the projection operator P, respectively. The theorem below shows that the orthogonal projection $P : H \mapsto U$ is a linear continuous self-adjoint operator.

Theorem 2.1.4 *The orthogonal projection $P : H \mapsto U$ defined from Hilbert space H onto the closed subspace $U \subset H$ is a linear continuous self-adjoint operator with $P^2 = P$ and $\|P\| = 1$, for $U \neq \{0\}$. Conversely, if $P : H \mapsto H$ is a linear continuous self-adjoint operator with $P^2 = P$, then it defines an orthogonal projection from H onto the closed subspace $\mathcal{R}(P)$.*

Proof It follows from (2.1.6) that

$$\|v\|^2 := \|u + w\|^2 = \|u\|^2 + \|w\|^2,$$

since $(u, w) = 0$, by $u \in U$ and $w \in U^\perp$. This implies $\|Pv\| \le \|v\|$, $Pv := u$ for all $v \in H$. In particular, for $v \in U \subset H$ we have $Pv = v$, which means $\|P\| = 1$. We prove now that P is self-adjoint. Let $v_k = u_k + w_k, u_k \in U, w_k \in U^\perp, k = 1, 2$. Multiplying both sides of $v_1 = u_1 + w_1$ by u_2 and both sides of $v_2 = u_2 + w_2$ by u_1, then taking into account $(u_k, v_m) = 0, k, m = 1, 2$, we conclude

$$(v_1, u_2) = (u_1, u_2), \ (v_2, u_1) = (u_2, u_1).$$

This implies $(u_1, v_2) = (u_1, u_2) = (v_1, u_2)$. Using the definition $Pv_k := u_k$ we deduce:

$$(Pv_1, v_2) := (v_1, Pv_2), \ \forall v_1, v_2 \in H,$$

i.e. the projection operator P is self-adjoint. Assuming now $v = u \in U \subset H$ in (2.1.6) we have: $u = u + 0$, where $0 \in U^\perp$. Then $Pu = u$, and for any $v \in H$ we have $P^2v = P(Pv) = Pu = u = Pv$. Hence $P^2v = Pv$, for all $v \in H$.

We prove now the second part of the theorem. Evidently, $\mathcal{R}(P) := \{Pv : v \in H\}$ is a linear subspace. We prove that the range of the projection operator $\mathcal{R}(P)$ is closed. Indeed, let $\{u_n\} \subset \mathcal{R}(P)$, such that $u_n \to u$, as $n \to \infty$. Then there exists such an element $v_n \in H$ that $u_n = Pv_n$. Together with the property $P^2 v_n = Pv_n$ this implies: $Pu_n = P^2 v_n = Pv_n = u_n$. Hence, $Pu_n = u_n$ for all $u_n \in \mathcal{R}(P)$. Letting to the limit and using the continuity of the operator P, we obtain:

$$u = \lim_{n \to \infty} u_n = \lim_{n \to \infty} Pu_n = Pu,$$

i.e. $Pu = u$, which means $u \in \mathcal{R}(P)$. Thus $\mathcal{R}(P)$ is a linear closed subspace of H and all its elements are fixed points of the operator P, that is, $Pu = u$ for all $u \in \mathcal{R}(P)$. On the other hand, P is a self-adjoint operator with $P^2 = P$, by the asumption. Then for any $v \in H$,

$$(Pv, (I - P)v) = (Pv, v) - (Pv, Pv) = (Pv, v) - (P^2 v, v) = 0.$$

Since $v \in H$ is an arbitrary element, the orthogonality $(Pv, (I - P)v) = 0$ means that $(I - P)v \in \mathcal{R}(P)^\perp$. Then the identity

$$v = Pv + (I - P)v, \ \forall v \in H$$

with Corollary 2.1.1 implies that P is a projection operator, since $Pv \in \mathcal{R}(P) \subset H$ and $(I - P)v \in \mathcal{R}(P)^\perp$. □

Remark 2.1.1 Based on the properties of the orthogonal projection $P : H \mapsto H$, we conclude that the operator $I - P : H \mapsto H$ is also an orthogonal projection. Indeed, $(I - P)^2 = I - 2P + P^2 = I - P$.

2.1 Best Approximation and Orthogonal Projection

Remark 2.1.2 Let us define the set $\mathcal{M} := \{u \in H : Pu = u\}$, i.e. the set of fixed points of the orthogonal projection P. It follows from the proof of Theorem 2.1.4 that $\mathcal{M} = \mathcal{R}(P)$. Similarly, $\mathcal{R}(I - P) = \mathcal{N}(P)$.

Some other useful properties of the projection operator $P : H \mapsto H$ are summarized in the following corollary.

Corollary 2.1.2 *Let $P : H \mapsto H$ be an orthogonal projection defined on a Hilbert space H. Then the following assertions hold:*
(p1) $\mathcal{N}(P)$ *and* $\mathcal{R}(P)$ *are closed linear subspaces of H.*
(p2) *Each element $v \in H$ can be written uniquely as the following decomposition:*

$$v = u + w, \; u \in \mathcal{R}(P), \; w \in \mathcal{N}(P). \tag{2.1.7}$$

Moreover,

$$\|v\|^2 = \|u\|^2 + \|w\|^2. \tag{2.1.8}$$

(p3) $\mathcal{N}(P) = \mathcal{R}(P)^\perp$ *and* $\mathcal{R}(P) = \mathcal{N}(P)^\perp$.

Proof The assertions **(p1)**−−**(p2)** follow from Corollary 2.1.1 and Theorem 2.1.4. We prove **(p3)**. Evidently, $\mathcal{N}(P) \perp \mathcal{R}(P)$ and $\mathcal{N}(P) \subset \mathcal{R}(P)^\perp$. Hence to prove the first part of the assertion **(p3)** we need to show that $\mathcal{R}(P)^\perp \subset \mathcal{N}(P)$. Let $v \in \mathcal{R}(P)^\perp \subset H$. Then, there exist such elements $u \in \mathcal{R}(P), w \in \mathcal{N}(P)$ that $v = u + w$, according to (2.1.7). Multiplying both sides by an arbitrary element $\tilde{v} \in \mathcal{R}(P)$ we obtain: $(v, \tilde{v}) = (u, \tilde{v}) + (w, \tilde{v})$. The left hand side is zero, since $v \in \mathcal{R}(P)^\perp$. Also, $(w, \tilde{v}) = 0$, by $\mathcal{N}(P) \perp \mathcal{R}(P)$. Thus, $(u, \tilde{v}) = 0$, for all $\tilde{v} \in \mathcal{R}(P)$. But $u \in \mathcal{R}(P)$. This implies $u = 0$, and as a result, $v = 0 + w$, where $w \in \mathcal{N}(P)$. Hence $v \in \mathcal{N}(P)$.

The second part of the assertion **(p3)** can be proved similarly. □

We illustrate the above results in the following example.

Example 2.1.2 Fourier Series and Orthogonal Projection.

Let $\{\varphi_n\}_{n=1}^\infty$ be an orthonormal basis of an infinite-dimensional real Hilbert space H. Then any element $u \in H$ can be written uniquely as the following convergent Fourier series:

$$u = \sum_{n=1}^\infty (u, \varphi_n)\varphi_n \equiv \sum_{n=1}^N (u, \varphi_n)\varphi_n + \sum_{n=N+1}^\infty (u, \varphi_n)\varphi_n.$$

Let us consider the finite system $\{\varphi_n\}_{n=1}^N$ which forms a basis for the finite-dimensional Hilbert subspace $H_N \subset H$. We define the operator $P : H \mapsto H_N$ as follows:

$$Pu := \sum_{n=1}^N (u, \varphi_n)\varphi_n, \; u \in H. \tag{2.1.9}$$

Evidently, P is a linear bounded operator. Moreover, $P^2 = P$. Indeed,

$$P^2 u := P\left(\sum_{n=1}^{N}(u,\varphi_n)\varphi_n\right) = \sum_{m=1}^{N}\left(\sum_{n=1}^{N}(u,\varphi_n)\varphi_n, \varphi_m\right)\varphi_m$$

$$= \sum_{m=1}^{N}(u,\varphi_m)\varphi_m = Pu,$$

by $(\varphi_n, \varphi_m) = \delta_{n,m}$. Thus, $P : H \mapsto H_N$, defined by (2.1.9), is a projection operator from the infinite-dimensional Hilbert space H *onto* the finite-dimensional Hilbert space $H_N \subset H$, with $\mathcal{R}(P) = H_N$. To show the orthogonality of $\mathcal{R}(P)$ and $\mathcal{N}(P)$, let $u \in \mathcal{R}(P)$ and $v \in \mathcal{N}(P)$ be arbitrary elements. Then $Pu = u$ and

$$(v, u) = (v, Pu) = \left(v, \sum_{n=1}^{N}(u,\varphi_n)\varphi_n\right) = \sum_{n=1}^{N}(u,\varphi_n)(v,\varphi_n)$$

$$= \left(\sum_{n=1}^{N}(v,\varphi_n)\varphi_n, u\right) = (Pv, u).$$

But $Pv = 0$, due to $v \in \mathcal{N}(P)$. Hence $(v, u) = 0$, for all $u \in \mathcal{R}(P)$ and $v \in \mathcal{N}(P)$, which implies $\mathcal{R}(P) \perp \mathcal{N}(P)$.

Now we show the projection error, defined as

$$u - Pu := \sum_{n=N+1}^{\infty}(u,\varphi_n)\varphi_n,$$

is orthogonal to H_N. Let $v \in H_N$ be any element. Then

$$(u - Pu, v) := \left(\sum_{n=N+1}^{\infty}(u,\varphi_n)\varphi_n, v\right) = \sum_{n=N+1}^{\infty}(u,\varphi_n)(\varphi_n, v)$$

and the right hand side tends to zero as $N \to \infty$, due to the convergent Fourier series. Thus $u - Pu \perp H_N$ for all $v \in H_N$.

Finally, we use the above result to estimate the approximation error $\|u - v\|_{H_N}$, where $v \in H_N$ is an arbitrary element. We have:

$$\|u - v\|_{H_N}^2 := (u - v, u - v) = (u - Pu + Pu - v, u - Pu + Pu - v)$$

$$= \|u - Pu\|_{H_N}^2 + \|v - Pu\|_{H_N}^2,$$

by (2.1.8). The right hand side has minimum value when $v = Pu$, i.e. $v \in H_N$ is a projection of $u \in H$. In this case we obtain:

2.1 Best Approximation and Orthogonal Projection

$$\inf_{v \in H_N} \|u - v\|_{H_N} = \|u - Pu\|_{H_N},$$

which is the best approximation problem. □

2.2 Range and Null-Space of Adjoint Operators

Relationships between the null-spaces and ranges of a linear operator and its adjoint play an important role in inverse problems. The results given below show that the range of a linear bounded operator can be derived via the null-space of its adjoint. Note that for the linear bounded operator $A : H \mapsto \tilde{H}$, defined between Hilbert spaces H and $\tilde{H}$, the *adjoint operator* $A^* : \tilde{H} \mapsto H$ is defined as follows:

$$(Au, v)_{\tilde{H}} = (u, A^*v)_H, \quad \forall u \in H, \ v \in \tilde{H}.$$

Theorem 2.2.1 *Let $A : H \mapsto \tilde{H}$ be a linear bounded operator, defined between Hilbert spaces H and $\tilde{H}$, and $A^* : \tilde{H} \mapsto H$ be its adjoint. Then*
(p1) $\mathcal{R}(A)^\perp = \mathcal{N}(A^*)$;
(p2) $\overline{\mathcal{R}(A)} = \mathcal{N}(A^*)^\perp$,

where $\mathcal{R}(A)$ and $\mathcal{N}(A^*)$ are the range and null-space of the operators A and A^*, correspondingly.

Proof Let $v \in \mathcal{N}(A^*)$. Then $A^*v = 0$, and for all $u \in H$ we have:

$$0 = (u, A^*v)_H = (Au, v)_{\tilde{H}}.$$

This implies that $v \in \mathcal{R}(A)^\perp$, i.e. $\mathcal{N}(A^*) \subset \mathcal{R}(A)^\perp$. Now suppose $v \in \mathcal{R}(A)^\perp$. Then $(Au, v)_{\tilde{H}} = 0$, for all $u \in H$. Hence $0 = (Au, v)_{\tilde{H}} = (u, A^*v)_H$, for all $u \in H$, which means $v \in \mathcal{N}(A^*)$. This implies $\mathcal{R}(A)^\perp \subset \mathcal{N}(A^*)$.

To prove **(p2)** let us assume first that $v \in \overline{\mathcal{R}(A)}$ is an arbitrary element. Then there exists such a sequence $\{v_n\} \in \overline{\mathcal{R}(A)}$ that

$$v_n = Au_n \text{ and } \lim_{n \to \infty} v_n = v.$$

Assuming $w \in \mathcal{N}(A^*)$ we conclude that $A^*w = 0$, so

$$(v_n, w)_{\tilde{H}} = (Au_n, w)_{\tilde{H}} = (u_n, A^*w)_H = 0.$$

Hence

$$|(v, w)| \leq |(v - v_n, w)| + |(v_n, w)| \leq \|(v - v_n, w)\| \|w\|.$$

The right hand side tends to zero as $n \to \infty$, which implies $(v, w) = 0$, for all $w \in \mathcal{N}(A^*)$, i.e. $v \in \mathcal{N}(A^*)^\perp$. Therefore $\overline{\mathcal{R}(A)} \subset \mathcal{N}(A^*)^\perp$. To prove $\mathcal{N}(A^*)^\perp \subset \overline{\mathcal{R}(A)}$, we need to prove that if $v \notin \overline{\mathcal{R}(A)}$, then $v \notin \mathcal{N}(A^*)^\perp$. Since $\overline{\mathcal{R}(A)}$ is a closed subspace of the Hilbert space H, by Corollary 2.1.1 there exists a unique decomposition of the above defined element $v \in H$:

$$v = v_0 + w_0, \quad v_0 \in \overline{\mathcal{R}(A)}, \quad w_0 \in \mathcal{R}(A)^\perp,$$

with $v_0 := Pv$. Then $(v, w_0) := (v_0 + w_0, w_0) = \|w_0\|^2 \neq 0$. But by (**p1**), $w_0 \in \mathcal{N}(A^*)$, so $(v, w_0) \neq 0$, which means that $v \notin \mathcal{N}(A^*)^\perp$. This completes the proof. □

Lemma 2.2.1 *Let $A : H \mapsto \tilde{H}$ be a bounded linear operator. Then $(A^*)^* = A$ and $\|A\|^2 = \|A^*\|^2 = \|AA^*\| = \|A^*A\|$.*

Proof We can easily show that $(A^*)^* = A$. Indeed, for all $u \in H$, $v \in \tilde{H}$,

$$\left(v, \left(A^*\right)^* u\right)_{\tilde{H}} = (A^*v, u)_H = \overline{(u, A^*v)_H} = \overline{(Au, v)_{\tilde{H}}} = (v, Au)_{\tilde{H}}.$$

Further, it follows from the definition $\|A^*v\|^2 := (A^*v, A^*v)_H$, $v \in \tilde{H}$, that $\|A^*v\|^2 = (AA^*v, v)_{\tilde{H}} \leq \|A\|\|A^*v\|\|v\|$, and hence $\|A^*v\| \leq \|A\|\|v\|$, which implies boundedness of the adjoint operator: $\|A^*\| \leq \|A\|$. In the same way we can deduce $\|A\| \leq \|A^*\|$, interchanging the roles of the operators A and A^*. Therefore, $\|A^*\| = \|A\|$.

To prove the second part of the lemma, we again use the definition $\|Au\|^2 := (Au, Au)_{\tilde{H}}$, $u \in H$. Then, $\|Au\|^2 = (A^*Au, u)_H \leq \|A^*Au\|\|u\|$, and we get $\|A\|^2 \leq \|A^*A\|$. On the other hand, $\|A^*A\| \leq \|A^*\|\|A\| = \|A\|^2$, since $\|A^*\| = \|A\|$. Thus $\|A\|^2 = \|A^*A\|$. □

Corollary 2.2.1 *Let conditions of Theorem 2.2.1 hold. Then*
(**c1**) $\mathcal{N}(A^*) = \mathcal{N}(AA^*)$;
(**c2**) $\overline{\mathcal{R}(A)} = \overline{\mathcal{R}(AA^*)}$.

Proof Let $v \in \mathcal{N}(A^*)$. Then $A^*v = 0$ and hence $AA^*v = 0$, which implies $v \in \mathcal{N}(AA^*)$, i.e. $\mathcal{N}(A^*) \subset \mathcal{N}(AA^*)$. Suppose now $v \in \mathcal{N}(AA^*)$. Then $AA^*v = 0$ and $\|A^*v\|_H^2 := (A^*v, A^*v)_H = (AA^*v, v)_{\tilde{H}} = 0$. This implies $A^*v = 0$, i.e. $v \in \mathcal{N}(A^*)$, which completes the proof of (**c1**). To prove (**c2**) we use the formula $(AA^*)^* := A^{**}A^* = AA^*$ and the second relationship (**p2**) in Theorem 2.2.1, replacing here A by AA^*. We have $\overline{\mathcal{R}(AA^*)} = \mathcal{N}(AA^*)^\perp$. Taking into account here (**c1**) we conclude $\overline{\mathcal{R}(AA^*)} = \mathcal{N}(A^*)^\perp$. With the relationship (**p1**) in Theorem 2.2.1, this completes the proof. □

Remark that if A is a bounded linear operator defined on a Hilbert space H, then AA^* and A^*A are positive.

Now we briefly show a crucial role of adjoint operators in studying the solvability of linear equations. Let $A : H \mapsto \tilde{H}$ be a bounded linear operator. Consider the linear operator equation

2.2 Range and Null-Space of Adjoint Operators

$$Au = f, \ u \in H, \ f \in \tilde{H}. \tag{2.2.1}$$

Denote by $v \in \tilde{H}$ a solution of the homogeneous adjoint equation $A^*v = 0$. Then we have:

$$(f, v)_{\tilde{H}} := (Au, v)_{\tilde{H}} = (u, A^*v)_H = 0.$$

Hence $(f, v) = 0$ for all $v \in \mathcal{N}(A^*)$, and by $\tilde{H} = \mathcal{N}(A^*)^\perp \oplus \mathcal{N}(A^*)$, this implies: $f \in \mathcal{N}(A^*)^\perp$. On the other hand, Fredholm alternative asserts that Eq. (2.2.1) has a (non-unique) solution if and only if $f \perp v$ for each solution $v \in \tilde{H}$ of the homogeneous adjoint equation $A^*v = 0$. This leads to the following result.

Proposition 2.2.1 *Let $A: H \mapsto \tilde{H}$ be a bounded linear operator on a Hilbert space H. Then a necessary condition for existence of a solution $u \in H$ of Eq. (2.2.1) is the condition*

$$f \in \mathcal{N}(A^*)^\perp. \tag{2.2.2}$$

Using Theorem 2.2.1 we can write H as the orthogonal (direct) sum

$$H = \overline{\mathcal{R}(A)} \oplus \mathcal{N}(A^*). \tag{2.2.3}$$

If the range $\mathcal{R}(A)$ of A is closed in H, that is, if $\overline{\mathcal{R}(A)} = \mathcal{R}(A)$, then using (2.2.3) and Proposition 2.2.1, we obtain the following necessary and sufficient condition for the solvability of Eq. (2.2.1).

Theorem 2.2.2 *Let $A: H \mapsto \tilde{H}$ be a bounded linear operator with closed range. Then Eq. (2.2.1) has a solution $u \in H$ if and only if condition (2.2.2) holds.*

This theorem provides a useful tool for proving existence of a solution of the closed range operator Eq. (2.2.1) via the null-space of the adjoint operator.

2.3 Moore-Penrose Generalized Inverse

Let $A: H \mapsto \tilde{H}$ be a bounded linear operator between the real Hilbert spaces H and $\tilde{H}$. Consider the operator Eq. (2.2.1). Evidently, a solution of (2.2.1) exists if and only if $f \in \mathcal{R}(A) \subset \tilde{H}$. This means that the first condition (**i1**) of Hadamard's Definition 1.1.1 holds. Assume now that $f \in \tilde{H}$ does not belong to the range $\mathcal{R}(A)$ of the operator A which usually appears in applications. It is natural to extend the notion of solution for this case, looking for an approximate (or generalized) solution of (2.2.1) which satisfies this equation as well as possible. For this aim we introduce the *residual* $\|f - Au\|_{\tilde{H}}$ and then look for an element $u \in H$, as in Definition 2.1.1, which minimizes this norm:

$$\|f - Au\|_{\tilde{H}} = \inf_{v \in H} \|f - Av\|_{\tilde{H}}. \tag{2.3.1}$$

The minimum problem (2.3.1) is called a *least squares problem* and accordingly, the best approximation $u \in H$ is called a *least squares solution* to (2.2.1).

Let us consider first the minimum problem (2.3.1) from differential calculus viewpoint. Introduce the functional

$$\mathcal{J}(u) = \frac{1}{2}\|Au - f\|_{\tilde{H}}^2, \ u \in H.$$

Using the identity

$$\mathcal{J}(u+h) - \mathcal{J}(u) = 2\left(A^*(Au - f), h\right) + \|Au\|_H^2, \ \forall h \in H,$$

we obtain the Fréchet differential

$$\left(\mathcal{J}'(u), h\right) = 2\left(A^*(Au - f), h\right), \ h \in H$$

of this functional. Hence the least squares solution $u \in H$ of the operator Eq. (2.2.1) is defined from the condition $\left(\mathcal{J}'(v), h\right) = 0$, for all $h \in H$, as follows: $(A^*(Au - f), h) = 0$, i.e. as a solution of the equation

$$A^*Au = A^*f. \tag{2.3.2}$$

This shows that least squares problem (2.3.1) is equivalent to Eq. (2.3.2) with the formal solution

$$u = \left(A^*A\right)^{-1} A^* f, \ f \in \tilde{H}. \tag{2.3.3}$$

Equation (2.3.2) is called the *normal equation*.

The normal equation plays an important role in studying ill-posed problems as we will see in next sessions. First of all, remark that the operator A in (2.2.1) may not be injective, which means non-uniqueness in view of Hadamard's definition. The first important property of the normal Eq. (2.3.2) is that the operator A^*A is injective on the range $\mathcal{R}(A^*)$ of the adjoint operator A^*, even if the bounded linear operator A is not injective. For this reason, the normal equation is the most appropriate one to obtain a least squares solution of an inverse problem.

Lemma 2.3.1 *Let $A : H \mapsto \tilde{H}$ be a bounded linear operator and $H, \tilde{H}$ Hilbert spaces. Then the operator $A^*A : \mathcal{R}(A^*) \subset H \mapsto H$ is injective.*

Proof Note, first of all, that $\overline{\mathcal{R}(A^*)} = \overline{\mathcal{R}(A^*A)}$, as it follows from Corollary 2.2.1 (replacing A by A^*). Let $u \in \mathcal{R}(A^*)$ be such an element that $A^*Au = 0$. Then $Au \in \mathcal{N}(A^*)$, by definition of the null-space. But $\mathcal{N}(A^*) = \mathcal{R}(A)^\perp$, due to Theorem 2.2.1. On the other hand, $Au \in \mathcal{R}(A)$, since Au is the image of the element $u \in H$ under the transformation A. Thus, $Au \in \mathcal{R}(A) \cap \mathcal{R}(A)^\perp$ which implies $Au = 0$. This in

2.3 Moore-Penrose Generalized Inverse

turn means that $u \in \mathcal{N}(A) = \mathcal{R}(A^*)^\perp$. With the above assumption $u \in \mathcal{R}(A^*)$, we conclude that $u \in \mathcal{R}(A^*) \cap \mathcal{R}(A^*)^\perp$, i.e. $u = 0$. Therefore $A^*Au = 0$ implies $u = 0$, which proves the injectivity of the operator A^*A. □

Let us explain now an interpretation of the "inverse operator" $(A^*A)^{-1}A^*$ in (2.3.3), in view of the orthogonal projection.

As noted above, $f \in \tilde{H}$ may not belong to the range $\mathcal{R}(A)$ of the operator A. So, we assume that $f \in \tilde{H} \setminus \mathcal{R}(A)$ is an arbitrary element and try *to construct a unique linear extension of an "inverse operator" from $\mathcal{R}(A)$ to the subspace $\mathcal{R}(A) \oplus \mathcal{R}(A)^\perp$*. Since the closure $\overline{\mathcal{R}(A)}$ of the range $\mathcal{R}(A)$ is a closed subspace of $\tilde{H}$, by Corollary 2.1.2, $\tilde{H} = \overline{\mathcal{R}(A)} \oplus \mathcal{R}(A)^\perp$ (note that $\overline{\mathcal{R}(A)}^\perp = \overline{\mathcal{R}(A)^\perp} = \mathcal{R}(A)^\perp$). Hence $\mathcal{R}(A) \oplus \mathcal{R}(A)^\perp$ is *dense* in $\tilde{H}$. By the same corollary, the projection Pf of the arbitrary element $f \in \mathcal{R}(A) \oplus \mathcal{R}(A)^\perp$ is in $\mathcal{R}(A)$. This means that there exists such an element $u \in H$ that

$$Au = Pf, \quad f \in \mathcal{R}(A) \oplus \mathcal{R}(A)^\perp. \tag{2.3.4}$$

By the Orthogonal Decomposition, the element Au being the projection of f onto $\overline{\mathcal{R}(A)}$, is an element of $\mathcal{R}(A)$. Furthermore, the element $u \in H$ is a least squares solution to (2.2.1), i.e. is a solution of the minimum problem (2.3.1). This implies that a least squares solution $u \in H$ of (2.2.1) exists if and only if f is an element of the *dense* in $\tilde{H}$ subspace $\mathcal{R}(A) \oplus \mathcal{R}(A)^\perp$.

On the other hand, for each element $f \in \overline{\mathcal{R}(A)} \oplus \mathcal{R}(A)^\perp$ the following (unique) decomposition holds:

$$f = Pf + h, \quad Pf \in \overline{\mathcal{R}(A)}, \ h \in \mathcal{R}(A)^\perp. \tag{2.3.5}$$

Hence for each projection $Pf \in \mathcal{R}(A)$ we have $f - Pf \in \mathcal{R}(A)^\perp$. Taking into account (2.3.4), we conclude from (2.3.5) that

$$f - Au \in \mathcal{R}(A)^\perp. \tag{2.3.6}$$

But $\mathcal{R}(A)^\perp = \mathcal{N}(A^*)$, by Theorem 2.2.1. Hence

$$f - Au \in \mathcal{N}(A^*). \tag{2.3.7}$$

By definition of the null-space, (2.3.7) implies that $A^*(f - Au) = 0$. Thus, again we arrive at the same result: $u \in H$ satisfies the normal Eq. (2.3.2).

Therefore we have constructed *a mapping $A^\dagger$ from $\mathcal{R}(A) \oplus \mathcal{R}(A)^\perp$ into H, which associates each element $f \in \mathcal{R}(A) \oplus \mathcal{R}(A)^\perp$ to the least squares solution $u \in H$ of the operator equation* (2.2.1). Furthermore, the domain $\mathcal{D}(A)^\dagger := \mathcal{R}(A) \oplus \mathcal{R}(A)^\perp$ of this mapping is obtained in a natural way.

This mapping is called the *Moore-Penrose (generalized) inverse* of the bounded linear operator A and is denoted by $A^\dagger$:

$$A^\dagger : \mathcal{R}(A) \oplus \mathcal{R}(A)^\perp \mapsto H. \tag{2.3.8}$$

Evidently, the generalized inverse is a densely defined linear operator, that is, $\overline{\mathcal{D}(A^\dagger)} = \tilde{H}$, since a least squares solution exists only if f is an element of the dense in $\tilde{H}$ subspace $\mathcal{R}(A) \oplus \mathcal{R}(A)^\perp$.

To complete this definition, let us answer the question: the operator $A^\dagger$ is an inverse of which operator? First of all, the normal Eq. (2.3.2) shows that a least squares solution exists if and only if $\mathcal{N}(A^*A) = \{0\}$, or equivalently, $\mathcal{N}(A) = 0$, due to $\mathcal{N}(A^*A) = \mathcal{N}(A)$, by Corollary 2.2.1. Since $A^*A : H \mapsto H$ is a self-adjoint operator, it follows from Theorem 2.2.1 and Corollary 2.2.1 that

$$\begin{aligned} H &:= \overline{\mathcal{R}(A^*A)} \oplus \mathcal{R}(A^*A)^\perp \\ &= \overline{\mathcal{R}(A^*)} \oplus \mathcal{N}(A^*A)^\perp \\ &= \mathcal{N}(A)^\perp \oplus \mathcal{N}(A) \end{aligned} \tag{2.3.9}$$

The last line of decompositions (2.3.9) shows that to ensure the existence, we need to restrict the domain of the linear operator $A : H \mapsto \tilde{H}$ from $\mathcal{D}(A)$ to $\mathcal{N}(A)^\perp$. By this way, we define this operator as follows:

$$\mathring{A} := A|_{\mathcal{N}(A)^\perp}, \quad \mathring{A} : \mathcal{N}(A)^\perp \subset H \mapsto \mathcal{R}(A) \subset \tilde{H}.$$

It follows from this construction that $\mathcal{N}(\mathring{A}) = \{0\}$ and $\mathcal{R}(\mathring{A}) = \mathcal{R}(A)$. Therefore, the inverse operator

$$\mathring{A}^{-1} : \mathcal{R}(A) \subset \tilde{H} \mapsto \mathcal{N}(A)^\perp \subset H \tag{2.3.10}$$

exists. However, the range $\mathcal{R}(\mathring{A}^{-1})$ of this inverse operator is in $\tilde{H}$ and does not contain the elements $f \in \tilde{H} \setminus \mathcal{R}(A)$. For this reason, at the second stage of the above construction, we extended this range from $\mathcal{R}(A)$ to $\mathcal{R}(A) \oplus \mathcal{R}(A)^\perp$, in order include those elements which may not belong to $\mathcal{R}(A)$.

Thus, following to [23], we can define the *Moore-Penrose inverse* $A^\dagger$ as follows.

Definition 2.3.1 The Moore-Penrose (generalized) inverse of a bounded linear operator A is the unique linear extension of the inverse operator (2.3.10) from $\mathcal{R}(A)$ to $\mathcal{R}(A) \oplus \mathcal{R}(A)^\perp$:

$$A^\dagger : \mathcal{R}(A) \oplus \mathcal{R}(A)^\perp \mapsto \mathcal{N}(A)^\perp \subset H, \tag{2.3.11}$$

with

$$\mathcal{N}(A^\dagger) = \mathcal{R}(A)^\perp. \tag{2.3.12}$$

The requirement (2.3.12) in this definition is due to (2.3.6) and (2.3.7). This requirement implies, in particular, that the Moore-Penrose inverse $A^\dagger$ is a linear operator.

2.3 Moore-Penrose Generalized Inverse

Corollary 2.3.1 *Let $f \in \mathcal{D}(A^\dagger)$. Then $u \in H$ is a least squares solution of the operator equation $Au = f$ if and only if it is a solution of the normal Eq. (2.3.2).*

Proof It follows from (2.3.1) that $u \in H$ is a least squares solution of $Au = f$ if and only if Au is the closest element to f in $\mathcal{R}(A)$. The last assertion is equivalent to (2.3.6), i.e. $f - Au \in \mathcal{R}(A)^\perp$. By Theorem 2.2.1, $\mathcal{R}(A)^\perp = \mathcal{N}(A^*)$. Hence $f - Au \in \mathcal{N}(A^*)$, which means $A^*(f - Au) = 0$ or $A^*Au = A^*f$. $\square$

The following theorem shows that the Moore-Penrose inverse of a bounded linear operator is a closed operator. Remark that a linear operator $L : H_1 \mapsto H_2$ is closed if and only if for any sequence $\{u_n\} \subset \mathcal{D}(L)$, satisfying

$$\lim_{n \to \infty} u_n = u, \text{ and } \lim_{n \to \infty} Lu_n = v,$$

the conditions hold:

$$u \in \mathcal{D}(L), \text{ and } v = Lu. \tag{2.3.13}$$

Theorem 2.3.1 *Let $A : H \mapsto \tilde{H}$ be a linear bounded operator from the Hilbert spaces H into $\tilde{H}$. Then the Moore-Penrose inverse $A^\dagger$, defined by (2.3.11) and (2.3.12), is a closed operator.*

Proof Let $\{f_n\} \subset \mathcal{D}(A^\dagger)$, $n = 1, 2, 3, \ldots$, $f_n \to f$, and $A^\dagger f_n \to u$, as $n \to \infty$. Denote by $u_n := A^\dagger f_n$ the unique solution of the normal Eq. (2.3.2) for each n, that is, $A^*Au_n = A^*f_n$, $u_n \in \mathcal{N}(A)^\perp$. Since $\mathcal{N}(A)^\perp$ is closed, $\{u_n\} \subset \mathcal{N}(A)^\perp$ and $u_n \to u$, as $n \to \infty$, which implies $u \in \mathcal{N}(A^\perp)$, i.e. $u \in \mathcal{R}(A^\dagger)$, by (2.3.11). Hence, the first condition of (2.3.13) holds. Now we prove that $u = A^\dagger f$. Due to the continuity of the operators A^*A and A^* we have:

$$A^*Au_n \to A^*Au, \text{ and } A^*Au_n = A^*f_n \to A^*f, \text{ as } n \to \infty.$$

The left hand sides are equal, so $A^*Au = A^*f$, i.e. $u \in \mathcal{N}(A)^\perp$ is the solution of the normal Eq. (2.3.3). By Corollary 2.3.1, $u = A^\dagger f$. This completes the proof. $\square$

Corollary 2.3.2 *Let conditions of Theorem 2.3.1 hold. Then the Moore-Penrose inverse $A^\dagger$ is continuous if and only if $\mathcal{R}(A)$ is closed.*

Proof Let $A^\dagger$, defined by (2.3.11) and (2.3.12), be a continuous operator. Assume that $\{f_n\} \subset \mathcal{R}(A)$ be a convergent sequence: $f_n \to f$, as $n \to \infty$. We need to prove that $f \in \mathcal{R}(A)$. Denote by $u_n := A^\dagger f_n$. Then $u_n \in \mathcal{N}(A)^\perp$, for all $n = 1, 2, 3, \ldots$. Since $A^\dagger$ is continuous and $\mathcal{N}(A)^\perp$ is closed we conclude:

$$u := \lim_{n \to \infty} u_n = \lim_{n \to \infty} A^\dagger f_n = A^\dagger f, \text{ and } u \in \mathcal{N}(A)^\perp.$$

On the other hand, $Au_n = f_n$ and $A : H \mapsto \tilde{H}$ is a continuous operator, so $Au = f$, where f is the above defined limit of the sequence $\{f_n\} \subset \mathcal{R}(A)$. But the element

Au, being the projection of f onto $\overline{\mathcal{R}(A)}$, is an element of $\mathcal{R}(A)$, by (2.3.4). Hence, $f \in \mathcal{R}(A)$, which implies that $\mathcal{R}(A)$ is closed.

To prove the second part of the corollary, we assume now $\mathcal{R}(A)$ is closed. By definition (2.3.11), $\mathcal{D}(A^\dagger) := \mathcal{R}(A) \oplus \mathcal{R}(A)^\perp$, which implies $\mathcal{D}(A^\dagger)$ is closed, since the orthogonal complement $\mathcal{R}(A)^\perp$ is a closed subspace. As a consequence, the graph $G_{A^\dagger} := \{(f, A^\dagger f) : f \in \mathcal{D}(A^\dagger)\}$ of the operator $A^\dagger$ is closed. Then, as a closed graph linear operator, $A^\dagger$ is continuous. □

Remember that in the case when A is a linear compact operator, the range $\mathcal{R}(A)$ is closed if and only if it is finite-dimensional.

Remark, finally, that the notion of generalized inverse has been introduced by E. H. Moore and R. Penrose [69, 80, 81]. For ill-posed problems this very useful concept has been developed in [23, 31].

2.4 Singular Value Decomposition

As we have seen already in the introduction, inverse problems with compact operators are an challenging case. Most inverse problems related to differential equations are represented by these operators. Indeed, all input-output operators corresponding to these inverse problems, are compact operators. Hence, *compactness of the operator A is a main source of ill-posedness* of the operator equation (2.2.1) and our interest will be directed towards the case $A : H \mapsto \tilde{H}$ in (2.2.1) is a linear compact operator. When A is a self-adjoint compact operator, i.e. for all $u \in H$ and $v \in \tilde{H}$, $(Au, v) = (u, Av)$, we may use the *spectral representation*

$$Au = \sum_{n=1}^{\infty} \lambda_n (u, u_n) u_n, \ \forall u \in H, \tag{2.4.1}$$

where λ_n, $n = 1, 2, 3, \ldots$ are nonzero real eigenvalues (repeated according to its multiplicity) and $\{u_n\} \subset H$ is the complete set of corresponding orthonormal eigenvectors u_n. The set $\{\langle \lambda_n, u_n \rangle\}$, consisting of all pairs of nonzero eigenvalues and corresponding eigenvectors, is defined an *eigensystem* of the self-adjoint operator A. It is also known from the spectral theory of self-adjoint compact operators that $\lambda_n \to 0$, as $n \to \infty$. If $\dim \mathcal{R}(A) = \infty$ then for any $\varepsilon > 0$ the index set $\{n \in \mathbb{N} : |\lambda_n| \geq \varepsilon\}$ is finite. Here and below $\mathbb{N}$ is the set of natural numbers. If the range $\mathcal{R}(A)$ of a compact operator is finite, $\lambda_n = 0$, for all $n > \dim \mathcal{R}(A)$.

However, *if A is not-self-adjoint*, there are no eigenvalues, hence no eigensystem. In this case, the notion *singular system* replaces the eigensystem. To describe this system we use the operators A^*A and AA^*. Both $A^*A : H \mapsto H$ and $AA^* : \tilde{H} \mapsto \tilde{H}$ are compact self-adjoint nonnegative operators. We denote the eigensystem of the self-adjoint operator A^*A by $\{\langle \mu_n, u_n \rangle\}$. Then $A^*Au_n = \mu_n u_n$, for all $u_n \in H$, which implies $(A^*Au_n, u_n) = \mu_n(u_n, u_n)$. Hence $\|Au_n\|_{\tilde{H}}^2 = \mu_n \|u_n\|_H^2 \geq 0$, which means all nonzero eigenvalues are positive: $\mu_n > 0$, $n \in \mathbf{N}$, where $\mathbf{N} := \{n \in \mathbb{N} : \mu_n \neq 0\}$

2.4 Singular Value Decomposition

(at most countable) index set of positive eigenvalues. We denote the square roots of the positive eigenvalues μ_n of the self-adjoint operator $A^*A : H \mapsto H$ by $\sigma_n := \sqrt{\mu_n}$, $n \in \mathbf{N}$. Below we will assume that these eigenvalues are ordered as follows: $\mu_1 \geq \mu_2 \geq \ldots \geq \mu_n \geq \ldots > 0$.

Definition 2.4.1 Let $A : H \mapsto \tilde{H}$ be a linear compact operator with adjoint $A^* : \tilde{H} \mapsto H$, H and $\tilde{H}$ be Hilbert spaces. The square root $\sigma_n := \sqrt{\mu_n}$ of the eigenvalue $\mu_n > 0$ of the self-adjoint operator $A^*A : H \mapsto H$ is called the singular value of the operator A.

Using the spectral representation (2.4.1) for the self-adjoint compact operator A^*A we have:

$$A^*Au = \sum_{n=1}^{\infty} \sigma_n^2 (u, u_n) u_n, \quad \forall u \in H. \qquad (2.4.2)$$

Let us introduce now the orthonormal system $\{v_n\}$ in $\tilde{H}$, via the orthonormal system $\{u_n\} \subset H$ as follows: $v_n := Au_n / \|Au_n\|$. Applying A^* to both sides we have: $A^*v_n = A^*Au_n / \|Au_n\|$. By the above definition $A^*Au_n = \sigma_n^2 u_n$ and $\|Au_n\| = \sigma_n$. This implies:

$$A^*v_n = \sigma_n u_n.$$

Act by the adjoint operator A^* now on both sides of the Fourier representation $v = \sum_{n=1}^{\infty}(v, v_n)v_n$, $v \in \tilde{H}$, where $v_n = Au_n / \|Au_n\|$. Taking into account the definition $A^*v_n = \sigma_n u_n$, we have:

$$A^*v = \sum_{n=1}^{\infty} \sigma_n (v, v_n) u_n, \quad v \in \tilde{H}. \qquad (2.4.3)$$

Applying A to both sides of (2.4.3) and using $\|Au_n\| = \sigma_n$ we obtain the spectral representation for the self-adjoint operator AA^*:

$$AA^*v = \sum_{n=1}^{\infty} \sigma_n^2 (v, v_n) v_n, \quad v \in \tilde{H}. \qquad (2.4.4)$$

It is seen from (2.4.2) and (2.4.4) that the eigenvalues $\sigma_n^2 > 0$, $n \in \mathbf{N}$, of the self-adjoint operators AA^* and A^*A are the same, as expected.

The representation (2.4.3) is called *singular value expansion* of the adjoint operator $A^* : \tilde{H} \mapsto H$. For the operator $A : H \mapsto \tilde{H}$ this expansion can be derived in a similar way:

$$Au = \sum_{n=1}^{\infty} \sigma_n (u, u_n) v_n, \quad u \in H. \qquad (2.4.5)$$

Substituting $v = v_n$ in (2.4.3) and $u = u_n$ in (2.4.5), we obtain the following formulae:

$$Au_n = \sigma_n v_n, \quad A^* v_n = \sigma_n u_n. \tag{2.4.6}$$

The triple $\{\sigma_n, u_n, v_n\}$ is called the *singular system for the non-self-adjoint operator* $A : H \mapsto \tilde{H}$.

As we will see in the next chapter, some input-output operators related to inverse source problems are self-adjoint. If $A : H \mapsto \tilde{H}$ is a self-adjoint operator with eigensystem $\{\langle \lambda_n, u_n \rangle\}$, then $\|Au_n\| = |\lambda_n|$ and

$$v_n := Au_n / \|Au_n\| = \lambda u_n / |\lambda_n|,$$

by the above construction. Therefore, *the singular system for the self-adjoint operator* $A : H \mapsto \tilde{H}$ is defined as the triple $\{\sigma_n, u_n, v_n\}$, with $\sigma = |\lambda_n|$ and $v_n = \lambda u_n / |\lambda_n|$.

Example 2.4.1 Singular values of a self-adjoint integral operator

Assuming $H = L^2(0, \pi)$, we define the non-self-adjoint integral operator $A : H \mapsto H$ as follows:

$$(Au)(x) := \int_x^\pi u(\xi) d\xi, \quad x \in (0, \pi), \ u \in H. \tag{2.4.7}$$

By using the integration by parts formula and the definition $(Au, v)_{L^2(0,\pi)} = (u, A^* v)_{L^2(0,\pi)}$, $\forall u, v \in H$, we can easily construct the adjoint operator $A^* : H \mapsto H$:

$$(A^* v)(x) = \int_0^x v(\xi) d\xi, \quad x \in (0, \pi), \ v \in H. \tag{2.4.8}$$

Evidently, both integral operators (2.4.7) and (2.4.8) are compact. Indeed, let $\{u_n\}_{n=1}^\infty$ be a bounded sequence in $L^2(0, \pi)$ with $\|u_n\|_{L^2(0,\pi)} \leq M$, $M > 0$. Then for any $x_1, x_2 \in [0, \pi]$, (2.4.7) implies:

$$|(Au_n)(x_1) - (Au_n)(x_2)| \leq \left| \int_{x_1}^{x_2} u_n(\xi) d\xi \right| \leq M |x_1 - x_2|^{1/2}.$$

This shows that $\{(Au)_n\}$ is an equicontinuous family of functions in $C[0, \pi]$. Hence, there exists a subsequence $\{(Au)_m\} \subset \{(Au)_n\}$ that converges uniformly in $C[0, \pi]$ to a continuous function v. Since uniform convergence implies convergence in $L^2[0, \pi]$, we conclude that the subsequence $\{(Tu)_m\}$ converges in $L^2[0, \pi]$. Therefore the integral operator defined by (2.4.7) is compact because the image of a bounded sequence always contains a convergent subsequence.

Now we define the self-adjoint integral operator $A^* A$:

2.4 Singular Value Decomposition

$$(A^*Au)(x) = \int_0^x \int_\xi^\pi u(\eta)d\eta d\xi, \quad u \in H. \tag{2.4.9}$$

To find the nonzero positive eigenvalues $\mu_n > 0$, $n \in \mathbb{N}$, of the self-adjoint integral operator A^*A, defined by (2.4.9), the eigenvalue problem should be solved for the integral equation

$$(A^*Au)(x) = \sigma^2 u(x), \quad x \in (0, \pi). \tag{2.4.10}$$

Differentiating both sides of (2.4.10) twice with respect to $x \in [0, \pi]$ we arrive at the Sturm-Liouville equation: $-u''(x) = \lambda u(x)$, $\lambda = 1/\sigma^2$. To derive the boundary conditions, we first substitute $x = 0$ in (2.4.10). Then we get $u(0) = 0$, by (2.4.9). Differentiating both sides of (2.4.10) and substituting $x = \pi$ we conclude $u'(\pi) = 0$. Hence, problem (2.4.10) is equivalent (in well-known sense) to the eigenvalue problem

$$\begin{cases} -u''(x) = \lambda u(x), & \text{a.e. } x \in (0, \pi), \ \lambda = 1/\sigma^2, \\ u(0) = u'(\pi) = 0, \end{cases} \tag{2.4.11}$$

for the self-adjoint positive-defined differential operator $Au := -u''$. The solution of this problem is in $\mathring{H}^2[0, \pi] := \{u \in H^2(0, \pi) : u(0) = 0\}$, where $H^2(0, \pi)$ is the Sobolev space.

Solving the eigenvalue problem (2.4.11) we find the eigenvalues $\lambda_n = (n - 1/2)^2$, $n \in \mathbb{N}$, and the corresponding normalized eigenvectors $u_n(x) = \sqrt{2/\pi} \sin(\sqrt{\lambda_n} x)$. Hence, the eigenvalues $\sigma_n^2 = 1/\lambda_n$ and the corresponding eigenvectors $u_n(x)$ of the self-adjoint integral operator A^*A are

$$\sigma_n^2 = (n - 1/2)^{-2}, \quad u_n(x) = \sqrt{2/\pi} \sin((n - 1/2)x).$$

By Definition 2.4.1, $\sigma_n = (n - 1/2)^{-1}$, $n \in \mathbb{N}$, are the eigenvalues of the non-self-adjoint integral operator A. The corresponding eigenvectors, given in equations (2.4.6) are

$$u_n(x) = \sqrt{2/\pi} \sin((n - 1/2)x), \quad v_n(x) = \sqrt{2/\pi} \cos((n - 1/2)x).$$

Thus, the singular system $\{\sigma_n, u_n, v_n\}$ for the non-self-adjoint integral operator (2.4.7) is defined as follows:

$$\{(n - 1/2)^{-1}, \sqrt{2/\pi} \sin((n - 1/2)x), \sqrt{2/\pi} \cos((n - 1/2)x)\}. \quad \square$$

Remark 2.4.1 The above example insights into the degree of ill-posedness of simplest integral equations $Au = f$ and $A^*Au = A^*f$, with the operators A and A^*A, defined by (2.4.7) and (2.4.9). In the first case one needs an operation differentiation (which is an ill-posed procedure) to find $u = A^{-1}f$. As a result, $\sigma_n = \mathcal{O}(n^{-1})$. In the second case the operator A^*A, defined by (2.4.8), contains two integration and hence one needs to differentiate twice to find $u = (A^*A)^{-1}A^*f$, which results in the singular values as $\mathcal{O}(n^{-2})$. We come back to this issue in the next chapter.

The above considerations lead to so-called *Singular Value Decomposition* (or normal form) of compact operators.

Theorem 2.4.1 (Picard) Let H and $\tilde{H}$ be Hilbert spaces and $A : H \mapsto \tilde{H}$ be a linear compact operator with the singular system $\{\sigma, u_n, v_n\}$. Then the equation $Au = f$ has a solution if and only if

$$f \in \mathcal{N}(A^*)^\perp \text{ and } \sum_{n=1}^{\infty} \frac{1}{\sigma_n^2}|(f, v_n)|^2 < +\infty. \qquad (2.4.12)$$

In this case

$$u := A^\dagger f = \sum_{n=1}^{\infty} \frac{1}{\sigma_n}(f, v_n)u_n \qquad (2.4.13)$$

is the solution of the equation $Au = f$.

Proof Let the equation $Au = f$ has a solution. Then f must be in $\overline{\mathcal{R}(A)}$. But by Theorem 2.2.1, $\overline{\mathcal{R}(A)} = \mathcal{N}(A^*)^\perp$. Hence $f \in \mathcal{N}(A^*)^\perp$ and the first part of (2.4.12) holds. To prove the second part of (2.4.12) we use the relation $A^*v_n = \sigma_n u_n$ in (2.4.6) to get

$$\sigma_n(u, u_n) = (u, A^*v_n) = (Au, v_n) = (f, v_n).$$

Hence, $(u, u_n) = (f, v_n)/\sigma_n$. Using this in

$$u = \sum_{n=1}^{\infty}(u, u_n)u_n, \ u \in H \qquad (2.4.14)$$

we obtain:

$$u = \sum_{n=1}^{\infty} \frac{1}{\sigma_n}(f, v_n)u_n.$$

But the orthonormal system $\{u_n\}$ is complete, so the Fourier series (2.4.14) is convergent. By the convergence criterion this implies the second condition of (2.4.12):

$$\sum_{n=1}^{\infty} \frac{1}{\sigma_n^2}|(f, v_n)|^2 = \sum_{n=1}^{\infty}|(u, u_n)|^2 < +\infty.$$

To prove the second part of the theorem, we assume now that conditions (2.4.12) hold. Then series (2.4.13) converges. Acting on both sides of this series by the

2.4 Singular Value Decomposition

operator A, using $f \in \mathcal{N}(A^*)^\perp = \overline{\mathcal{R}(A)}$ and $Au_n = \sigma_n v_n$ we get:

$$Au = \sum_{n=1}^{\infty} \frac{1}{\sigma_n}(f, v_n) Au_n$$

$$\sum_{n=1}^{\infty}(f, v_n)v_n = f.$$

This completes the proof. □

Since $\mu_n > 0$, $n \in \mathbb{N}$ are eigenvalues of the self-adjoint operator A^*A (as well as AA^*) and $\sigma_n := \sqrt{\mu_n}$, we have: $\sigma_n \to 0$, as $n \to \infty$, if $\dim \mathcal{R}(A) = \infty$. Then it follows from formulae (2.4.12)-(2.4.13) that $A^\dagger$ is an unbounded operator. Indeed, for any fixed eigenvector v_k, with $\|v_k\| = 1$, we have:

$$\|A^\dagger v_k\| = \frac{1}{\sigma_k} \to \infty, \text{ as } n \to \infty.$$

The second condition (2.4.12), called *Picard criterion*, shows that the best approximate solution of the equation $Au = f$ exists if only the Fourier coefficients (f, v_n) of f decay faster than the singular values σ_n. Concrete examples related to this issue will be given in the next chapter.

As noted in Remark 2.4.1, singular value decomposition reflects the ill-posedness of the equation $Au = f$ with a compact operator A between the *infinite dimensional* Hilbert spaces H and $\tilde{H}$. Indeed, the decay rate of the non-increasing sequence $\{\sigma_n\}_{n=1}^{\infty}$ characterizes the *degree of ill-posedness* of an ill-posed problem. In particular, the amplification factors of a measured data errors in nth Fourier component of the series (2.4.13), corresponding to the integral operators (2.4.7) and (2.4.9), increase as n and n^2, respectively, due to the factor $1/\sigma_n$. In terms of corresponding problems $Au = f$ and $A^*Au = A^*f$ this means that the second problem is more ill-posed than the first one. Hence, solving numerically the second ill-posed problem is more difficult than the first one.

These considerations motivate the following definition of ill-posedness of problems governed by compact operators, proposed in [43].

Definition 2.4.2 Let $A : H \mapsto \tilde{H}$ be a linear compact operator between the infinite dimensional Hilbert spaces H and $\tilde{H}$. If there exists a constant $C > 0$ and a real number $s \in (0, \infty)$ such that

$$\sigma_n \geq \frac{C}{n^s}, \text{ for all } n \in \mathbf{N}, \quad (2.4.15)$$

then the equation $Au = f$ is called moderately ill-posed of degree at most s. If for any $\epsilon > 0$, condition (2.4.15) does not hold with s replaced by $s - \epsilon > 0$, then the equation $Au = f$ is called moderately ill-posed of degree s. If no such number

$s \in (0, \infty)$ exists such that condition (2.4.15) holds, then the equation $Au = f$ is called severely ill-posed.

Typical behavior of severe ill-posedness is exponential decay of the singular values of the compact operator A. As we will show in the next chapter, classical backward parabolic problem is severely ill-posed, whereas the final data inverse source problems related to parabolic and hyperbolic equations are only moderately ill-posed. Remark that some authors use more detailed classification, distinguishing between *mildly ill-posedness* ($s \in (0, 1)$) and moderately ill-posedness ($s \in (1, \infty)$).

In applications, to obtain an approximation of $A^\dagger f$ one can truncate the series (2.4.13):

$$u^N := \sum_{n=1}^{N} \frac{1}{\sigma_n}(f, v_n)u_n. \tag{2.4.16}$$

This method of obtaining the *approximate solution* u^N is called the *truncated singular value decomposition* (TSVD).

To understand the role of the *cutoff parameter* N, we assume that the right hand side $f \in H$ of the equation $Au = f$ is given with some *measurement error* $\delta > 0$, i.e. $\|f - f^\delta\| \leq \delta$, where f^δ is a *noisy data*. Then

$$u^{N,\delta} := \sum_{n=1}^{N} \frac{1}{\sigma_n}(f^\delta, v_n)u_n \tag{2.4.17}$$

is an approximate solution of the equation $Au = f^\delta$ corresponding to the noisy data f^δ. Let us estimate the norm $\|u^N - u^{N,\delta}\|$, i.e. the difference between the approximate solutions corresponding to the noise free (f) and noisy (f^δ) data. From (2.4.16)-(2.4.17) we deduce the estimate:

$$\|u^N - u^{N,\delta}\|^2 = \sum_{n=1}^{N} \frac{1}{\sigma_n^2}\left|(f - f^\delta, v_n)\right|^2$$

$$\leq \frac{1}{\sigma_N^2} \sum_{n=1}^{N} \left|(f - f^\delta, v_n)\right|^2 \leq \frac{\delta^2}{\sigma_N^2}.$$

Using this estimate we can find the *accuracy error* $\|u^{N,\delta} - A^\dagger f\|$, i.e. the difference between the best approximate solution $A^\dagger f$, corresponding to the noise free data f, and the approximate solution $u^{N,\delta}$, obtained by TSVD and corresponding to the noisy data f^δ:

$$\|u^{N,\delta} - A^\dagger f\| \leq \|u^N - A^\dagger f\| + \|u^N - u^{N,\delta}\|$$

$$\leq \|u^N - A^\dagger f\| + \frac{\delta}{\sigma_N}. \tag{2.4.18}$$

2.4 Singular Value Decomposition

The first term $\|u^N - A^\dagger f\|$ on the right hand side of estimate (2.4.18) depends only on the cutoff parameter N and does not depend on the measurement error $\delta > 0$. This term is called the *regularization error*. The second term $\|u^N - u^{N,\delta}\|$ on the right hand side of (2.4.18) depends not only on the cutoff parameter N, but also on the measurement error $\delta > 0$. This term is called the *data error*. This term exhibits some very distinctive features of a solution of the ill-posed problems. Namely, the approximation error $\|u^{N,\delta} - A^\dagger f\|$ decreases with $\delta > 0$, for a fixed value of the cutoff parameter N, on one hand. On the other hand, for a given $\delta > 0$ this error tends to infinity, as $N \to \infty$, since $\sigma_N := \sqrt{\mu_N} \to 0$. Hence, the parameter $N = N(\delta)$ needs to be chosen depending on $\delta > 0$ such that

$$\frac{\delta}{\sigma_{N(\delta)}} \to 0, \text{ as } \delta \to 0. \tag{2.4.19}$$

We use the right hand side of (2.4.17) to introduce the operator $R_{N(\delta)} : \tilde{H} \mapsto H$,

$$R_{N(\delta)} f^\delta := \sum_{n=1}^{N(\delta)} \frac{1}{\sigma_n} (f^\delta, v_n) u_n. \tag{2.4.20}$$

It follows from estimate (2.4.18) that if condition (2.4.19) holds, then

$$\|R_{N(\delta)} f^\delta - A^\dagger f\| \to 0, \ \delta \to 0.$$

Hence, the basic idea of TSVD in solving ill-posed problems is finding a finite dimensional approximation of the unbounded operator $A^\dagger$. A class of such finite dimensional operators defined by (2.4.20) and approximating the unbounded operator $A^\dagger$ can be defined as *regularization method* or *regularization strategy*. The cutoff parameter $N(\delta)$ plays role of the parameter of regularization [35].

2.5 Regularization Strategy. Tikhonov Regularization

Let $A : H \mapsto \tilde{H}$ be a linear injective bounded operator between infinite-dimensional real Hilbert spaces H and $\tilde{H}$. Consider the linear ill-posed operator equation

$$Au = f, \ u \in H, \ f \in \mathcal{R}(A). \tag{2.5.1}$$

By the condition $f \in \mathcal{R}(A)$, the operator Eq. (2.5.1) is ill-posed in the sense that a solution $u \in H$ exists, but doesn't depend continuously on the data $f \in \mathcal{R}(A)$. In practice this data always contains a random noise. We denote by $f^\delta \in \tilde{H}$ the noisy data and assume that

$$\|f^\delta - f\|_{\tilde{H}} \leq \delta, \ f \in \mathcal{R}(A), \ f^\delta \in \tilde{H}, \ \delta > 0. \tag{2.5.2}$$

Then the exact equality in the equation $Au = f^\delta$ can not be satisfied due to the noisy data f^δ and we may only consider the minimization problem

$$J(u) = \inf_{v \in H} J(v) \qquad (2.5.3)$$

for the *Tikhonov functional*

$$J(u) = \frac{1}{2} \|Au - f^\delta\|_{\tilde{H}}^2, \ u \in H, \ f^\delta \in \tilde{H}, \qquad (2.5.4)$$

where $\|Au - f^\delta\|_{\tilde{H}}^2 := (Au - f^\delta, Au - f^\delta)_{\tilde{H}}$.

A solution $u \in H$ of the minimization problem (2.5.3)–(2.5.4) is called *quasi-solution or least squares solution* of the ill-posed problem (2.5.1). If, in addition, this solution is defined as the minimum-norm solution, i.e. if

$$\|u\|_H = \inf \{\|w\|_H : w \in H \text{ is a least squares solution of } (1.5.1)\},$$

then this solution is called *best approximate solution* of (2.5.1). Note that the concept of quasi-solution has been introduced in [49].

Since H is infinite-dimensional and A is compact, the minimization problem for the functional (2.5.4) is ill-posed, the functional $J(u)$ doesn't depend continuously on the data $f^\delta \in \mathcal{R}(A)$. One of the possible ways of stabilizing the functional is to add the penalty term $\alpha \|u - u^0\|_H^2$, as in Optimal Control Theory, and then consider the minimization problem for the *regularized Tikhonov functional*

$$J_\alpha(u) := \frac{1}{2} \|Au - f^\delta\|_{\tilde{H}}^2 + \frac{1}{2} \alpha \|u - u^0\|_H^2, \ u \in H, \ f^\delta \in \tilde{H}. \qquad (2.5.5)$$

Here $\alpha > 0$ is the parameter of regularization and $u^0 \in H$ is an initial guess. Usually $u^0 \in H$ is one of the possible good approximations to the exact solution $u \in H$, but if such an initial guess is not known, we may take $u^0 = 0$. Below we assume that $u^0 = 0$.

This approach is defined as *Tikhonov regularization* [95, 96] or *Tikhonov-Phillips regularization* [83].

Theorem 2.5.1 *Let $A : H \mapsto \tilde{H}$ be a linear injective bounded operator between real Hilbert spaces H and $\tilde{H}$. Then the regularized Tikhonov functional (2.5.5) has a unique minimum $u_\alpha^\delta \in H$, for all $\alpha > 0$. This minimum is the solution of the linear equation*

$$\left(A^*A + \alpha I\right) u_\alpha^\delta = A^* f^\delta, \ u_\alpha^\delta \in H, \ f^\delta \in \tilde{H}, \ \alpha > 0 \qquad (2.5.6)$$

and has the form

$$u_\alpha^\delta = \left(A^*A + \alpha I\right)^{-1} A^* f^\delta. \qquad (2.5.7)$$

2.5 Regularization Strategy. Tikhonov Regularization

Moreover, the operator $A^*A + \alpha I$ is boundedly invertible, hence the solution u_α^δ continuously depends on f^δ.

Proof First of all, note that the Fréchet differentiability of the functional (2.5.5) follows from the identity:

$$J_\alpha(u+v) - J_\alpha(u) = \left(A^*(Au - f^\delta) + \alpha u, v\right) + \frac{1}{2}\|Av\|^2 + \frac{1}{2}\alpha\|v\|^2, \quad \forall u, v \in H,$$

where $A^* : \tilde{H} \mapsto H$ is the adjoint operator of A. This identity implies:

$$\begin{cases} (J'_\alpha(u), v) = (Au - f^\delta, Av) + \alpha(u, v), & \forall v \in H; \\ J''_\alpha(u; v, v) = \|Av\|^2 + \alpha\|v\|^2, & \forall v \in H. \end{cases} \quad (2.5.8)$$

Formula (2.5.8) for the second Fréchet derivative $J''_\alpha(u; v, v)$ shows that the regularized Tikhonov functional $J_\alpha(u)$ defined on a real Hilbert H space is strictly convex, since $\alpha > 0$, and $\lim_{\|u\|\to+\infty} J_\alpha(u) = +\infty$. Then it has a unique minimizer $u_\alpha^\delta \in H$ and this minimum is characterized by the following necessary and sufficient condition

$$(J'_\alpha(u), v) = 0, \quad \forall v \in H, \quad (2.5.9)$$

where $J'_\alpha(u)$ is the first Fréchet derivative of the regularized Tikhonov functional. Thus, condition (2.5.9) with formula (2.5.8) implies that the minimum $u_\alpha^\delta \in H$ of the regularized Tikhonov functional is the solution of the linear Eq. (2.5.6). This solution is defined by (2.5.7), since the operator $A^*A + \alpha I$ is boundedly invertible. This follows from the Lax-Milgram lemma and the positive definiteness of the operator $A^*A + \alpha I$:

$$\left((A^*A + \alpha I)v, v\right) = \|Av\|^2 + \alpha\|v\|^2 \geq \alpha\|v\|^2, \quad \forall v \in H, \; \alpha > 0.$$

Evidently, the operator $A^*A + \alpha I$ is one-to-one for each positive α. Indeed, multiplication of the homogeneous equation $(A^*A + \alpha I)v = 0$ by $v \in H$ implies: $(A^*Av, v) + \alpha(v, v) = (Av, Av) + \alpha(v, v) = 0$. This holds if and only if $v = 0$. This completes the proof. □

From (2.5.8) we deduce the *gradient formula for the regularized Tikhonov functional*.

Corollary 2.5.1 *For the Fréchet gradient $J'_\alpha(u)$ of the regularized Tikhonov functional (2.5.4) the following formula holds:*

$$J'_\alpha(u) = A^*\left(Au - f^\delta\right) + \alpha u, \quad u \in H. \quad (2.5.10)$$

The main consequence of the Picard's Theorems 2.4.1 and 2.5.1 is that the solution u_α^δ of the normal Eq. (2.5.6) corresponding to the noisy data f^δ can be represented by the following series:

$$u_\alpha^\delta = \sum_{n=1}^{\infty} \frac{q(\alpha; \sigma_n)}{\sigma_n} (f^\delta, v_n) u_n, \quad \alpha > 0, \qquad (2.5.11)$$

where

$$q(\alpha; \sigma) = \frac{\sigma^2}{\sigma^2 + \alpha} \qquad (2.5.12)$$

is called the *filter function*.

Corollary 2.5.2 *Let conditions of Theorem 2.4.1 hold and $f^\delta \in \mathcal{N}(A^*)^\perp$. Then the unique regularized solution $u_\alpha^\delta \in H$, given by (2.5.7), can be represented as the convergent series (2.5.11).*

Proof It follows from the normal Eq. (2.5.6) that $\alpha u_\alpha^\delta = A^* f^\delta - A^* A u_\alpha^\delta$. By Corollary 2.2.1, $\overline{\mathcal{R}(A^*A)} = \overline{\mathcal{R}(A^*)}$, so this implies that $u_\alpha^\delta \in \overline{\mathcal{R}(A^*)}$. But due to Theorem 2.2.1, $\overline{\mathcal{R}(A^*)} = \mathcal{N}(A)^\perp$ and we conclude that $u_\alpha^\delta \in \mathcal{N}(A)^\perp$. The orthonormal system $\{u_m\}$ spans $\mathcal{N}(A)^\perp$. Hence

$$u_\alpha^\delta = \sum_{m=1}^{\infty} c_m u_m. \qquad (2.5.13)$$

To find the unknown parameters c_m we substitute (2.5.13) into the normal Eq. (2.5.6):

$$\sum_{m=1}^{\infty} (\sigma_m^2 + \alpha) c_m u_m = A^* f^\delta.$$

Multiplying both sides by u_n we get:

$$(\sigma_n^2 + \alpha) c_n = (A^* f^\delta, u_n).$$

But $(A^* f^\delta, u_n) = (f^\delta, A u_n) = \sigma_n (f^\delta, v_n)$. Therefore

$$(\sigma_n^2 + \alpha) c_n = \sigma_n (f^\delta, v_n)$$

and the unknown parameters are defined as follows:

$$c_m = \frac{\sigma_m}{\sigma_m^2 + \alpha} (f^\delta, v_m)$$

Using this in (2.5.13) we obtain:

$$u_\alpha^\delta = \sum_{m=1}^{\infty} \frac{\sigma_m}{\sigma_m^2 + \alpha} (f^\delta, v_m) u_n.$$

2.5 Regularization Strategy. Tikhonov Regularization

With formula (2.5.12), this implies (2.5.11). □

Remark that the filter function has the following properties:

$$q(\alpha; \sigma) \leq \min\left\{\frac{\sigma}{2\sqrt{\alpha}}; \frac{\sigma^2}{\alpha}\right\}, \quad \forall \alpha > 0. \tag{2.5.14}$$

These properties follow from the obvious inequalities:

$$\sigma^2 + \alpha \equiv (\sigma - \sqrt{\alpha})^2 + 2\sigma\sqrt{\alpha} \geq 2\sigma\sqrt{\alpha},$$

$$\frac{\sigma^2}{\sigma^2 + \alpha} \equiv \frac{\sigma^2/\alpha}{\sigma^2/\alpha + 1} \leq \frac{\sigma^2}{\alpha}, \quad \forall \alpha > 0, \ \sigma > 0.$$

The linear Eq. (2.5.6) is defined as a *regularized form of the normal equation* $A^*Au = A^*f^\delta$ [97].

If A^*A is invertible, then for $\alpha = 0$ formula (2.5.7) implies that

$$u_0^\delta = (A^*A)^{-1}A^*f^\delta =: A^\dagger f^\delta, \quad \alpha = 0 \tag{2.5.15}$$

is the solution of the *normal equation*

$$A^*Au^\delta = A^*f^\delta, \quad u^\delta \in H, \ f^\delta \in \tilde{H}. \tag{2.5.16}$$

Evidently $u_0^\delta \in \mathcal{N}(A)^\perp$, by Definition 2.3.1 of Moore-Penrose inverse $A^\dagger$.

It follows from (2.5.15) that Moore-Penrose inverse $A^\dagger := (A^*A)^{-1}A^*$ arises naturally as a result of minimization of Tikhonov functional. Remark that the provisional replacement of the compact operators AA^* or A^*A by the non-singular operators $AA^* + \alpha I$ or $A^*A + \alpha I$, is the main idea of Tikhonov's regularization procedure. This procedure has originally been introduced by Tikhonov in [94] for uniform approximations of solutions of Fredholm's equation of the first kind. However, Tikhonov does not point to the relation of his ideas to Moore-Penrose inverse.

Let us assume now that $\alpha > 0$. The right hand side of (2.5.7) defines family of continuous operators

$$R_\alpha := (A^*A + \alpha I)^{-1} A^* : \tilde{H} \mapsto H, \ \alpha > 0, \tag{2.5.17}$$

depending on the parameter of regularization $\alpha > 0$. Obviously, when data is noise free, then the regularized solution $R_\alpha f$ should converge (in some sense) to $A^\dagger f$, as $\alpha \to 0$, that is, $R_\alpha f \to A^\dagger f$, for all $f \in \mathcal{D}(A^\dagger)$. In practice this data is always noisy and we may only assume that it is known up to some error $\delta > 0$, that is, $\|f - f^\delta\| \leq \delta$. Hence the parameter of regularization $\alpha > 0$ depends on the noise level $\delta > 0$ and the noise data $f^\delta \in \tilde{H}$, and should be chosen appropriately, keeping an error $\|u_\alpha^\delta - u\|$ as small as possible. A strategy of choosing the parameter of regularization $\alpha = \alpha(\delta, f^\delta)$ is called a *parameter choice rule*. These considerations lead to the following definition of *regularization strategy*.

Definition 2.5.1 Let $A : H \mapsto \tilde{H}$ be an injective compact operator. Assume that $f, f^\delta \in \tilde{H}$ be noise free and noisy data respectively, that is, $\|f - f^\delta\| \leq \delta < \|f^\delta\|$, $\delta > 0$. A family $\{R_{\alpha(\delta, f^\delta)}\}_\alpha$ of bounded linear operators is called a regularization strategy or a convergent regularization method if for all $f \in \mathcal{D}(A^\dagger)$,

$$\limsup_{\delta \to 0^+} \left\{ \|R_{\alpha(\delta, f^\delta)} f^\delta - A^\dagger f\|_H : f^\delta \in \tilde{H}, \ \|f - f^\delta\| \leq \delta \right\} = 0, \quad (2.5.18)$$

with $\alpha : \mathbb{R}_+ \times \tilde{H} \mapsto \mathbb{R}_+$, such that

$$\limsup_{\delta \to 0^+} \left\{ \alpha(\delta, f^\delta) : f^\delta \in \tilde{H}, \ \|f - f^\delta\| \leq \delta \right\} = 0. \quad (2.5.19)$$

If the parameter of regularization $\alpha = \alpha(\delta, f^\delta)$ depends only on the noise level $\delta > 0$, the rule $\alpha(\delta, f^\delta)$ is called *a-priori parameter choice rule*. Otherwise, this rule is called *a-posteriori parameter choice rule*. Tikhonov regularization is a typical example of a priori parameter choice rule, since the choice of the parameter of regularization $\alpha > 0$ is made a priori, i.e. before computations, as we will see in Theorem 2.5.2. The Morozov's Discrepancy Principle is a regularization strategy with a-posteriori parameter choice rule, since the choice of the parameter of regularization $\alpha > 0$ is made during the process of computing. Regarding the iterative methods, the number of iterations plays here the role of the regularization parameter. As we will show in the next chapter, Landweber's Method and Conjugate Gradient Method together with appropriate parameter choice rule are also a regularization strategy in sense of Definition 2.5.1. We remark finally that, as in the case of Tikhonov regularization, some widely used regularization operators R_α *are linear*. In particular, the Morozov's Discrepancy Principle and Landweber's Method can be formulated as linear regularization methods. However, the Conjugate Gradient Method is a *nonlinear regularization method*, since the right hand side of the equation $Au = f$ does not depend linearly on the parameter of regularization α.

Above definition, with Theorem 2.5.1, implies that the family of operators $\{R_\alpha\}_{\alpha > 0}$, $R_\alpha : \tilde{H} \mapsto H$, defined as

$$R_\alpha := (A^*A + \alpha I)^{-1} A^*, \quad \alpha > 0, \quad (2.5.20)$$

is the regularization strategy corresponding to Tikhonov regularization. That is, the regularization operators R_α, $\alpha > 0$, approximate the unbounded inverse $A^\dagger$ of the operator A on $\mathcal{R}(A)$. On the other hand, Corollary 2.5.1, implies that the singular value expansion of this operator is

$$R_\alpha f^\delta := \sum_{n=1}^\infty \frac{q(\alpha; \sigma_n)}{\sigma_n} (f^\delta, v_n) u_n, \quad \alpha > 0. \quad (2.5.21)$$

2.5 Regularization Strategy. Tikhonov Regularization

It is easy to prove that the family of operators $\{R_\alpha\}$ are not uniformly bounded, i.e. there exists a sequence $\{\alpha_m\}_{m=1}^\infty$, $\alpha_m > 0$, such that $\|R_{\alpha_m}\| \to \infty$, as $\alpha_m \to 0$. Indeed, taking $f_\alpha^\delta = v_m$ in (2.5.21), where v_m is any fixed eigenvector, with $\|v_m\| = 1$, we have:

$$\|R_{\alpha_m} v_m\| = \frac{\sigma_m}{\sigma_m^2 + \alpha_m}.$$

For the sequence $\{\alpha_m\}$ satisfying the conditions

$$\alpha_m \to 0 \text{ and } \alpha_m/\sigma_m \to 1, \text{ as } m \to \infty,$$

we conclude:

$$\|R_{\alpha_m} v_m\| = \frac{1}{\sigma_m + \alpha_m/\sigma_m} \to \infty, \text{ as } m \to \infty.$$

The regularization strategy R_α possesses this property in general case as well.

Lemma 2.5.1 *Let R_α be regularization strategy corresponding to the compact operator $A : H \mapsto \tilde{H}$. Then the operators $\{R_\alpha\}_{\alpha>0}$, $R_\alpha : \tilde{H} \mapsto H$, are not uniformly bounded, i.e. there exists a sequence $\{\alpha_m\}_{m=1}^\infty$, $\alpha_m > 0$, such that $\|R_{\alpha_m}\| \to \infty$, as $\alpha_m \to 0$.*

Proof Assume, in contrary, that there exists a constant $M > 0$, independent on $\alpha > 0$, such that $\|R_\alpha\| \leq M$, for all $\alpha > 0$. Then for any $f \in \mathcal{R}(A) \subset \tilde{H}$ we have:

$$\|A^\dagger f\| \leq \|A^\dagger f - R_\alpha f\| + \|R_\alpha f\| \leq \|A^\dagger f - R_\alpha f\| + \|R_\alpha\| \|f\|$$
$$\leq \|A^\dagger f - R_\alpha f\| + M\|f\|.$$

The first norm on the right-hand side tends to zero as $\alpha \to 0$, by definition (2.5.18). Then passing to the limit we get: $\|A^\dagger f\| \leq M\|f\|$, for all $f \in \mathcal{R}(A)$, which implies $\|A^\dagger\| \leq M$, i.e. boundedness of the generalized inverse $A^\dagger$. This contradiction completes the proof. □

Therefore, one needs to find a bounded approximation of the unbounded operator $A^\dagger$. This approximation will evidently depends on both, the parameter of regularization $\alpha > 0$ and the noisy data f^δ. So, the main problem of regularization strategy is to find such a bounded approximation, which will be convergent to the best approximate solution $A^\dagger f$ corresponding to the exact data $f \in \mathcal{N}(A^*)^\perp$, as $\alpha \to 0$ and $\delta \to 0$.

The following theorem gives an answer to this issue.

Theorem 2.5.2 *Let $A : H \mapsto \tilde{H}$ be a linear injective bounded operator between Hilbert spaces H and $\tilde{H}$. Assume that conditions of Theorem 2.4.1 hold. Denote by $f^\delta \in \mathcal{N}(A^*)^\perp$ the noisy data: $\|f - f^\delta\|_{\tilde{H}} \leq \delta$, $\delta > 0$. Suppose that the conditions hold:*

$$\alpha(\delta) \to 0 \quad \text{and} \quad \frac{\delta^2}{\alpha(\delta)} \to 0, \quad \text{as } \delta \to 0. \tag{2.5.22}$$

Then the solution $u_\alpha^\delta := R_{\alpha(\delta)} f^\delta$ of the regularized form of the normal Eq. (2.5.6) converges to the best approximate solution $u := A^\dagger f$ of Eq. (2.5.1) in the norm of the space H, that is,

$$\|R_{\alpha(\delta)} f^\delta - A^\dagger f\|_H \to 0, \quad \delta \to 0. \tag{2.5.23}$$

Proof Let us estimate the norm $\|R_{\alpha(\delta)} f^\delta - A^\dagger f\|_H$ using (2.4.13) and (2.5.21). We have

$$\|R_{\alpha(\delta)} f^\delta - A^\dagger f\|_H^2 := \sum_{n=1}^{\infty} \left\|\left(\frac{q(\alpha(\delta); \sigma_n)}{\sigma_n}(f^\delta, v_n) - \frac{1}{\sigma_n}(f, v_n)\right) u_n\right\|^2$$

$$= \sum_{n=1}^{\infty} \left\|\left(\frac{q(\alpha(\delta); \sigma_n)}{\sigma_n}(f^\delta - f, v_n) + \left(\frac{q(\alpha(\delta); \sigma_n)}{\sigma_n} - \frac{1}{\sigma_n}\right)(f, v_n)\right) u_n\right\|^2$$

$$\leq 2 \sum_{n=1}^{\infty} \frac{q^2(\alpha(\delta); \sigma_n)}{\sigma_n^2} |(f^\delta - f, v_n)|^2 + 2 \sum_{n=1}^{\infty} \frac{\alpha^2(\delta)}{\sigma_n^2(\sigma_n^2 + \alpha(\delta))^2} |(f, v_n)|^2. \tag{2.5.24}$$

Denote by S_1 and S_2 the first and the second summands on the right-hand side of (2.5.24), respectively.

We use from (2.5.14) the property $q^2(\alpha; \sigma_n)/\sigma_n^2 \leq 1/(4\alpha)$ of the filter function and Parseval's identity to estimate the term S_1:

$$S_1 \leq \frac{1}{2\alpha(\delta)} \sum_{n=1}^{\infty} |(f^\delta - f, v_n)|^2 \leq \frac{1}{2\alpha(\delta)} \|f^\delta - f\|^2 \leq \frac{\delta^2}{2\alpha(\delta)}. \tag{2.5.25}$$

For estimating the term S_2 we rewrite it in the following form:

$$S_2 = 2 \sum_{n=1}^{N} \frac{\alpha^2(\delta)}{\sigma_n^2(\sigma_n^2 + \alpha(\delta))^2} |(f, v_n)|^2 + 2 \sum_{n=N+1}^{\infty} \frac{\alpha^2(\delta)}{\sigma_n^2(\sigma_n^2 + \alpha(\delta))^2} |(f, v_n)|^2$$

$$=: S_{2N} + R_{2N}. \tag{2.5.26}$$

To estimate the Nth partial sum S_{2N} we use the properties $\sigma_1 \geq \sigma_2 \geq \ldots \geq \sigma_n \geq \ldots > 0$ and $\sigma_n \to 0$, as $n \to \infty$, of the singular values. For any small value $\alpha = \alpha(\delta) > 0$ of the parameter of regularization, depending on the noise level $\delta > 0$, there exists such a positive integer $N = N(\alpha(\delta)) \equiv N(\delta)$ that

$$\min_{1 \leq n \leq N(\delta)} \sigma_n^2 = \sigma_{N(\delta)}^2 \geq \sqrt{\alpha(\delta)} > \sigma_{N(\delta)+1}^2, \tag{2.5.27}$$

2.5 Regularization Strategy. Tikhonov Regularization

where $N = N(\alpha(\delta)) \to \infty$, as $\delta \to 0$. For such $N = N(\delta)$ we estimate the Nth partial sum S_{2N} in (2.5.26) as follows:

$$S_{2N} \leq \frac{2\alpha^2(\delta)}{(\sigma_{N(\delta)}^2 + \alpha(\delta))^2} \sum_{n=1}^{N(\delta)} \frac{1}{\sigma_n^2} |(f, v_n)|^2$$

$$= \frac{2\alpha(\delta)}{(\sigma_{N(\delta)}^2/\sqrt{\alpha(\delta)} + \sqrt{\alpha(\delta)})^2} \sum_{n=1}^{N(\delta)} \frac{1}{\sigma_n^2} |(f, v_n)|^2$$

By the condition (2.5.27), $\sigma_{N(\delta)}^2/\sqrt{\alpha(\delta)} \geq 1$. Hence

$$S_{2N} \leq \frac{2\alpha(\delta)}{(1 + \sqrt{\alpha(\delta)})^2} \sum_{n=1}^{N(\delta)} \frac{1}{\sigma_n^2} |(f, v_n)|^2. \quad (2.5.28)$$

Let us estimate now the series remainder term R_{2N} defined in (2.5.26). We have

$$R_{2N} = 2 \sum_{n=N(\delta)+1}^{\infty} \frac{1}{(\sigma_n^2/\alpha(\delta) + 1)^2} \frac{1}{\sigma_n^2} |(f, v_n)|^2$$

$$\leq 2 \sum_{n=N(\delta)+1}^{\infty} \frac{1}{\sigma_n^2} |(f, v_n)|^2.$$

Taking into account this estimate with estimates (2.5.25) and (2.5.28) in (2.5.24) we finally deduce:

$$\|R_{\alpha(\delta)} f^\delta - A^\dagger f\|_H^2 \leq \frac{\delta^2}{2\alpha} + \frac{2\alpha(\delta)}{(1+\sqrt{\alpha(\delta)})^2} \sum_{n=1}^{N(\delta)} \frac{1}{\sigma_n^2} |(f, v_n)|^2$$

$$+ 2 \sum_{n=N(\delta)+1}^{\infty} \frac{1}{\sigma_n^2} |(f, v_n)|^2. \quad (2.5.29)$$

The first right hand side term tends to zero, as $\delta \to 0$, by the second condition of (2.5.22). The factor before the partial sum of the second right hand side term tends to zero, since $\alpha(\delta) \to 0$, as $\delta \to 0$, by the first condition of (2.5.22), and this partial sum is finite due to the convergence condition (2.4.12) of the Picard's Theorem 2.4.1. The third right hand side term also tends to zero, as $\delta \to 0$, since in this case $N(\delta) \to \infty$, and, as a result, the series remainder term tends to zero, by the same convergence condition. This implies (2.5.23). □

Estimate (2.5.29) clearly shows the role of all parameters $\delta > 0$, $\alpha(\delta) > 0$ and $N = N(\delta)$ in the regularization strategy.

As we have seen in the previous section, the number N of first terms in singular value expansion of a compact operator can also be considered as a regularization parameter. As noted above in iterative methods, the number of iterations also plays role of the regularization parameter, that is, $\alpha \sim 1/N$. For more detailed analysis of regularization methods for ill-posed problems we refer to the books [23, 54, 90].

Remark finally that besides the Tikhonov regularization, there are other regularization methods, such as Lavrentiev regularization, asymptotic regularization, local regularization, etc. Since zero is the only accumulation point of the singular values of a compact operator, the underlying idea in all these regularization techniques is modifying the smallest singular values, shifting all singular values by $\alpha > 0$. In other words, the idea is to approximate the compact operator A or A^*A by a family of operators $A + \alpha I$ or $A^*A + \alpha I$. The first one corresponds to *Lavrentiev regularization*. Specifically, while Tikhonov regularization is based on the normal equation $A^*Au = A^*f^\delta$, Lavrentiev regularization is based on the original equation $Au = f^\delta$. The main advantage of the first approach over the second one is that the operator A^*A is always injective, due to Lemma 2.3.1, even if the operator $A : \mapsto \tilde{H}$ is not injective. The second advantage of Tikhonov regularization is that the class of admissible values of the parameter of regularization $\alpha > 0$ for convergent regularization method is larger than the same class in Lavrentiev regularization. Specifically, when $\alpha = \delta$ the condition $\delta^2/\alpha(\delta) \to 0$, as $\delta \to 0$, for convergent regularization strategy in Tikhonov regularization holds, but does not hold in Lavrentiev regularization, as we will see below.

Corollary 2.5.2 can be adopted to the case of *Lavrentiev regularization* when the linear bounded injective operator $A : H \mapsto H$ is self-adjoint and positive semidefinite.

Corollary 2.5.3 *Let conditions of Theorem 2.4.1 hold and $f^\delta \in \mathcal{R}(A)$. Then the unique solution $u_\alpha^\delta \in H$ of the regularized equation*

$$(A + \alpha I) u_\alpha^\delta = f^\delta \qquad (2.5.30)$$

can be represented as the series

$$u_\alpha^\delta = \sum_{n=1}^\infty \frac{\tilde{q}(\alpha; \sigma_n)}{\sigma_n} (f^\delta, u_n) u_n, \quad \alpha > 0, \qquad (2.5.31)$$

where

$$\tilde{q}(\alpha; \sigma) = \frac{\sigma}{\sigma + \alpha}. \qquad (2.5.32)$$

Proof We use the singular system $\{\sigma_n, u_n, u_n\}$ for the non-self-adjoint operator $A : H \mapsto \tilde{H}$, that is, $Au_n = \sigma_n u_n$ and the representation

$$u_\alpha^\delta = \sum_{m=1}^\infty c_m u_m. \qquad (2.5.33)$$

2.5 Regularization Strategy. Tikhonov Regularization

Substituting this into the normal Eq. (2.5.30) we obtain:

$$\sum_{m=1}^{\infty}(\sigma_m + \alpha)c_m u_m = f^\delta.$$

Multiplying both sides by u_n we find the unknown parameters c_m:

$$c_m = \frac{1}{\sigma_m + \alpha}(f^\delta, u_m)$$

Using this in (2.5.33) we find:

$$u_\alpha^\delta = \sum_{m=1}^{\infty} \frac{1}{\sigma_m + \alpha}\left(f^\delta, u_m\right) u_m.$$

By (2.5.32), this is exactly the required expansion (2.5.31). □

Now the question we seek to answer here is that under which conditions it is possible to construct a convergent regularization strategy for Lavrentiev regularization. The following theorem, which is an analogue of Theorem 2.5.2, answers this question.

Theorem 2.5.3 *Let the linear bounded injective operator $A : H \mapsto \tilde{H}$ be a self-adjoint and positive semi-definite. Assume that conditions of Theorem 2.4.1 hold. Denote by $f^\delta \in \mathcal{R}(A)$ the noisy data: $\|f - f^\delta\|_{\tilde{H}} \leq \delta$, $\delta > 0$. Suppose that the following conditions hold:*

$$\alpha(\delta) \to 0 \text{ and } \frac{\delta}{\alpha(\delta)} \to 0, \text{ as } \delta \to 0. \tag{2.5.34}$$

Then the operator $R_{\alpha(\delta)}$ defined by the right hand side of (2.5.31) is a convergent regularization strategy, that is,

$$R_{\alpha(\delta)} f^\delta := \sum_{n=1}^{\infty} \frac{\tilde{q}(\alpha(\delta); \sigma_n)}{\sigma_n}(f^\delta, u_n)u_n, \quad \alpha > 0, \tag{2.5.35}$$

that is, the solution $u_\alpha^\delta := R_{\alpha(\delta)} f^\delta$ of the regularized form of the normal Eq. (2.5.30) converges to the best approximate solution $u := A^\dagger f$ of Eq. (2.5.1) in the norm of the space H.

Proof The proof is similar to the proof of Theorem 2.5.2. Here we first use the following property

$$\frac{\sigma}{\sigma + \alpha} \leq \frac{\sigma}{\alpha} \tag{2.5.36}$$

of the filter function (2.5.32) in (2.5.24) to obtain the estimate

$$\|R_{\alpha(\delta)}f^\delta - A^\dagger f\|_H^2 := \sum_{n=1}^{\infty}\left\|\left(\frac{\tilde{q}(\alpha(\delta);\sigma_n)}{\sigma_n}(f^\delta,u_n) - \frac{1}{\sigma_n}(f,u_n)\right)u_n\right\|^2$$

$$\leq 2\sum_{n=1}^{\infty}\frac{\tilde{q}^2(\alpha(\delta);\sigma_n)}{\sigma_n^2}|(f^\delta - f, u_n)|^2 + 2\sum_{n=1}^{\infty}\frac{\alpha^2(\delta)}{\sigma_n^2(\sigma_n + \alpha(\delta))^2}|(f,u_n)|^2 \quad (2.5.37)$$

Denote by S_1 and S_2 the first and the second right-hand side summands of (2.5.37), respectively. By (2.5.36) we conclude $\tilde{q}^2(\alpha;\sigma_n)/\sigma_n^2 \leq 1/\alpha^2$. Using this inequality in (2.5.37) we obtain the estimate

$$S_1 := 2\sum_{n=1}^{\infty}\frac{\tilde{q}^2(\alpha(\delta);\sigma_n)}{\sigma_n^2}|(f^\delta - f, u_n)|^2 \leq \frac{1}{\alpha^2(\delta)}\|f^\delta - f\|^2 \leq \frac{\delta^2}{\alpha^2(\delta)} \quad (2.5.38)$$

Second, we rewrite the term S_2 in the following form:

$$S_2 = 2\sum_{n=1}^{N}\frac{\alpha^2(\delta)}{\sigma_n^2(\sigma_n + \alpha(\delta))^2}|(f,u_n)|^2 + 2\sum_{n=N+1}^{\infty}\frac{\alpha^2(\delta)}{\sigma_n^2(\sigma_n + \alpha(\delta))^2}|(f,u_n)|^2$$

$$=: S_{2N} + R_{2N}. \quad (2.5.39)$$

For estimating the first right hand side term S_2 in (2.5.39), we use the properties $\sigma_1 \geq \sigma_2 \geq \ldots \geq \sigma_n \geq \ldots > 0$, $\sigma_n \to 0$, as $n \to \infty$, of the eigenvalues $\lambda_n = \sigma_n$ of self-adjoint positive semidefinite operator A. For any small values $\alpha = \alpha(\delta) > 0$ of the parameter of regularization, depending on the noise level $\delta > 0$, there exists such a positive integer $N = N(\alpha(\delta)) \equiv N(\delta)$ that

$$\min_{1\leq n\leq N(\delta)}\sigma_n^2 = \sigma_{N(\delta)}^2 \geq \alpha(\delta) > \sigma_{N(\delta)+1}^2, \quad (2.5.40)$$

where $N = N(\alpha(\delta)) \to \infty$, as $\delta \to 0$. For such $N = N(\delta)$ we get the following estimate for S_{2N} in (2.5.39):

$$S_{2N} \leq \frac{2\alpha^2(\delta)}{(\sigma_{N(\delta)} + \alpha(\delta))^2}\sum_{n=1}^{N(\delta)}\frac{1}{\sigma_n^2}|(f,v_n)|^2$$

$$= \frac{2\alpha(\delta)}{(\sigma_{N(\delta)}/\sqrt{\alpha(\delta)} + \sqrt{\alpha(\delta)})^2}\sum_{n=1}^{N(\delta)}\frac{1}{\sigma_n^2}|(f,v_n)|^2$$

By the condition (2.5.40), $\sigma_{N(\delta)}/\sqrt{\alpha(\delta)} \geq 1$, which implies:

2.5 Regularization Strategy. Tikhonov Regularization

$$S_{2N} \leq 2\alpha(\delta) \sum_{n=1}^{N(\delta)} \frac{1}{\sigma_n^2} |(f, u_n)|^2. \tag{2.5.41}$$

The factor before the above partial sum, which is finite by the convergence condition (2.4.12) of the Picard's Theorem 2.4.1, tends to zero, as $\alpha(\delta) \to 0$. Hence, $S_{2N} \to 0$, as $\alpha(\delta) \to 0$.

Let us estimate now the series remainder term R_{2N} defined in (2.5.39). We have

$$R_{2N} = 2 \sum_{n=N(\delta)+1}^{\infty} \frac{1}{(\sigma_n/\alpha(\delta)+1)^2} \frac{1}{\sigma_n^2} |(f, v_n)|^2$$

$$\leq 2 \sum_{n=N(\delta)+1}^{\infty} \frac{1}{\sigma_n^2} |(f, v_n)|^2.$$

Substituting this estimate with (2.5.38) and (2.5.41) into (2.5.37) we conclude:

$$\|R_{\alpha(\delta)} f^\delta - A^\dagger f\|_H^2 \leq \\ \frac{\delta^2}{\alpha(\delta)^2} + 2\alpha(\delta) \sum_{n=1}^{N(\delta)} \frac{1}{\sigma_n^2} |(f, u_n)|^2 + 2 \sum_{n=N(\delta)+1}^{\infty} \frac{1}{\sigma_n^2} |(f, v_n)|^2 . \tag{2.5.42}$$

It is easy to verify that under the conditions (2.5.34) all three right hand side terms tend to zero, as $\delta \to 0$. $\square$

Comparing the convergence conditions in Theorems 2.5.2 and 2.5.3 we first observe that, in Tikhonov regularization the class of admissible values of the parameter of regularization $\alpha > 0$ for convergent regularization method is larger than the same class in Lavrentiev regularization. While, for example, in Tikhonov regularization the values $\alpha = \delta$ and $\alpha = \sqrt{\delta}$ of the parameter of regularization are admissible for convergent regularization method, as the second condition of (2.5.22) shows, in Lavrentiev regularization they are not admissible by the condition (2.5.34). Moreover, the dependence on the parameter of regularization $\alpha > 0$ of the numbers $N = N(\alpha(\delta))$, defined in proofs of these theorems and corresponding to these regularizations, are different. Thus, if $\sigma_n = \mathcal{O}(1/n^2)$, then condition (2.5.27) of Theorem 2.5.2 implies that there exists the constants $c_2 > c_1 > 0$ such that

$$\frac{c_2}{N^4} \geq \sqrt{\alpha(\delta)} > \frac{c_1}{(N+1)^4}.$$

For the same $\sigma_n = \mathcal{O}(1/n^2)$, the condition (2.5.40) of Theorem 2.5.3 implies:

$$\frac{\tilde{c}_2}{N^4} \geq \alpha(\delta) > \frac{\tilde{c}_1}{(N+1)^4}, \quad \tilde{c}_2 > \tilde{c}_1 > 0.$$

As a result, $N(\alpha(\delta)) = \mathcal{O}\left(\alpha(\delta)^{-1/8}\right)$ in Tikhonov regularization and $N(\alpha(\delta)) = \mathcal{O}\left(\alpha(\delta)^{-1/4}\right)$ in Lavrentiev regularization.

Note, finally that the topic of Tikhonov regularization is very broad and here we described it for only linear inverse problems. We refer the reader to the books [48, 90] on the mathematical theory of regularization methods related to nonlinear inverse problems.

2.6 Morozov's Discrepancy Principle

In applications the right hand side $f \in H$ of the equation $Au = f$ always contains a noise. Instead of this equation one needs to solve the equation $Au^\delta = f^\delta$, where f^δ is a noisy data: $\|f - f^\delta\| \leq \delta$, where $\delta > 0$. This, in particular, implies that in the ideal case the residual or *discrepancy* $\|Au^\delta - f^\delta\|$ can only be at most in the order of δ. On the other hand, in order to construct a bounded approximation $R_{\alpha(\delta)}$ of the unbounded operator $A^\dagger$ one needs to choose the parameter of regularization $\alpha > 0$ depending on the noise level $\delta > 0$. Thus, the parameter of regularization needs to be chosen by a compromise between the residual $\|Au^\delta - f^\delta\|$ and the given bound $\delta > 0$ for the noise level. This is the main criteria of so-called *Morozov's Discrepancy Principle* due to Morozov [67, 68]. This principle is now one of the simplest tools and most widely used regularization method for ill-posed problems.

Definition 2.6.1 Let $f^\delta \in \tilde{H}$ be a noisy data with an arbitrary given noise level $\delta > 0$, that is, $\|f^\delta - f\| \leq \delta$, where $f \in \mathcal{R}(A)$ is a noise free (exact) data. If there exists such a value $\alpha = \alpha(\delta)$ of the parameter of regularization, depending on $\delta > 0$, that the corresponding solution $u^\delta_{\alpha(\delta)} \in H$ of the equation $Au = f^\delta$ satisfies the condition

$$\beta_1 \delta \leq \|Au^\delta_{\alpha(\delta)} - f^\delta\| \leq \beta_2 \delta, \quad \beta_2 \geq \beta_1 \geq 1, \qquad (2.6.1)$$

then the parameter of regularization $\alpha = \alpha(\delta)$ is said to be chosen according to Morozov's Discrepancy Principle.

Let us assume that the size $\delta := \|\delta f\| = \|f^\delta - f\| > 0$ of the noise is known (although we do not know the random perturbation δf). Denote by $u \in \mathcal{N}(A)^\perp$ the solution of the equation $Au = f$ with a noise free (exact) data $f \in \mathcal{N}(A^*)^\perp = \overline{\mathcal{R}(A)}$. Then

$$\|Au - f^\delta\| = \|Au - f - \delta f\| = \|\delta f\| =: \delta, \quad \delta > 0.$$

Thus, if u is the exact solution, corresponding to the exact data f, and f^δ is the noisy data, then

$$\|Au - f^\delta\| = \delta. \qquad (2.6.2)$$

Now, having the size $\delta = \|f - f^\delta\| > 0$ of noise, we want to use Tikhonov regularization to find the solution u^δ_α defined by (2.5.7) and corresponding to the

2.6 Morozov's Discrepancy Principle

noisy data $f^\delta := f + \delta f$. Then, as it follows from (2.6.2), *the best that we can require from the parameter of regularization $\alpha > 0$ is the residual (or discrepancy)* $\|Au^\delta_{\alpha(\delta)} - f^\delta\| = \delta$.

We prove that there exists such a value of the parameter of regularization $\alpha = \alpha(\delta)$ which satisfies the following conditions:

$$\|Au^\delta_{\alpha(\delta)} - f^\delta\| = \|f - f^\delta\| = \delta, \quad \delta > 0. \tag{2.6.3}$$

Theorem 2.6.1 *Let conditions of Theorem 2.4.1 hold. Denote by $f^\delta \in \tilde{H}$ the noisy data with $\|f^\delta\| > \delta > 0$. Then there exists a unique value $\alpha = \alpha(\delta)$ of the parameter of regularization satisfying conditions* (2.6.3).

Proof By the unique decomposition, for any $f^\delta \in \tilde{H} = \mathcal{R}(A) \oplus \mathcal{R}(A)^\perp$ we have

$$f^\delta = \sum_{n=1}^{\infty} (f^\delta, v_n) v_n + Pf^\delta, \tag{2.6.4}$$

where $Pf^\delta \in \mathcal{R}(A)^\perp$ is the projection of the noisy data on $\mathcal{R}(A)^\perp$.

Now we rewrite the solution u^δ_α, given by (2.5.11), of the normal Eq. (2.5.6) in the following form:

$$u^\delta_\alpha = \sum_{n=1}^{\infty} \frac{\sigma_n}{\sigma_n^2 + \alpha} (f^\delta, v_n) u_n, \quad \alpha > 0. \tag{2.6.5}$$

Acting on this solution by the operator A and using the relation $Au_n = \sigma_n v_n$ we deduce:

$$Au^\delta_\alpha = \sum_{n=1}^{\infty} \frac{\sigma_n^2}{\sigma_n^2 + \alpha} (f^\delta, v_n) v_n.$$

With (2.6.4) this yields:

$$f^\delta - Au^\delta_\alpha = \sum_{n=1}^{\infty} \frac{\alpha}{\sigma_n^2 + \alpha} (f^\delta, v_n) v_n + Pf^\delta.$$

Then we obtain the discrepancy:

$$\|Au^\delta_\alpha - f^\delta\|^2 = \sum_{n=1}^{\infty} \frac{\alpha^2}{(\sigma_n^2 + \alpha)^2} |(f^\delta, v_n)|^2 + \|Pf^\delta\|^2. \tag{2.6.6}$$

It can be verified that $g(\alpha) := \|Au^\delta_\alpha - f^\delta\|$ is a monotonically increasing function for $\alpha > 0$. To complete the proof of the theorem we need to show that the equation $g(\alpha) = \delta$, $\alpha \in (0, +\infty)$, has a unique solution. For $\alpha \to 0^+$ we use $Pf = 0$ for the noise free data $f \in \mathcal{R}(A)$ to get

$$\lim_{\alpha \to 0^+} g(\alpha) = \|Pf^\delta\| = \|P(f^\delta - f)\| \leq \|f^\delta - f\| = \delta, \qquad (2.6.7)$$

by (2.6.3). Note that the sum in (2.6.6) tends to zero, as $\alpha \to 0^+$.
For $\alpha \to +\infty$ we use the following limit

$$\lim_{\alpha \to +\infty} \frac{\alpha^2}{(\sigma_n^2 + \alpha)^2} = \lim_{\alpha \to +\infty} \frac{1}{(\sigma_n^2/\alpha^2 + 1)^2} = 1$$

and the identity (2.1.8) to get

$$\lim_{\alpha \to +\infty} g(\alpha) = \left(\sum_{n=1}^{\infty} |(f^\delta, v_n)|^2 + \|Pf^\delta\|^2 \right)^{1/2} =: \|f^\delta\|,$$

by (2.6.6).

By the assumption $\|f^\delta\| > \delta$ we conclude:

$$\lim_{\alpha \to +\infty} g(\alpha) > \delta.$$

This implies with (2.6.7) that the equation $g(\alpha) = \delta$ has a unique solution for $\alpha \in (0, +\infty)$. □

Remark that $\|f^\delta\| > \delta > 0$ is a natural condition in the theorem. Otherwise, i.e. if $\|f^\delta\| < \delta$, then $u^\delta_{\alpha(\delta)} = 0$ can be assigned as a regularized solution.

The theorem below shows that Morozov's Discrepancy Principle provides a convergent regularization strategy. First we need the following notion.

Definition 2.6.2 Let $A : H \mapsto \tilde{H}$ be a linear injective compact operator between Hilbert spaces H and $\tilde{H}$, and $A^* : \tilde{H} \mapsto H$ its adjoint. Denote by $u \in H$ the solution $u = A^\dagger f$ of the equation $Au = f$ with the noise free (exact) data $f \in \mathcal{R}(A)$. If there exists such an element $v \in \tilde{H}$ with $\|v\|_{\tilde{H}} \leq M$ that $u = A^* v$, then we say that $u \in H$ satisfies the source condition.

Remark that besides of the classical concept of source condition, in recent years different new concepts, including approximate source conditions, have been developed [44].

Theorem 2.6.2 Let $A : H \mapsto \tilde{H}$ be a linear injective compact operator between Hilbert spaces H and $\tilde{H}$. Assume that the solution $u = A^\dagger f$ of the equation $Au = f$ with the noise free data $f \in \mathcal{R}(A)$ satisfies the source condition. Suppose that the parameter of regularization $\alpha = \alpha(\delta)$, $\delta > 0$, is defined according to conditions (2.6.3). Then the regularization method $R_{\alpha(\delta)}$ is convergent, that is,

$$\|R_{\alpha(\delta)} f^\delta - A^\dagger f\|_H \to 0, \quad \delta \to 0, \qquad (2.6.8)$$

2.6 Morozov's Discrepancy Principle

where $f^\delta \in \tilde{H}$ is the noisy data and $R_{\alpha(\delta)} f^\delta =: u^\delta_{\alpha(\delta)}$ is the regularized solution.

Proof $u^\delta_{\alpha(\delta)}$ furnishes a minimum to the regularized Tikhonov functional,

$$J_\alpha(v) := \frac{1}{2}\|Av - f^\delta\|^2_{\tilde{H}} + \frac{1}{2}\alpha\|v\|^2_H, \ v \in H, \ f^\delta \in \tilde{H}. \tag{2.6.9}$$

Hence $J_\alpha(u^\delta_{\alpha(\delta)}) \leq J_\alpha(v)$, for all $v \in H$. In particular, for the unique solution $u \in \mathcal{N}(A)^\perp$ of the equation $Au = f$ with the noise free data $f \in \mathcal{R}(A)$ we have:

$$J_\alpha(u^\delta_{\alpha(\delta)}) \leq J_\alpha(u). \tag{2.6.10}$$

According to (2.6.3), $\|Au^\delta_{\alpha(\delta)} - f^\delta\| = \delta$ and $\|Au - f^\delta\| = \delta$, $\delta > 0$. Taking into account this in (2.6.9) we conclude:

$$J_\alpha(u^\delta_{\alpha(\delta)}) = \frac{1}{2}\delta^2 + \frac{1}{2}\alpha\|u^\delta_{\alpha(\delta)}\|^2_H,$$

$$J_\alpha(u) = \frac{1}{2}\delta^2 + \frac{1}{2}\alpha\|u\|^2_H.$$

This implies that with (2.6.10) that

$$\|u^\delta_{\alpha(\delta)}\|_H \leq \|u\|_H, \ \forall \delta > 0, \ \alpha > 0, \tag{2.6.11}$$

i.e. the set $\{u^\delta_{\alpha(\delta)}\}_{\delta>0}$ is uniformly bounded in H by the norm of the solution $u \in \mathcal{N}(A)^\perp$ of the equation $Au = f$.

Having the uniform boundedness of $\{u^\delta_{\alpha(\delta)}\}_{\delta>0}$ we estimate now the difference between the regularized and exact solutions:

$$\|u^\delta_{\alpha(\delta)} - u\|^2 = \|u^\delta_{\alpha(\delta)}\|^2 - 2Re(u^\delta_{\alpha(\delta)}, u) + \|u\|^2$$
$$\leq 2\left(\|u\|^2 - Re(u^\delta_{\alpha(\delta)}, u)\right) = 2Re(u - u^\delta_{\alpha(\delta)}, u).$$

Since $u = A^*v$, we transform this estimate as follows:

$$\|u^\delta_{\alpha(\delta)} - u\|^2 \leq 2Re(u - u^\delta_{\alpha(\delta)}, A^*v) = 2Re(Au - Au^\delta_{\alpha(\delta)}, v)$$
$$= 2Re(f - Au^\delta_{\alpha(\delta)}, v) \leq 2Re(f - f^\delta, v) + 2Re(f^\delta - Au^\delta_{\alpha(\delta)}, v)$$
$$\leq 2\|v\|\left[\|f - f^\delta\| + \|f^\delta - Au^\delta_{\alpha(\delta)}\|\right].$$

Using conditions (2.6.3) and $\|v\|_{\tilde{H}} \leq M$ on the right hand side we finally get:

$$\|u^\delta_{\alpha(\delta)} - u\| \leq 2\sqrt{M\delta}. \tag{2.6.12}$$

The right hand side tends to zero as $\delta \to 0^+$, which is the desired result. □

Remark 2.6.1 Estimate (2.6.12) depends on the norm of the element $v \in \tilde{H}$ which image $u = A^*v$ under the adjoint operator $A^* : \tilde{H} \mapsto H$ is the solution of the equation $Au = f$ with the noise free data $f \in \mathcal{R}(A)$. This element $v \in \tilde{H}$ is called a sourcewise element. In this we say that the exact solution $u = A^\dagger f$ satisfies the source condition $u = A^*v, v \in \tilde{H}$.

In applications, it is not necessary to satisfy the condition (2.6.3) exactly. Instead, the relaxed form condition

$$\beta_* \delta \leq \|Au^\delta_{\alpha(\delta)} - f^\delta\| \leq \beta^* \delta, \quad \beta^* > \beta_* > 0, \, \delta > 0 \qquad (2.6.13)$$

can be used.

Therefore, if the noise level $\delta > 0$ is known, then in the iteration algorithm one of the forms of condition (2.6.13) is used as a *stopping rule*, according to Morozov's Discrepancy Principle. Iteration is terminated if

$$\|Au^{\delta,n(\delta)}_{\alpha(\delta)} - f^\delta\| \leq \tau_M \delta < \|Au^{\delta,n(\delta)-1}_{\alpha(\delta)} - f^\delta\|, \quad \tau_M > 1, \, \delta > 0, \qquad (2.6.14)$$

that is, if for the first time the condition

$$\|Au^{\delta,n(\delta)}_{\alpha(\delta)} - f^\delta\| \leq \tau_M \delta \quad \tau_M > 1, \, \delta > 0 \qquad (2.6.15)$$

holds. Here $\tau_M > 1$ is a fixed parameter.

Chapter 3
Inverse Source Problems with Final Overdetermination

Inverse source problems for evolution PDEs $u_t = Au + F, t \in (0, T_f]$, represent a well-known area in inverse problems theory and has many engineering applications. These problems play a key role in providing estimations of unknown and inaccessible source terms involved in the associated mathematical model, using some measured data. An inverse problem with the final overdetermination $u_T := u(T), T > 0$, for one-dimensional heat equation has first been considered by A.N. Tikhonov in study of geophysical problems [93]. In this work the heat equation with prescribed lateral and final data is studied in half-plane and the uniqueness of the bounded solution is proved. For parabolic equations in a bounded domain, when in addition to usual initial and boundary conditions, a solution is given at the final time, well-posedness of inverse source problem has been proved by Isakov [46, 47].

In this chapter we study inverse source problems for one-dimensional linear evolution equations with final overdetermination. The main reason of considering of this class of problems is to demonstrate an understanding of major concepts of inverse problems. First, we discuss the most widely studied and classical inverse source problem with final overdetermination for one-dimensional heat equation. Along with the other objectives of this chapter, we will show how the unique regularized solution obtained by Tikhonov regularization applied to the input-output operator $\Phi : H \mapsto \tilde{H}$ is related to the Singular Value Decomposition (SDV) of the Moore-Penrose inverse $\Phi^\dagger := (\Phi^*\Phi)^{-1}\Phi^*$, of this operator. For this aim, an adjoint problem, corresponding to each considered inverse source problem is introduced. This allows to derive a gradient formula for the Fréchet derivative $J'_\alpha(F)$ of the regularized Tikhonov functional $J_\alpha(F)$ via the weak solution of the adjoint problem. Using the gradient formula and solving the equation $J'_\alpha(F) = 0$ with respect to the unknown source F, a quasi-solution of the inverse problem is obtained as a singular value expansion of the Moore-Penrose inverse $\Phi^\dagger$. This approach allows to show that final data inverse source problems for parabolic and hyperbolic equations, as well as backward parabolic problem, have the same filter function $q(\sigma_n, \alpha)$, but with different singular values σ_n.

3.1 Inverse Source Problem for Heat Equation

In this section we demonstrate Tikhonov regularization method and Singular Value Decomposition of Compact Operators for heat source identification problem with final data measurement. In the variable coefficient case we will develop an adjoint problem approach based on weak solution theory for PDEs. This approach allows not only to derive an explicit gradient formula for the Fréchet derivative of the regularized Tikhonov functional, but also permits to prove the Lipschitz continuity of this gradient [38]. In the constant coefficient case, we will prove that the solution of the nonlinear equation $J'_\alpha(F) = 0$ is the singular value expansion of the unique minimizer of the regularized Tikhonov functional.

Consider one dimensional heat conduction in a finite homogeneous rod, occupying the interval $(0, l)$, where the source term has the following form: $F(x)G(x, t)$. Assume that this process is governed by the following mixed initial-boundary value problem:

$$\begin{cases} u_t = (k(x)u_x)_x + F(x)G(x, t), & (x, t) \in \Omega_T; \\ u(x, 0) = f(x), & x \in (0, l); \\ u(0, t) = 0, \ u_x(l, t) = 0, & t \in (0, T], \end{cases} \quad (3.1.1)$$

where $\Omega_T := \{0 < x < l, \ 0 < t \leq T\}$ and $T > 0$ is the final time and $k(x) > 0$ is the thermal conductivity. The initial and boundary conditions are assumed to be homogeneous without loss of generality, since the parabolic problem is linear.

It is assumed in the considered model (3.1.1) that the *spacewise-dependent source* $F(x)$ *is unknown and needs to be identified from the measured temperature*

$$u_T(x) := u(x, T), \quad x \in (0, l), \quad (3.1.2)$$

at the final time $T > 0$. The *final time measured output* u_T is assumed to be non-smooth, that is $u_T \in L^2(0, l)$ and can contain a random noise.

The problem of determining the pair $\langle u(x, t), F(x)\rangle$ in (3.1.1) and (3.1.2) will be defined as an *inverse source problem with final data for heat equation*. For a given function $F(x)$ from some class of admissible sources, the problem (3.1.1) will be referred as a *direct (or forward) problem*.

An analysis of inverse problems given in this chapter will be based on weak solution theory for PDEs, since in practice, input and output data may not be smooth. Assuming that these data satisfy the following conditions

$$\begin{cases} k \in L^\infty(0, l), \ 0 < k_0 \leq k(x) \leq k_1 < \infty, \\ f, F \in L^2(0, l), \ G \in L^2(0, T; L^\infty(0, l)), \end{cases} \quad (3.1.3)$$

where the norm in the Banach space $L^2(0, T; L^\infty(0, l))$ is defined as follows:

3.1 Inverse Source Problem for Heat Equation

$$\|G\|_{L^2(0,T;L^\infty(0,l))} := \left(\int_0^T \|G\|^2_{L^\infty(0,l)} dt\right)^{1/2},$$

We will define the *weak solution of the direct problem* (3.1.1) as a function $u \in L^2(0, T; \mathcal{V}(0, l))$, with $u_t \in L^2(0, T; H^{-1}(0, l))$, satisfying the integral identity

$$\int_0^l (u_t v + k(x)u_x v_x)dx = \int_0^l F(x)G(x,t)v(x)dx, \text{ a.e. } t \in (0, T), \quad (3.1.4)$$

and the condition $u(x, 0) = f(x)$, $x \in (0, l)$, for every $v \in \mathcal{V}(0, l)$, where $\mathcal{V}(0, l) := \{v \in H^1(0, l) : v(0) = 0\}$. Here and below $H^1(0, l)$ is the Sobolev space of functions which consists of all integrable real functions $v : (0, l) \mapsto \mathbb{R}$ such that $v, v_x \in L^2(0, l)$ and $H^{-1}(0, l)$ is the dual space of $H^1(0, l)$.

Remark that the definition $u \in L^2(0, T; \mathcal{V}(0, l))$ of the weak solution does not directly imply the continuity of this solution with respect to the time variable $t \in (0, T]$. Hence, it is not clear in this definition the pointwise values $u(x, 0)$ and $u(x, T)$ at the initial and final times. As it is follows from the theorem below, the conditions $u \in L^2(0, T; \mathcal{V}(0, l))$ and $u_t \in L^2(0, T; H^{-1}(0, l))$ imply $u \in C([0, T]; L^2(0, l))$, which means that values $u(x, 0)$ and $u(x, T)$ are well defined (see [24], Chap. 5.9, Theorem 3).

Theorem 3.1.1 *Let $u \in L^2(0, T; \mathcal{V}(0, l))$ and $u_t \in L^2(0, T; H^{-1}(0, l))$.*
(i) Then $u \in C([0, T]; L^2(0, l))$.
(ii) The real valued function $t \mapsto \|u(\cdot, t)\|^2_{L^2(0,l)}$ is weakly differentiable and

$$\frac{1}{2}\frac{d}{dt}\int_0^l u^2(x,t)dx = \int_0^l u_t(x,t)u(x,t)dx, \text{ for a.e. } t \in (0, T);$$

(iii) There exists a constant $C_0 = C_0(T) > 0$ such that the following estimate holds:

$$\max_{t \in [0,T]} \|u\|_{L^2(0,l)} \leq C_0 \left(\|u\|_{L^2(0,T;\mathcal{V}(0,l))} + \|u_t\|_{L^2(0,T;H^{-1}(0,l))}\right). \quad (3.1.5)$$

Theorem 3.1.1 with estimate (3.1.5) shows, in particular, that as a function of $t \in [0, T]$, $u(x, t)$ can be identified with its continuous representative and the initial and final time values $u(x, 0)$ and $u(x, T)$, $x \in [0, l]$, make sense.

The above defined weak solution satisfies the following estimate (see [24], Chap. 7.1, Theorem 2):

$$\max_{t \in [0,T]} \|u\|_{L^2(0,l)} + \|u\|_{L^2(0,T;\mathcal{V}(0,l))} + \|u_t\|_{L^2(0,T;H^{-1}(0,l))} \leq$$
$$C_1 \left[\|FG\|_{L^2(0,T;L^2(0,l))} + \|f\|_{L^2(0,l)}\right], \quad (3.1.6)$$

where the constant $C_1 > 0$ depends $T, l > 0$ and the coefficient $k(x)$.

In subsequent analysis of the inverse problem (3.1.1) and (3.1.2), we need to write the norm of the unknown spacewise-dependent source $F(x)$ in a separate term on the right hand side of (3.1.5). For this aim, we use the condition in (3.1.3) for the function $G(x, t)$ and the estimate:

$$\|FG\|_{L^2(0,T;L^2(0,l))} := \left(\int_0^T \|FG\|^2_{L^2(0,l)} dt\right)^{1/2} \le$$
$$\|F\|_{L^2(0,l)} \left(\int_0^T \|G\|^2_{L^\infty(0,l)} dt\right)^{1/2} =: \|F\|_{L^2(0,l)} \|G\|_{L^2(0,T;L^\infty(0,l))}. \quad (3.1.7)$$

Using this estimate in (3.1.6) we get:

$$\max_{t\in[0,T]} \|u\|_{L^2(0,l)} + \|u\|_{L^2(0,T;\mathcal{V}(0,l))} + \|u_t\|_{L^2(0,T;H^{-1}(0,l))} \le$$
$$C_1 \left(\|F\|_{L^2(0,l)} \|G\|_{L^2(0,T;L^\infty(0,l))} + \|f\|_{L^2(0,l)}\right). \quad (3.1.8)$$

If the coefficient $k(x)$ is smooth and the initial data $f(x)$ is more regular, then the regularity of the above defined weak solution increases. Specifically, if, in addition to condition (3.1.3), the conditions

$$k \in C^1(0, l), \quad f \in \mathcal{V}(0, l), \quad G_t \in L^2(0, T; L^2(0, l)) \quad (3.1.9)$$

also hold, then we can define the *regular weak solution of the direct problem* (3.1.1). As the above weak solution, this solution is defined initially in $L^2(0, T; H^2(0, l)) \cap L^\infty(0, T; \mathcal{V}(0, l))$, with $u_t \in L^2(0, T; L^2(0, l))$. Then, in order to get a mapping to a better space, it can be proved that if $u \in L^2(0, T; H^2(0, l))$, with $u_t \in L^2(0, T; L^2(0, l))$, then $u \in C([0, T]; H^1(0, l))$ (see Theorem 4 [24], Chap. 5.9). This result is given by the following extension of Theorem 3.1.1.

Theorem 3.1.2 *Let $u \in L^2(0, T; H^2(0, l))$ and $u_t \in L^2(0, T; L^2(0, l))$.*
(i) Then $u \in C([0, T]; H^1(0, l))$, and hence, $u \in C([0, T] \times (0, l))$.
(ii) Furthermore, the following estimate holds:

$$\max_{t\in[0,T]} \|u\|_{H^1(0,l)} \le C_2 \left(\|u\|_{L^2(0,T;H^2(0,l))} + \|u_t\|_{L^2(0,T;L^2(0,l))}\right), \quad (3.1.10)$$

where the constant $C_2 > 0$ depends only on $T > 0$ and $l > 0$.

This is an adopted version of Theorem 4 (see [24], Chap. 5.9) to the considered case. This theorem with estimate (3.1.10), implies that, for a fixed $t \in [0, T]$ the function $u(x, t)$, as a one dimensional function of $x \in (0, l)$, is an element of the space $H^1(0, l)$.

3.1 Inverse Source Problem for Heat Equation

For the regular weak solution the following estimate holds (see [24], Chap. 7.1, Theorem 5):

$$\operatorname*{ess\,sup}_{t \in [0,T]} \|u\|_{\mathcal{V}(0,l)} + \|u\|_{L^2(0,T;H^2(0,l))} + \|u_t\|_{L^2(0,T;L^2(0,l))} \leq$$
$$C_3 \left[\|F\|_{L^2(0,l)} \|G\|_{L^2(0,T;L^\infty(0,l))} + \|f\|_{\mathcal{V}(0,l)} \right], \quad (3.1.11)$$

where the constant $C_3 > 0$ depends on the final time $T > 0$ and the constants $l, k_0, k_1 > 0$.

Details of the weak solution theory for evolution PDEs can be found in [24].

3.1.1 Compactness of Input-Output Operator and Fréchet Gradient

Let $u(x, t; F)$ be the weak solution of (3.1.1) for a given $F \in L^2(0, l)$. We introduce the *input-output operator* $\Phi : L^2(0, l) \mapsto L^2(0, l)$, defined as $(\Phi F)(x) := u(x, T; F)$, $x \in (0, l)$, that is, operator Φ transforms each admissible *input* $F(x)$ to the *output* $u(x, T; F)$. Then the inverse source problem, defined by (3.1.1) and (3.1.2), can be reformulated as the following operator equation:

$$\Phi F = u_T, \quad u_T \in L^2(0, l). \quad (3.1.12)$$

Let us analyze the compactness of the input-output operator $\Phi : L^2(0, l) \mapsto L^2(0, l)$. Due to linearity of the problem, we may assume here that the initial data is zero: $f(x) = 0$.

Lemma 3.1.1 *Let conditions (3.1.3) and (3.1.9) hold. Then the input-output operator*

$$\Phi : F \in L^2(0, l) \mapsto u(x, T; F) \in L^2(0, l) \quad (3.1.13)$$

is a linear compact operator.

Proof Let $\{F_m\} \subset L^2(0, l)$ be a bounded sequence. Then the sequence of corresponding regular weak solutions $\{u(x, t; F_m)\}$ of the direct problem (3.1.1) is bounded, due to estimate (3.1.11). In particular, the sequence $\{u_m\}$, $u_m := u(x, T; F_m)$, is bounded in the norm of $\mathcal{V}(0, l)$. By the compact imbedding $\mathcal{V}(0, l) \hookrightarrow L^2(0, l)$, there is a subsequence $\{u_n\}$ of $\{u_m\}$ which converges in L^2-norm. Hence the input-output operator transforms any bounded in $L^2(0, l)$ sequence $\{F_m\}$ to the compact sequence $\{u_n\}$ in $L^2(0, l)$. □

Remark 3.1.1 If the initial data is not zero, then assuming that condition (3.1.9) also hold, we obtain that the input-output operator is an affine compact operator.

It follows from Lemma 3.1.1 that the inverse problem (3.1.12) or (3.1.1 and 3.1.2) is ill-posed. Then we may use Tikhonov regularization

$$J_\alpha(F) = J(F) + \frac{1}{2}\alpha \|F\|^2_{L^2(0,l)}, \tag{3.1.14}$$

introducing the Tikhonov functional

$$J(F) = \frac{1}{2}\|u(\cdot, T; F) - u_T\|^2_{L^2(0,l)}, \quad F \in L^2(0,l) \tag{3.1.15}$$

and the parameter of regularization $\alpha > 0$. Since Φ is a linear compact operator, by Theorem 2.1.1 the unique minimum $F_\alpha \in L^2(0,l)$ of the regularized functional (3.1.14) exists and is the solution of the regularized normal equation

$$(\Phi^*\Phi + \alpha I)F_\alpha = \Phi^* u_T, \tag{3.1.16}$$

where $\Phi^* : L^2(0,l) \mapsto L^2(0,l)$ is the adjoint operator. This unique solution can be written as follows:

$$F_\alpha = \mathcal{R}_\alpha u_T, \quad \mathcal{R}_\alpha := (\Phi^*\Phi + \alpha I)^{-1}\Phi^* : L^2(0,l) \mapsto L^2(0,l), \tag{3.1.17}$$

where the operator $\mathcal{R}_\alpha : L^2(0,1) \mapsto L^2(0,l)$, $\alpha > 0$, is a regularization strategy.

Consider first the non-regularized case, when $\alpha = 0$. The following lemma, which establishes the *gradient formula for Tikhonov functional* (3.1.15), allows to derive the Fréchet gradient $J'(F)$ of Tikhonov functional (3.1.15) via the solution of corresponding (unique) adjoint problem solution $\psi(x,t)$ and known source $G(x,t)$.

Lemma 3.1.2 *Let conditions (3.1.3) hold. Denote by $u(x,t;F)$ is the weak solution of the parabolic problem (3.1.1) corresponding to a given $F \in L^2(0,l)$. Then Tikhonov functional (3.1.15) is Fréchet differentiable and for the Fréchet gradient $J'(F)$ the following gradient formula holds:*

$$J'(F)(x) = (\psi(x,\cdot;F), G(x,\cdot))_{L^2(0,T)} := \int_0^T \psi(x,t;F)G(x,t)dt, \tag{3.1.18}$$

for a.e. $x \in (0,l)$, where $\psi(x,t;F)$ is the weak solution of the following adjoint problem:

$$\begin{cases} \psi_t = -(k(x)\psi_x)_x, & (x,t) \in \Omega_T, \\ \psi(x,T) = u(x,T;F) - u_T(x), & x \in (0,l), \\ \psi(0,t) = 0, \; \psi_x(l,t) = 0, & t \in [0,T]. \end{cases} \tag{3.1.19}$$

3.1 Inverse Source Problem for Heat Equation

Proof Assuming that $F, \ F + \delta F \in L^2(0, l)$, we calculate the increment $\delta J(F) := J(F + \delta F) - J(F)$ of functional (3.1.15):

$$\delta J(F) = \int_0^l [u(x, T; F) - u_T(x)]\delta u(x, T; F)dx$$
$$+ \frac{1}{2}\int_0^l [\delta u(x, T; F)]^2 dx, \qquad (3.1.20)$$

where $\delta u(x, t; F) := u(x, t; F + \delta F) - u(x, t; F)$ is the weak solution of the following parabolic problem

$$\begin{cases} \delta u_t = (k(x)\delta u_x)_x + \delta F(x)G(x, t), \ (x, t) \in \Omega_T, \\ \delta u(x, 0) = 0, \ x \in (0, l), \\ \delta u(0, t) = 0, \ \delta u_x(l, t) = 0, \ t \in (0, T]. \end{cases} \qquad (3.1.21)$$

We transform the first integral on the right hand side of (3.1.20), assuming that $\psi(x, t; F)$ and $\delta u(x, t; F)$ are the solutions of problems (3.1.19) and (3.1.21), accordingly. We have:

$$\int_0^l [u(x, T; F) - u_T(x)]\delta u(x, T; F)dx = \int_0^l \psi(x, T; F)\delta u(x, t; F)dx$$
$$= \int_0^l \left\{\int_0^T (\psi(x, t; F)\delta u(x, t; F))_t \, dt\right\} dx$$
$$= \int\int_{\Omega_T} \{\psi_t(x, t; F)\delta u(x, t; F) + \psi(x, t; F)\delta u_t(x, t; F)\} dxdt$$
$$= \int\int_{\Omega_T} \{-(k(x)\psi_x)_x \delta u(x, t; F) + \psi(x, t; F)(k(x)\delta u_x)_x\} dxdt$$
$$+ \int\int_{\Omega_T} \delta F(x)G(x, t)\psi(x, t; F)dxdt$$
$$= \int_0^T \{-k(x)\psi_x\delta u + k(x)\delta u_x\psi\}_{x=0}^{x=l} dt +$$
$$\int\int_{\Omega_T} \delta F(x)G(x, t)\psi(x, t; F)dxdt.$$

Taking into account here the boundary conditions given in (3.1.19) and (3.1.21) we obtain the following integral identity:

$$\int_0^l [u(x, T; F) - u_T(x)]\delta u(x, T; F)dx = \int\int_{\Omega_T} \psi(x, t; F)\delta F(x)G(x, t)dxdt,$$

for all $F, \delta F \in L^2(0, l)$. With formula (3.1.20) this implies:

$$\delta J(F) = \int_0^l \left(\int_0^T \psi(x, t; F) G(x, t) dt \right) \delta F(x) dx +$$
$$\frac{1}{2} \int_0^l [\delta u(x, T; F)]^2 dx. \quad (3.1.22)$$

The last right hand side integral is of the order $\mathcal{O}\left(\|\delta F\|_{L^2(0,l)}^2\right)$ by estimate (3.1.8). This completes the proof of lemma. □

One of most important issues in numerical solution of inverse problems is the Lipschitz continuity of the Fréchet gradient. Lemma 3.4.4 in Sect. 3.4.3 shows that if $\{F^{(n)}\}$ is the a sequence of iterations defined by the gradient algorithm, then $J(F^{(n)})$ is a monotone decreasing sequence.

Lemma 3.1.3 *Let conditions (3.1.3) hold. Then Fréchet gradient of Tikhonov functional (3.1.15) is Lipschitz continuous,*

$$\|J'(F + \delta F) - J'(F)\|_{L^2(0,l)} \leq L_1 \|\delta F\|_{L^2(0,l)}, \ \forall F, \delta F \in L^2(0, l), \quad (3.1.23)$$

with the Lipschitz constant $L_1 > 0$, depending on $T, k_0, l > 0$ and the norm $M_G = \|G\|_{L^2(0,T;L^\infty(0,l))}$ of the given source term $G(x, t)$.

Proof By definition,

$$\|J'(F + \delta F) - J'(F)\|_{L^2(0,l)} = \left(\int_0^l \left(\int_0^T \delta\psi(x, t; F) G(x, t) dt \right)^2 dx \right)^{1/2},$$

where $\delta\psi(x, t; F) := \psi(x, t; F + \delta F) - \psi(x, t; F)$ is the weak solution of the problem:

$$\begin{cases} \delta\psi_t = -(k(x)\delta\psi_x)_x, \ (x, t) \in \Omega_T; \\ \delta\psi(x, T) = \delta u(x, T; F), \ x \in (0, l); \\ \delta\psi(0, t) = 0, \ \delta\psi_x(l, t) = 0, \ t \in [0, T]. \end{cases} \quad (3.1.24)$$

Applying the Hölder inequality we deduce that

$$\|J'(F + \delta F) - J'(F)\|_{L^2(0,l)} \leq M_G \|\delta\psi(\cdot, \cdot; F)\|_{L^2(0,T;L^2(0,l))}. \quad (3.1.25)$$

Hence, we need to prove that the right hand side norm is of the order $\mathcal{O}\left(\|\delta F\|_{L^2(0,l)}\right)$. Multiplying both sides of Eq. (3.1.24) by $\delta\psi(x, t; F)$, integrating on $[0, l]$, applying the integration by parts formula and then using the homogeneous boundary conditions in (3.1.24), we obtain the following identity:

3.1 Inverse Source Problem for Heat Equation

$$\frac{1}{2}\frac{d}{dt}\int_0^l (\delta\psi(x,t;F))^2\,dx = \int_0^l k(x)\,(\delta\psi_x(x,t;F))^2\,dx,\ t\in [0,T].$$

Integrating this identity on $[0,T]$ using the condition $\delta\psi(x,T) = \delta u(x,T;F)$ in (3.1.24) and applying then Poincaré inequality $\|\psi_x\|_{L^2(0,l)}^2 \geq (2/l^2)\|\psi\|_{L^2(0,l)}^2$ to the right hand side integral, we conclude:

$$\int_0^T\int_0^l (\delta\psi(x,t;F))^2\,dx \leq \frac{l^2}{4k_0}\int_0^l (\delta u(x,T;F))^2\,dx$$

where $k_0 > 0$ is the constant in (3.1.3). The last integral is of the order $\mathcal{O}\left(\|\delta F\|_{L^2(0,l)}^2\right)$ by estimate (3.1.8), since $\delta u(x,t;F)$ is the solution of problem (3.1.21). With estimate (3.1.25) this implies the desired result. □

An implementation of the gradient formula (3.1.18) in Conjugate Gradient Algorithm will be discussed in subsequent sections. Obviously, it is an implicit formula, since the solution $\psi(x,t;F)$ of the adjoint problem (3.1.19) depends on the unknown source. We may use now formula (3.1.18) in (3.1.14) to obtain the gradient formula for the regularized functional.

Corollary 3.1.1 *For the Fréchet gradient of the regularized functional $J_\alpha(F)$, $\alpha > 0$, defined by (3.1.14) the following formula holds:*

$$J'_\alpha(F)(x) = \int_0^T \psi(x,t;F)G(x,t)\,dt + \alpha F(x),\ \text{a.e. } x\in(0,l). \quad (3.1.26)$$

Taking into account Corollary 2.5.1 in Sect. 2.5, we may rewrite the gradient formula (3.1.26) for the regularized functional via the input-output mapping $\Phi : L^2(0,l) \mapsto L^2(0,l)$ as follows:

$$J'_\alpha(F) = \Phi^*(\Phi F - u_T) + \alpha F,\ F\in L^2(0,l). \quad (3.1.27)$$

With the necessary condition $J'_\alpha(F) = 0$, formula (3.1.26) allows to derive the following representation for the unique regularized solution of the inverse source problem with final data.

Corollary 3.1.2 *The unique regularized solution $F_\alpha \in L^2(0,l)$ of the inverse problem (3.1.1) and (3.1.2) can be represented as follows:*

$$F_\alpha(x) = -\frac{1}{\alpha}\int_0^T \psi(x,t;F_\alpha)G(x,t)\,dt,\ \text{a.e. } x\in(0,l),\ \alpha > 0. \quad (3.1.28)$$

This representation is an analogue of the representation

$$u_\alpha^\delta = -\frac{1}{\alpha}A^*(Au_\alpha^\delta - f^\delta),\ f^\delta \in \tilde{H},$$

for the solution of the regularized normal equation $(A^*A + \alpha I)u_\alpha^\delta = f^\delta$, introduced in Sect. 2.5.

Remark 3.1.2 Let the source term in the heat equation (3.1.1) has the separable form $F(x)G(t)$. As it was shown above, the regularized solution $F_\alpha \in L^2(0, l)$ of the inverse problem (3.1.1) and (3.1.2) is defined uniquely for all $G \in L^2(0, T)$, as a solution of the normal equation (3.1.16). However, the non-regularized solution $F \in L^2(0, l)$ of the inverse problem (3.1.1) and (3.1.2) may not be unique for all $G \in L^2(0, T)$. Specifically, the time dependent source $G \in L^2(0, T)$ must ensure the fulfillment of the condition $\sigma_n \neq 0$, for all $n \in \mathbb{N}$, as we will see in Sect.3.1.2. For example, if the function $G(t)$ satisfies the conditions $G(t) > 0$ and $G \in L^2(0, T)$, then $\sigma_n \neq 0$, for all $n \in \mathbb{N}$ and the non-regularized solution is unique. In this case the gradient formula (3.1.18) has the form:

$$J'(F)(x) = \int_0^T \psi(x, t; F)G(t)dt \equiv \Phi^*(\Phi F - u_T)(x), \qquad (3.1.29)$$

a.e. for all $x \in (0, l)$.

3.1.2 Singular Value Decomposition of Input-Output Operator

In this subsection we will assume that the source term in the heat equation (3.1.1) has the separable form $F(x)G(t)$, the time dependent source function $G(t)$ satisfies the conditions

$$\|G\|_{L^2(0,l)} > 0, \quad G(t) \geq 0, \quad t \in [0, T] \qquad (3.1.30)$$

and the initial data is zero: $f(x) = 0$. Specifically, consider the problem of determining the unknown space-wise dependent source $F \in L^2(0, l)$ in

$$\begin{cases} u_t = (k(x)u_x)_x + F(x)G(t), & (x, t) \in \Omega_T; \\ u(x, 0) = 0, & x \in (0, l); \\ u(0, t) = 0, \quad u_x(l, t) = 0, & t \in (0, T]; \\ u_T(x) := u(x, T), & x \in (0, l). \end{cases} \qquad (3.1.31)$$

where $u_T \in L^2(0, l)$ is a noise free output satisfying the following *consistency conditions*:

$$u_T(0) = 0, \quad u'(l) = 0. \qquad (3.1.32)$$

3.1 Inverse Source Problem for Heat Equation

First we will prove that the input-output operator $\Phi : L^2(0, l) \mapsto L^2(0, l)$ corresponding to the inverse problem (3.1.31) is a self-adjoint operator. Then we will use the singular value decomposition (SVD) of compact operators on a Hilbert space given in Sect. 2.4, to obtain singular value expansion of the regularized solution $F_\alpha \in L^2(0, l)$ of this inverse problem.

Lemma 3.1.4 *Let conditions (3.1.3) and (3.1.30) hold. Then the input-output operator Φ, defined by (3.1.13) and corresponding to the inverse problem (3.1.31), is self-adjoint and positive defined. Moreover,*

$$(\Phi \varphi_m)(x) = \kappa_m \varphi_m(x), \quad (3.1.33)$$

that is, $\{\kappa_m, \varphi_m\}$ is the eigensystem of the input-output operator Φ, where $\{\varphi_n\}_{n=1}^\infty$ are orthonormal eigenvectors of the differential operator $\ell : \mathcal{V}(0, l) \mapsto L^2(0, l)$ defined by

$$\begin{cases} (\ell \varphi)(x) := -(k(x)\varphi'(x))' = \lambda \varphi(x), \quad x \in (0, l); \\ \varphi(0) = 0, \quad \varphi'(l) = 0, \end{cases} \quad (3.1.34)$$

corresponding to eigenvalues $\{\lambda_n\}_{n=1}^\infty$, and

$$\kappa_n = \int_0^T \exp(-\lambda_n(T - t))G(t)dt, \quad n = 1, 2, 3, \ldots . \quad (3.1.35)$$

Proof Evidently, the differential operator $\ell : \mathcal{V}(0, l) \mapsto L^2(0, l)$ defined by (3.1.34) is self-adjoint and positive defined. Hence there exists an infinite number of positive eigenvalues $\{\lambda_n\}_{n=1}^\infty$, $0 < \lambda_1 < \lambda_2 < \lambda_3 \ldots$, repeated according to their (finite) multiplicity that $\lambda_n \to \infty$, as $n \to \infty$.

We assume, without loss of generality, that the eigenvectors $\{\varphi_n\}_{n=1}^\infty$, corresponding to the eigenvalues λ_n, are normalized (dividing the both sides of (3.1.34) by $\|\varphi_n\|_{L^2(0,l)}$). Then, these eigenvectors $\{\varphi_n\}_{n=1}^\infty$ form an orthonormal basis in $L^2(0, l)$ for the operator ℓ, defined by (3.1.34). Indeed, the identities $-(k(x)\varphi_n'(x))' = \lambda \varphi_n(x)$ and $-(k(x)\varphi_m'(x))' = \lambda \varphi_m(x)$ imply:

$$(\lambda_n - \lambda_m)\varphi_n(x)\varphi_m(x) = \frac{d}{dx}\left[k(x)\varphi_n'(x)\varphi_m(x) - k(x)\varphi_n(x)\varphi_m'(x)\right].$$

Integrate this identity on $[0, l]$ and use the boundary conditions (3.1.34):

$$(\lambda_n - \lambda_m)\varphi_n(x)\varphi_m(x) = 0, \quad \forall n, m = 1, 2, 3, \ldots .$$

Therefore, for $\lambda_n \neq \lambda_m$,

$$(\varphi_n(x), \varphi_m(x))_{L^2(0,l)} := \int_0^l \varphi_n(x)\varphi_m(x)dx = 0, \quad n \neq m,$$

which means that $(\varphi_n, \varphi_m)_{L^2(0,l)} = \delta_{nm}$, $n, m = 1, 2, 3, \ldots$. Here and below δ_{nm} is the Kronecker delta.

With this orthonormal basis we use the Fourier series expansion

$$u(x, t) = \sum_{n=1}^{\infty} u_n(t)\varphi_n(x) \tag{3.1.36}$$

of the solution of the initial boundary value problem given by the first three equations of (3.1.31). Here

$$u_n(t) = F_n \int_0^t \exp(-\lambda_n(t - \tau))G(\tau)d\tau, \quad t \in (0, T] \tag{3.1.37}$$

is the solution of the Cauchy problem

$$\begin{cases} u'_n(t) + \lambda_n u_n(t) = F_n G(t), \quad t \in (0, T], \\ u_n(0) = 0, \end{cases}$$

for each $n = 1, 2, 3, \ldots$ and $F_n := (F, \varphi_n)_{L^2(0,l)}$ is the Fourier coefficient of the function $F \in L^2(0, l)$.

Now we can use (3.1.36) and (3.1.37) to obtain the Fourier series expansion of the input-output operator Φ, defined as $(\Phi F)(x) := u(x, T; F)$:

$$(\Phi F)(x) = \sum_{n=1}^{\infty} \kappa_n F_n \varphi_n(x), \tag{3.1.38}$$

where κ_n is defined by (3.1.35).

Remark that, $\{\varphi_n\}_{n=1}^{\infty}$ are eigenvectors of the input-output operator Φ, corresponding to eigenvalues $\{\kappa_n\}_{n=1}^{\infty}$. To show this, we replace $F(x)$ by $\varphi_m(x)$ in (3.1.38). Then we get (3.1.33):

$$(\Phi\varphi_m)(x) = \sum_{n=1}^{\infty} \kappa_n (\varphi_m, \varphi_n)_{L^2(0,l)} \varphi_n(x) = \kappa_m \varphi_m(x). \tag{3.1.39}$$

Furthermore, $\Phi : L^2(0, l) \mapsto L^2(0, l)$ is a self-adjoint operator, that is,

$$(\Phi F, \tilde{F})_{L^2(0,l)} = (F, \Phi\tilde{F})_{L^2(0,l)}, \quad \forall F, \tilde{F} \in L^2(0, l).$$

3.1 Inverse Source Problem for Heat Equation

Indeed,

$$(\Phi F, \tilde{F})_{L^2(0,l)} := \left(\sum_{n=1}^{\infty} \kappa_n F_n \varphi_n(x), \sum_{m=1}^{\infty} \tilde{F}_m \varphi_m(x) \right)_{L^2(0,l)}$$

$$= \sum_{n=1}^{\infty} \kappa_n F_n \tilde{F}_n = (F, \Phi \tilde{F})_{L^2(0,l)},$$

where $\tilde{F}_n := (\tilde{F}, \varphi_n)_{L^2(0,l)}$ is the Fourier coefficient of the function $\tilde{F} \in L^2(0, l)$. Hence, $\Phi = \Phi^*$, where $\Phi^* : L^2(0, l) \mapsto L^2(0, l)$ is the adjoint operator. □

Thus, the input-output operator $\Phi : L^2(0, l) \mapsto L^2(0, l)$ is a self-adjoint operator with eigenvectors $\{\varphi_n\}_{n=1}^{\infty}$, corresponding to different eigenvalues $\{\kappa_n\}_{n=1}^{\infty}$. It follows from (3.1.39) that $\|\Phi\|_{L^2(0,l)} = \kappa_1$. Hence,

$$\|\Phi\|_{L^2(0,l)} = \kappa_1 > \kappa_2 > \kappa_3 > \dots . \quad (3.1.40)$$

Further, formula (3.1.38) implies that

$$(\Phi^* \Phi F)(x) = \sum_{n=1}^{\infty} \kappa_n^2 F_n \varphi_n(x). \quad (3.1.41)$$

By Definition 2.4.1 in Sect. 2.4, the square root of eigenvalues κ_n^2 of the self-adjoint operator $\Phi^* \Phi$ is defined as *the singular values of the input-output operator* Φ, that is, $\sigma_n := \kappa_n > 0$, $n = 1, 2, 3, \dots$. Hence, *the singular system* $\{\sigma_n, u_n, v_n\}$ *for the self-adjoint operator input-output operator* Φ is $\{\kappa_n, \varphi_n, \varphi_n\}$, according to the definition in Sect. 2.4. Indeed, $u_n = \varphi_n$ and $v_n := \Phi u_n / \|\Phi u_n\| = \Phi \varphi_n / \|\Phi \varphi_n\|$. But $\Phi \varphi_n = \kappa_n \varphi_n$, by (3.1.33), and $\|\Phi \varphi_n\| = \kappa_n$. This implies that $v_n = \kappa_n \varphi_n / |\kappa_n|$. Since $\kappa_n > 0$, we have $v_n = \varphi_n$.

Therefore, if conditions (3.1.30) hold, then $\kappa_n > 0$, for all $n = 1, 2, 3, \dots$, as formula (3.1.35) shows. In this case, *the singular system* $\{\kappa_n, \varphi_n, \varphi_n\}$ *for the self-adjoint operator input-output operator* Φ *is uniquely determined by the eigensystem* $\{\kappa_n, \varphi_n\}$ *of the differential operator* (3.1.34).

The following theorem gives a series representation of the unique solution of the regularized form (3.1.16) of the normal equation.

Theorem 3.1.3 *Let conditions* (3.1.3) *hold. Assume that* $u_T \in L^2(0, l)$ *is a noise free measured output data defined in* (3.1.31). *Then for the unique minimum* $F_\alpha \in L^2(0, l)$ *of the regularized functional* (3.1.14) *the following singular value expansion holds:*

$$F_\alpha(x) = \sum_{n=1}^{\infty} \frac{q(\alpha; \kappa_n)}{\kappa_n} u_{T,n} \varphi_n(x), \quad x \in (0, l), \quad (3.1.42)$$

where

$$q(\alpha; \kappa) = \frac{\kappa^2}{\kappa^2 + \alpha} \tag{3.1.43}$$

is the filter function, $\alpha > 0$ is the parameter of regularization, κ_n, $n = 1, 2, 3, \ldots$, defined by formula (3.1.35) are the eigenvalues of the input-output operator Φ, $u_{T,n} := (u_T, \varphi_n)$ is the nth Fourier coefficient of $u_T(x)$ and $\varphi_n(x)$ are the normalized eigenfunctions corresponding to the eigenvalues λ_n of the operator $\ell : \mathcal{V}(0, l) \mapsto L^2(0, l)$ defined by (3.1.34).

Proof Let $u(x, t; F)$ be the weak solution of the direct problem defined by the first three equations of (3.1.31), for a given admissible $F \in L^2(0, l)$. Then for the output $u(x, T; F)$ the expansion (3.1.38) holds:

$$u(x, T; F) = \sum_{n=1}^{\infty} \kappa_n F_n \varphi_n(x), \tag{3.1.44}$$

We use this with the nth Fourier coefficients $u_{T,n} := (u_T, \varphi)_{L^2(0,l)}$ of the noise free measured output datum $u_T \in L^2(0, l)$ in (3.1.14) and (3.1.15). Then we have:

$$J_\alpha(F) = \frac{1}{2} \sum_{n=1}^{\infty} \left[(\kappa_n F_n - u_{T,n})^2 + \alpha F_n^2 \right],$$

Transforming the nth term under the sum we get:

$$J_\alpha(F) = \frac{1}{2} \sum_{n=1}^{\infty} \left[(\kappa_n^2 + \alpha) \left(F_n - \frac{\kappa_n}{\kappa_n^2 + \alpha} u_{T,n} \right)^2 + \frac{\alpha}{\kappa_n^2 + \alpha} u_{T,n}^2 \right],$$

The regularized functional achieves minimum value if

$$F_n - \frac{\kappa_n}{\kappa_n^2 + \alpha} u_{T,n} = 0,$$

as the right hand side shows. This defines the nth Fourier coefficient of the unique minimum $F_\alpha \in L^2(0, l)$ of the regularized functional (3.1.14):

$$F_{\alpha,n} = \frac{\kappa_n}{\kappa_n^2 + \alpha} u_{T,n}.$$

3.1 Inverse Source Problem for Heat Equation

Substituting this into the Fourier series expansion

$$F_\alpha(x) = \sum_{n=1}^{\infty} F_{\alpha,n} \varphi_n(x) \qquad (3.1.45)$$

of the function $F_\alpha \in L^2(0,l)$ we arrive at the required expansion (3.1.42). □

The representation formula

$$F_\alpha(x) = -\frac{1}{\alpha} \int_0^T \psi(x,t; F_\alpha) G(t) dt, \quad \text{a.e. } x \in (0,l), \; \alpha > 0, \qquad (3.1.46)$$

which follows from (3.1.28) for the solution of the inverse problem (3.1.31) (the case $G(x,t) \equiv G(t)$), shows the dependence of the solution $F_\alpha \in L^2(0,l)$ of this inverse problem on the solution of the adjoint problem (3.1.19). The same dependence can be observed in the gradient formulae (3.1.18) and (3.1.29). Hence it is useful to illustrate a relationship between the singular value expansion (3.1.45) and the representation formula (3.1.46). Specifically, the following example shows their equivalence [40].

Example 3.1.1 The relationship between the singular value expansion (3.1.42) and the representation formula (3.1.46)

We assume here that $k(x) \equiv k = const > 0$. Then the above defined eigenvalues λ_n and normalized eigenfunctions $\varphi_n(x)$ of the operator $(\ell\varphi)(x) := -k\varphi''(x)$, defined by (3.1.34), are

$$\lambda_n = [(n-1/2)\pi/l]^2, \quad \varphi_n(x) = \sqrt{2/l}\sin(\sqrt{\lambda_n}x), \; n = 1,2,3,\ldots. \qquad (3.1.47)$$

Replacing in formula (3.1.35) λ_n with $k\lambda_n$, we get:

$$\kappa_n = \int_0^T \exp(-k\lambda_n(T-t)) G(t) dt, \quad n = 1,2,3,\ldots. \qquad (3.1.48)$$

Let us apply the Fourier method to the adjoint problem (3.1.19) using the above defined normalized eigenfunctions $\varphi_n(x)$ corresponding to the eigenvalues λ_n. We have:

$$\psi(x,t; F) = \sum_{n=1}^{\infty} \psi_n(t; F) \varphi_n(x), \qquad (3.1.49)$$

where the $\psi_n(t)$ is the solution of the backward Cauchy problem

$$\begin{cases} \psi_n'(t) = \lambda_n \psi_n(t), & t \in [0,T), \\ \psi_n(T) = u_n(T; F_n) - u_{T,n}. \end{cases}$$

The solution of this problem is

$$\psi_n(t; F_n) = [u_n(T; F_n) - u_{T,n}] \exp(-k\lambda_n(T-t)), \quad n = 1, 2, 3, \ldots . \quad (3.1.50)$$

Now we use the Fourier series expansions (3.1.45) and (3.1.49) in the representation formula (3.1.46) to obtain the formula

$$F_{\alpha,n} = -\frac{1}{\alpha} \int_0^T \psi_n(t; F_{\alpha,n}) G(t) dt, \quad \text{a.e. } x \in (0, l), \ \alpha > 0, \quad (3.1.51)$$

for the nth Fourier coefficient of the unique minimum $F_\alpha \in L^2(0, l)$ of the regularized functional (3.1.14). On the other hand, assuming $F(x) = F_\alpha(x)$ in (3.1.50), multiplying it both sides by $-G(t)/\alpha \neq 0$, integrating on $[0, T]$ and then using formula (3.1.35) for κ_n, we obtain:

$$-\frac{1}{\alpha} \int_0^T \psi_n(t; F_{\alpha,n}) G(t) dt = -\frac{\kappa_n}{\alpha} [u_n(T; F_\alpha) - u_{T,n}].$$

Comparing this formula with (3.1.51) we deduce:

$$F_{\alpha,n} = -\frac{\kappa_n}{\alpha} [u_n(T; F_{\alpha,n}) - u_{T,n}]. \quad (3.1.52)$$

Further, it follows from the expansions (3.1.36) and (3.1.44) that between the nth Fourier coefficient $F_{\alpha,n}$ of the unique minimum $F_\alpha \in L^2(0, l)$ and nth Fourier coefficient $u_n(T; F_\alpha)$ of the output data $u(x, T; F\alpha)$ the following relationship holds:

$$u_n(T; F_\alpha) = \kappa_n F_{\alpha,n}. \quad (3.1.53)$$

Using this formula in (3.1.52) we conclude:

$$\frac{u_n(T; F_{\alpha,n})}{\kappa_n} = -\frac{\kappa_n}{\alpha} [u_n(T; F_{\alpha,n}) - u_{T,n}],$$

which yields:

$$u_n(T; F_\alpha) = q(\alpha, \kappa_n) u_{T,n}, \quad n = 0, 1, 2, 3, \ldots . \quad (3.1.54)$$

This is a relationship between the Fourier coefficients $u_n(T; F_\alpha)$ and $u_{T,n}$ of the *output data* $u(x, T; F_\alpha)$ and the *measured output data* $u_T(x)$, via the filter function $q(\alpha, \kappa)$, defined by formula (3.1.43).

Formulae (3.1.53) and (3.1.54) permit to derive the nth Fourier coefficient $F_{\alpha,n}$ of the unique minimum $F_\alpha \in L^2(0, l)$ of the regularized functional (3.1.14) and nth Fourier coefficient $u_{T,n}$ of the measured output data $u_T(x)$ via the filter function $q(\alpha, \kappa)$:

3.1 Inverse Source Problem for Heat Equation

$$F_{\alpha,n} = \frac{q(\alpha, \kappa_n)}{\kappa_n} u_{T,n}, \quad n = 0, 1, 2, 3, \ldots. \tag{3.1.55}$$

The singular value expansion (3.1.42) for the unique minimum $F_\alpha \in L^2(0, l)$ follows from the expansion (3.1.45) and formula (3.1.55). □

3.1.3 Picard Criterion and Regularity of Input/Output Data

We deduce here several important corollaries from Theorem 3.1.3.

Let us first assume that $\alpha = 0$. Then, by formula (3.1.43), $q(\alpha, \sigma) = 1$, and expansion (3.1.42) becomes:

$$F(x) = \sum_{n=1}^{\infty} \frac{1}{\kappa_n} u_{T,n}\, \varphi_n(x), \quad x \in (0, l), \tag{3.1.56}$$

This expansion exactly coincides with the singular value expansion (2.4.13) given in Picard's Theorem 2.4.1 in Sect. 2.4. Since $F_0 = \Phi^\dagger u_T$ is the solution of the normal equation (3.1.16) with $\alpha = 0$ and $\Phi^\dagger : L^2(0, l) \mapsto L^2(0, l)$ is Moore-Penrose inverse of the input-output operator, we conclude that (3.1.56) is a singular value expansion of the solution of the normal equation, when $\alpha = 0$.

The Picard criterion given in Theorem 2.4.1 of Sect. 2.4 and applied to the inverse problem (3.1.31) implies that the series (3.1.56) converges if and only if $u_T \in \mathcal{N}(\Phi^*)^\perp$ and the following condition holds:

$$\sum_{n=1}^{\infty} \frac{u_{T,n}^2}{\kappa_n^2} < \infty. \tag{3.1.57}$$

Remember that $\{\varphi_n(x)\}$ forms a complete orthonormal system in $L^2(0, l)$ and therefore for the self-adjoint input-output operator $\Phi : L^2(0, l) \mapsto L^2(0, l)$ we deduce that $\mathcal{N}(\Phi) = \mathcal{N}(\Phi^*) = \{0\}$. On the other hand, $\mathcal{D}(\Phi^\dagger) := \mathcal{R}(\Phi) \oplus \mathcal{R}(\Phi)^\perp$, by the definition of the Moore-Penrose inverse. Since $\mathcal{N}(\Phi^*) = \mathcal{R}(\Phi)^\perp$, by Theorem 2.2.1, this implies that $\mathcal{D}(\Phi^\dagger) = \mathcal{R}(\Phi)$ and $\mathcal{D}(\Phi^\dagger)$ is dense in $L^2(0, l)$.

The fulfilment of the Picard criterion (3.1.57) depends on two factors: κ_n and $u_{T,n}$. As formula (3.1.35) shows, an asymptotic behavior of the first factor κ_n depends on a class where the input $G(t)$ is defined. The second factor is the Fourier coefficient $u_{T,n} := (u_T, \varphi_n)_{L^2(0,l)}$ of the measured output data $u_T(x)$. As a consequence, the convergence of the series (3.1.56) depends on the input data $G(t)$ and the measured output data $u_T(x)$. Based on this observation, we will analyze here the relationship between the convergence of the series (3.1.56) and the smoothness/regularity of the input $G \in L^2(0, T)$ and the measured output $u_T \in L^2(0, l)$ data.

The begin the following result which shows a necessary condition for solvability of the inverse problem (3.1.31) with the noise free measured output data $u_T \in \mathcal{N}(\Phi^*)^\perp \equiv \overline{\mathcal{R}(\Phi)}$.

Corollary 3.1.3 *Let conditions (3.1.3) hold. Assume that $u_T \in L^2(0, l)$ is a noise free measured output defined in (3.1.31). If the time dependent function $G \in L^2(0, T)$ satisfies conditions (3.1.30), then*

$$\sum_{n=1}^{\infty} n^2 u_{T,n}^2 < \infty, \ u_{T,n} := (u_T, \varphi_n)_{L^2(0,l)}. \tag{3.1.58}$$

is the necessary condition for solvability of the inverse problem (3.1.31).

Proof Using formula (3.1.35) for κ_n, we can estimate the singular values $\sigma_n := |\kappa_n|$ as follows:

$$0 < \kappa_n = \left| \int_0^T \exp(-\lambda_n(T-t))G(t)dt \right|$$
$$\leq \left(\int_0^T \exp(-2\lambda_n(T-t))dt \right)^{1/2} \|G\|_{L^2[0,T]} =$$
$$\frac{1}{\sqrt{2\lambda_n}}[1 - \exp(-2\lambda_n T)]^{1/2} \|G\|_{L^2[0,l]}. \tag{3.1.59}$$

It is known that the eigenvalues λ_n of the differential operator $\ell : \mathcal{V}(0, l) \mapsto L^2(0, l)$, defined in (3.1.34), are of order $\lambda_n = \mathcal{O}(n^2)$. Then we conclude from the above estimate that, $\kappa_n = \mathcal{O}(1/n)$. Using this property in the Picard criterion (3.1.57) we arrive at the condition (3.1.58). □

To find out *what means the necessary condition (3.1.58) in terms of the measured output* $u_T \in L^2(0, l)$, we need some auxiliary results from Fourier series theory. It is known that if $\{\varphi_n(x)\}_{n=1}^\infty$ is an *orthonormal* basis of $L^2(0, l)$ then the Fourier series of $f \in L^2(0, l)$ with respect to this basis converges in L^2-norm if and only if Parseval's identity

$$\|f\|_{L^2(0,l)}^2 = \sum_{n=1}^{\infty} f_n^2, \tag{3.1.60}$$

holds. Here and below, $f_n := (f, \varphi_n)_{L^2(0,l)}$ is the nth Fourier coefficient of the element $f \in L^2(0, l)$. Let now assume that $\{\varphi_n(x)\}_{n=1}^\infty \subset L^2(0, l)$ is the orthonormal basis defined by the eigenfunctions of the Sturm-Liouville problem (3.1.34). Then the theorem below shows that in this case Parseval's identity holds.

3.1 Inverse Source Problem for Heat Equation

Theorem 3.1.4 *Let $\{\varphi_n(x)\}_{n=1}^{\infty} \subset L^2(0, l)$ be the orthonormal basis defined by the eigenfunctions of the Sturm-Liouville problem (3.1.34). Then Parseval's identity (3.1.60) holds for $f \in L^2(0, l)$ with respect the basis $\{\varphi_n(x)\}$ and the Fourier series*

$$f(x) = \sum_{n=1}^{\infty} f_n \varphi_n(x) \tag{3.1.61}$$

of an element $f \in L^2(0, l)$ with respect to this basis converges in L^2-norm.

Proof of this theorem can be found in [30].

Now, replacing in this theorem $f \in L^2(0, l)$ by $u_T \in L^2(0, l)$ and reformulating the above assertion in terms of the measured output data we conclude that if $u_T \in L^2(0, l)$, then Parseval's identity holds:

$$\|u_T\|_{L^2(0,l)}^2 = \sum_{n=1}^{\infty} u_{T,n}^2, \tag{3.1.62}$$

More precisely, *having only the condition $u_T \in L^2(0, l)$ we can not guarantee the fulfilment of the Picard criterion (3.1.58)*. In other words, the more smoothness of u_T is required in order to fulfil the Picard criterion. To find out such a class of functions for which the Picard criterion (3.1.58) holds, we need the following auxiliary result.

Lemma 3.1.5 *Let conditions (3.1.3) hold. Assume that $\{\varphi_n(x)\}_{n=1}^{\infty} \subset L^2(0, l)$ is the orthonormal basis defined by the eigenfunctions corresponding to the eigenvalues $\{\lambda_n\}_{n=1}^{\infty}$ of the Sturm-Liouville problem (3.1.34). Then the following assertions hold true:*
(i) The system $\{\varphi_n(x)/\sqrt{\lambda_n}\}_{n=1}^{\infty}$ forms an orthonormal basis of $\mathcal{V}(0, l) := \{v \in H^1(0, l) : v(0) = 0\}$ with the inner product

$$(\ell w, v)_{L^2(0,l)} := \int_0^l k(x) w'(x) v'(x) dx. \tag{3.1.63}$$

(ii) If $v \in \mathcal{V}(0, l)$, then the Fourier series

$$v(x) = \sum_{n=1}^{\infty} v_n \varphi_n(x), \tag{3.1.64}$$

with the Fourier coefficients $v_n = (v, \varphi_n)_{L^2(0,l)}$, converges in the norm of the space $\mathcal{V}(0, l) \subset H^1(0, l)$.

Proof Multiplying the both sides of $\ell \varphi_n(x) = \lambda_n \varphi_n(x)$ by $\varphi_m(x)$ we conclude:

$$\begin{cases} (\ell \varphi_n, \varphi_n)_{L^2(0,l)} = \lambda_n \|\varphi_n\|_{L^2(0,l)}^2 = \lambda_n, \ m = n, \\ (\ell \varphi_n, \varphi_m)_{L^2(0,l)} = \lambda_n (\varphi_n, \varphi_m)_{L^2(0,l)} = 0, \ m \neq m. \end{cases} \tag{3.1.65}$$

Hence

$$(\ell\varphi_n/\sqrt{\lambda_n}, \varphi_m/\sqrt{\lambda_m})_{L^2(0,l)} = \begin{cases} 1, & m = n, \\ 0, & m \neq m. \end{cases}$$

This implies that $\{\varphi_n(x)/\sqrt{\lambda_n}\}_{n=1}^\infty$ forms an orthonormal subset of $\mathcal{V}(0, l)$. To prove that this subset is an orthonormal basis of $\mathcal{V}(0, l)$, it is sufficient to show that $(\ell\varphi_n, v)_{L^2(0,l)} = 0$ implies $v \equiv 0$, for any $v \in \mathcal{V}(0, l)$. Indeed,

$$(\ell\varphi_n, v)_{L^2(0,l)} = \lambda_n(\varphi_n, v)_{L^2(0,l)} = 0.$$

Since $\{\varphi_n(x)\}_{n=1}^\infty$ forms an orthonormal basis of $L^2(0, l)$, so, $\lambda_m(\varphi_n, u)_{L^2(0,l)} = 0$ implies $v \equiv 0$. Therefore, $\{\varphi_n(x)/\sqrt{\lambda_n}\}_{n=1}^\infty$ forms an orthonormal basis of $\mathcal{V}(0, l)$ with the inner product (3.1.63).

To prove the second assertion (ii), first we remark that the norms $\|v\|_{\mathcal{V}(0,l)}$ and $\|v'\|_{L^2(0,l)}$ are equivalent, due to the Dirichlet boundary condition $\varphi_n(0) = 0$ in (3.1.34). Furthermore, conditions $0 < k_0 \leq k(x) \leq k_1 < \infty$ imply an equivalence of the energy norm $\|v\|_\ell := (\ell v, v)_{L^2(0,l)}^{1/2}$, defined by the inner product (3.1.63), and the norm $\|v'\|_{L^2(0,l)}$.

Consider now the series

$$v(x) = \sum_{n=1}^\infty \mu_n \frac{\varphi_n(x)}{\sqrt{\lambda_n}}, \qquad (3.1.66)$$

with the Fourier coefficients $\mu_n = (\ell v, \varphi_n(x)/\sqrt{\lambda_n})_{L^2(0,l)}$ (with respect to the basis $\{\varphi_n(x)/\sqrt{\lambda_n}\}$). Since the system $\{\varphi_n(x)/\sqrt{\lambda_n}\}_{n=1}^\infty$ forms an orthonormal basis of $\mathcal{V}(0, l)$, the series (3.1.66) converges in the norm of $\mathcal{V}(0, l)$. Comparing the series (3.1.64) and (3.1.66) we deduce that $\mu_n = \sqrt{\lambda_n} v_n$. This implies that the series (3.1.64) also converges in the norm of $\mathcal{V}(0, l)$. □

Corollary 3.1.4 *Let conditions of Corollary 3.1.3 hold. Assume that the output $u_T(x)$ satisfies the consistency conditions (3.1.32). Then the necessary condition (3.1.58) for solvability of the inverse problem (3.1.31) is equivalent to the condition $u_T \in \mathcal{V}(0, l)$.*

Proof It follows from (3.1.65) and (3.1.66), applied to the measured output data $u_T \in \mathcal{V}(0, l)$, that

$$\|u_T\|_\ell^2 := (\ell u_T, u_T)_{L^2(0,l)} = \sum_{n=1}^\infty \lambda_n u_{T,n}^2, \quad u_{T,n} := (u_T, \varphi_n)_{L^2(0,l)}. \qquad (3.1.67)$$

Since the norms $\|u_T\|_\ell := \left((\ell u_T, u_T)_{L^2(0,l)}\right)^{1/2}$ and $\|u_T\|_{\mathcal{V}(0,l)}$ are equivalent and $\lambda_n = \mathcal{O}(n^2)$, the series (3.1.67) convergences if and only if condition (3.1.58) holds. □

3.1 Inverse Source Problem for Heat Equation

It is important to note that the Picard criterion (3.1.57) implicitly includes also the requirement

$$\kappa_n \neq 0, \text{ for all } n = 1, 2, 3 \ldots . \tag{3.1.68}$$

This condition coincides with unique solvability of the inverse problem (3.1.31). Indeed, it follows from the Fourier series expansion (3.1.38) of the input-output operator Φ that if $\kappa_m = 0$ for some m, then the mth Fourier coefficient $F_m := (F, \varphi_m)_{L^2(0,l)}$ of the unknown function $F \in L^2(0, l)$ can not be determined uniquely. More precisely, the solution $F \in L^2(0, l)$ of the inverse problem (3.1.31) can only be determined up to the additive term $C_m \varphi_m(x)$, where C_m is an arbitrary constant. In this context, (3.1.30) are the sufficient conditions ensuring the fulfilment of condition (3.1.68).

Remark also that if $\kappa_n = 0$, for some n, then the input-output operator Φ is not bijective, as the Fourier series expansion (3.1.38) shows. Hence in this case the bijectivity condition of Theorem 2.5.1 in Sect. 2.5 is not satisfied.

The following result shows that the conditions $0 < G_0 \leq G(t) \leq G_1 < \infty$, for all $t \in [0, T]$ can be considered as the sufficient condition ensuring the fulfilment of the requirement (3.1.68).

Corollary 3.1.5 *Let conditions (3.1.3) hold. Assume that $u_T \in L^2(0, l)$ is a noise free measured output defined in (3.1.31). If the time dependent function $G \in L^2(0, T)$ satisfies the conditions*

$$0 < G_0 \leq G(t) \leq G_1 < \infty, \ t \in [0, T], \tag{3.1.69}$$

then the inverse problem (3.1.31) is uniquely solvable if and only if

$$\sum_{n=1}^{\infty} n^4 u_{T,n}^2 < \infty. \tag{3.1.70}$$

Proof By using conditions (3.1.69), we can easily derive the estimate

$$0 < \frac{G_0}{\lambda_n}[1 - \exp(-\lambda_n T)] \leq \kappa_n \leq \frac{G_1}{\lambda_n}[1 - \exp(-\lambda_n T)], \ n = 1, 2, 3 \ldots . \tag{3.1.71}$$

It follows from estimate (3.1.71) that $0 < \kappa_n = \mathcal{O}(1/n^2)$, due to $\lambda_n = \mathcal{O}(n^2)$. Then the Picard criterion (3.1.57) becomes to the condition (3.1.70). □

Corollary 3.1.6 *The unique solvability (3.1.70) is equivalent to the conditions $u_T \in H^2(0, l)$ and $u_T(0) = u'_T(l) = 0$.*

We leave the proof of this corollary as an exercise for the reader. □

Summarizing the results given in Corollaries 3.1.3, 3.1.4, 3.1.5 and 3.1.6 we conclude that *an increase of smoothness of an input $G(t)$ forces an increase of smoothness of an output u_T*. Specifically, there exists a direct relationship between the smoothness of an input and output, for unique solvability of an inverse problem: smooth input requires an appropriate smooth output.

3.1.4 The Regularization Strategy by SVD. Truncated SVD

Let us assume now that $\alpha \neq 0$. Then, by Definition 2.5.1 in Sect. 2.5, the operator

$$\mathcal{R}_\alpha := \left(\Phi^*\Phi + \alpha I\right)^{-1} \Phi^* : L^2(0, l) \mapsto L^2(0, l), \; \alpha > 0, \quad (3.1.72)$$

is the regularization strategy. According to (3.1.42), the following singular value expansion holds:

$$(\mathcal{R}_\alpha u_T)(x) = \sum_{n=1}^{\infty} \frac{q(\alpha; \kappa_n)}{\kappa_n} u_{T,n} \varphi_n(x), \; \alpha \geq 0, \quad (3.1.73)$$

where $u_T \in L^2(0, l)$ is the noise free measured output.

Comparing the damping parameters $1/\kappa_n$ and $q(\alpha; \kappa)/\kappa$ in the singular value expansions (3.1.56) and (3.1.73), we find that Tikhonov regularization is reflected in the expansion (3.1.73) of the regularized solution $F_\alpha(x)$ as the factor $q(\alpha; \kappa)$. For this reason, the function $q(\alpha; \kappa)$, $0 < q(\alpha; \kappa) \leq 1$, $\alpha > 0$, given by (3.1.43) is also called a *regularizing filter*, corresponding to Tikhonov regularization. Remark that in Sect. 2.5 the factor $q(\alpha; \sigma)$ was defined as the filter function.

As noted above, the measured output $u_T \in L^2(0, l)$ in the inverse source problem (3.1.31) is never known as an exact datum. Instead the noisy data $u_T^\delta \in L^2(0, l)$ with a given noise level $\delta > 0$ is usually available:

$$\|u_T^\delta - u_T\|_{L^2(0,l)} \leq \delta, \; \delta > 0. \quad (3.1.74)$$

In this case the inverse source problem (3.1.31) with the noisy output can be reformulated as the following operator equation:

$$\Phi F^\delta = u_T^\delta, \; u_T^\delta \in L^2(0, l). \quad (3.1.75)$$

Applying Theorem 3.1.3 to this inverse problem and using the expansion (3.1.73), we can define *formally* the regularization strategy as follows:

$$(\mathcal{R}_{\alpha(\delta)} u_T^\delta)(x) = \sum_{n=1}^{\infty} \frac{q(\alpha(\delta); \kappa_n)}{\kappa_n} u_{T,n}^\delta \varphi_n(x),$$
$$q(\alpha(\delta); \kappa) = \frac{\kappa^2}{\kappa^2 + \alpha(\delta)}, \quad (3.1.76)$$

3.1 Inverse Source Problem for Heat Equation

where $u^\delta_{T,n} := (u^\delta_T, \varphi_n)_{L^2(0,l)}$ is the nth Fourier coefficient of the noisy output. The left hand side defines the regularized solution $F^\delta_\alpha(x) := (\mathcal{R}_{\alpha(\delta)} u^\delta_T)(x)$ of the inverse problem, i.e. the unique solution of the regularized form of the normal equation with noisy output $u^\delta_T \in L^2(0,l)$:

$$(\Phi^*\Phi + \alpha(\delta)I) F^\delta_{\alpha(\delta)} = \Phi^* u^\delta_T. \tag{3.1.77}$$

The following theorem shows that if the parameter of regularization $\alpha(\delta) > 0$, depending on the noise level $\delta > 0$, is chosen properly, then the regularizing filter $q(\alpha; \kappa)$, given in (3.1.76), generates a *convergent regularization strategy* for the of inverse source problem (3.1.31).

Theorem 3.1.5 *Let conditions (3.1.3) hold. Assume that the time dependent source $G(t)$ in (3.1.31) satisfies (3.1.69). Suppose that $u^\delta_T \in L^2(0,l)$ is the noisy data given by (3.1.74). If the parameter of regularization is chosen so that*

$$\alpha(\delta) \to 0 \text{ and } \frac{\delta^2}{\alpha(\delta)} \to 0, \text{ as } \delta \to 0, \tag{3.1.78}$$

then the regularized solution $F^\delta_\alpha(x) := (\mathcal{R}_{\alpha(\delta)} u^\delta_T)(x)$ given by (3.1.76) converges in L^2-norm to the unique solution $F = \Phi^\dagger u_T$ given by (3.1.56), of the operator equation (3.1.12), as $\delta \to 0$, that is,

$$\|\mathcal{R}_{\alpha(\delta)} u^\delta_T - F\|_{L^2(0,l)} \to 0, \text{ as } \delta \to 0. \tag{3.1.79}$$

Proof Let us estimate the above norm using the singular value expansions (3.1.56) and (3.1.76). We have:

$$\|\mathcal{R}_{\alpha(\delta)} u^\delta_T - F\|^2_{L^2(0,l)} = \sum_{n=1}^\infty \left| \left(u^\delta_{T,n} \frac{q(\kappa_n, \alpha(\delta))}{\kappa_n} - u_{T,n} \frac{1}{\kappa_n} \right) \right|^2$$

$$= \sum_{n=1}^\infty \left| (u^\delta_{T,n} - u_{T,n}) \frac{q(\kappa_n, \alpha(\delta))}{\kappa_n} + u_{T,n} \left(\frac{q(\kappa_n, \alpha(\delta))}{\kappa_n} - \frac{1}{\kappa_n} \right) \right|^2$$

$$\leq 2 \sum_{n=1}^\infty (u^\delta_{T,n} - u_{T,n})^2 \frac{q^2(\kappa_n, \alpha(\delta))}{\kappa_n^2} + 2 \sum_{n=1}^\infty u^2_{T,n} \frac{\alpha(\delta)^2}{(\kappa_n^2 + \alpha(\delta))^2 \kappa_n^2}. \tag{3.1.80}$$

By the property $(q(\kappa, \alpha)/\kappa)^2 \leq 1/(4\alpha)$ of the filter function, given by estimate (2.5.14) in Sect. 2.5, the first summand on the right hand side of (3.1.80) can be estimated as follows:

$$2\sum_{n=1}^{\infty}(u_{T,n}^{\delta}-u_{T,n})^2\left(\frac{q(\kappa_n,\alpha(\delta))}{\kappa_n}\right)^2$$
$$\leq \frac{1}{2\alpha(\delta)}\sum_{n=1}^{\infty}(u_{T,n}^{\delta}-u_{T,n})^2 = \frac{\delta^2}{2\alpha(\delta)}. \tag{3.1.81}$$

The right hand side tends to zero, by condition (3.1.78) of the theorem.

To estimate the second right hand summand in (3.1.80), we rewrite it in the following form

$$2\sum_{n=1}^{\infty} u_{T,n}^2 \frac{\alpha(\delta)^2}{(\kappa_n^2+\alpha(\delta))^2 \kappa_n^2}$$
$$= \sum_{n=1}^{N} \frac{2\alpha(\delta)^2}{(\kappa_n^2+\alpha(\delta))^2} \frac{u_{T,n}^2}{\kappa_n^2} + \sum_{n=N+1}^{\infty} \frac{2\alpha(\delta)^2}{(\kappa_n^2+\alpha(\delta))^2} \frac{u_{T,n}^2}{\kappa_n^2}, \tag{3.1.82}$$

and estimate each right hand side terms separately.

It follows from estimate (3.1.71) that $\kappa_n = \mathcal{O}(1/n^2)$. This implies that no matter how small the parameter of regularization $\alpha(\delta) > 0$ was selected, there exists such a natural number $N = N(\delta)$ that

$$\min_{1\leq n\leq N}\kappa_n^2 = \kappa_N^2 \geq \sqrt{\alpha(\delta)} > \kappa_{N+1}^2. \tag{3.1.83}$$

Then $N(\delta) = \mathcal{O}(\alpha(\delta)^{-1/8}) \to \infty$, as $\delta \to 0$. This and (3.1.83) allow to estimate the partial sum on right hand side of (3.1.82) as follows:

$$\sum_{n=1}^{N} \frac{2\alpha(\delta)^2}{(\kappa_n^2+\alpha(\delta))^2} \frac{u_{T,n}^2}{\kappa_n^2} \leq \frac{2\alpha(\delta)}{(1+\sqrt{\alpha(\delta)})^2} \sum_{n=1}^{N(\delta)} \frac{u_{T,n}^2}{\kappa_n^2}. \tag{3.1.84}$$

By the solvability condition (3.1.57), the sum is finite and the right hand side tends to zero, as $\delta \to 0$, by the first condition of (3.1.78).

The last right hand side term in (3.1.82) can easily be estimated by using the inequality $\alpha^2/(\kappa_n^2+\alpha)^2 < 1$, for all $n > N(\delta)+1$:

$$\sum_{n=N(\delta)+1}^{\infty} \frac{2\alpha(\delta)^2}{(\kappa_n^2+\alpha(\delta))^2} \frac{u_{T,n}^2}{\kappa_n^2} \leq 2\sum_{n=N(\delta)+1}^{\infty} \frac{u_{T,n}^2}{\kappa_n^2}.$$

Again, by the solvability condition (3.1.57), the right hand side tends to zero, as $\delta \to 0$. This, with (3.1.80), (3.1.81), (3.1.82) and (3.1.84) implies

3.1 Inverse Source Problem for Heat Equation

$$\|R_{\alpha(\delta)}u_T^\delta - F\|_{L^2(0,l)} \leq$$

$$\frac{\delta^2}{2\alpha(\delta)} + \frac{2\alpha(\delta)}{(1+\sqrt{\alpha(\delta)})^2}\sum_{n=1}^{N(\delta)}\frac{u_{T,n}^2}{\kappa_n^2} + 2\sum_{n=N(\delta)+1}^{\infty}\frac{u_{T,n}^2}{\kappa_n^2}. \quad (3.1.85)$$

This estimate yields (3.1.79), since all the right hand side terms tend to zero, as $\delta \to 0$, by conditions (3.1.78). This completes the proof of the theorem. □

As it was introduced in Sect. 2.4, the Singular Value Decomposition (SVD) can also be used to obtain an approximation

$$F^N(x) = \sum_{n=1}^{N}\frac{1}{\kappa_n}u_{T,n}\,\varphi_n(x), \quad x \in (0, l), \quad (3.1.86)$$

of the solution (3.1.56) of the normal equation (3.1.16) without regularization ($\alpha = 0$). The proof scheme of Theorem 3.1.5 shows that the *in the case when $\alpha > 0$ SVD can also be used to obtain a regularization strategy in finite dimensional space* $L_N^2(0,l) \subset L^2(0,l)$. This strategy is called Truncated Singular Value Decomposition (TSVD), with Tikhonov regularization. Indeed, let the cutoff parameter $N = N(\delta)$ be defined as in the proof of the theorem, by (3.1.83): $N(\delta) = \mathcal{O}(\alpha(\delta)^{-1/8})$. We define the Nth partial sum

$$\begin{cases} \left(\mathcal{R}_{\alpha(\delta)}^{N(\delta)}u_T^\delta\right)(x) := \sum_{n=1}^{N(\delta)}\frac{q(\alpha;\kappa_n)}{\kappa_n}u_{T,n}^\delta\varphi_n(x), \; q(\alpha;\kappa) = \frac{\kappa^2}{\kappa^2+\alpha}, \\ \alpha \geq 0, \; \kappa_n = \int_0^T \exp(\lambda_n(T-t))G(t)dt, \; n = 1,2,3,\ldots \end{cases} \quad (3.1.87)$$

of the series (3.1.76) as an approximation of the regularization strategy. We first estimate the approximation error $\|\mathcal{R}_{\alpha(\delta)}^{N(\delta)}u_T^\delta - F\|$:

$$\|\mathcal{R}_{\alpha(\delta)}^{N(\delta)}u_T^\delta - F\|_{L^2(0,l)}^2 \leq 2\left\|\mathcal{R}_{\alpha(\delta)}^{N(\delta)}u_T^\delta - F^N\right\|_{L^2(0,l)}^2 + 2\left\|F^N - F\right\|_{L^2(0,l)}^2.$$

Following (3.1.80) we get:

$$\|\mathcal{R}_{\alpha(\delta)}^{N(\delta)}u_T^\delta - F\|_{L^2(0,l)}^2 \leq 4\sum_{n=1}^{N(\delta)}(u_{T,n}^\delta - u_{T,n})^2\frac{q^2(\kappa_n,\alpha(\delta))}{\kappa_n^2}$$

$$+ 4\sum_{n=1}^{N(\delta)}u_{T,n}^2\frac{\alpha(\delta)^2}{(\kappa_n^2+\alpha(\delta))^2\kappa_n^2} + 2\sum_{N(\delta)+1}^{\infty}\frac{u_{T,n}^2}{\kappa_n^2}$$

For estimating the first and the second right hand side norms we use estimates (3.1.81) and (3.1.84). Then deduce the following *error estimate* for this approximation:

$$\|\mathcal{R}_{\alpha(\delta)}^{N(\delta)} u_T^\delta - F\|_{L^2(0,l)}^2 \leq$$

$$\frac{\delta^2}{\alpha(\delta)} + \frac{4\alpha(\delta)}{(1+\sqrt{\alpha(\delta)})^2} \sum_{n=1}^{N(\delta)} \frac{u_{T,n}^2}{\kappa_n^2} + 2 \sum_{N(\delta)+1}^{\infty} \frac{u_{T,n}^2}{\kappa_n^2}. \quad (3.1.88)$$

Remark that the approximation error estimate (3.1.88) is almost the same as estimate (3.1.85) in Theorem 3.1.5.

In the numerical implementations it is convenient to use the special case $\alpha(\delta) \sim \delta$. Evidently, in this case conditions (3.1.78) of Theorem 3.1.5 hold.

In the numerical example below we demonstrate the role of the above theoretical results in the TSVD algorithm with Tikhonov regularization applied to the problem of determining the unknown space-wise dependent source $F \in L^2(0,l)$ in (3.1.31). In the numerical analysis we will consider the simplest version of the heat equation, assuming that $k(x) = 1$, that is, only Simpson's rule in the numerical integration is responsible for increase of an error.

Example 3.1.2 Implementation of TSVD: identification of an unknown space-wise dependent source in (3.1.31)

The function $u(x,t) = \sin(\pi x/2)\left(1 - \exp(-\pi^2 t/2)\right)$ $(x,t) \in [0,1] \times [0,1]$ is the exact solution of the heat equation $u_t = u_{xx} + F(x)G(t)$ with the source functions $F(x) = \sin(\pi x/2)$, $G(t) = (\pi^2/4)\left[1 + \exp(-\pi^2 t/2)\right]$, and with homogeneous boundary and initial conditions: $u(0,t) = u_x(1,t) = 0$, $u(x,0) = 0$. The synthetic noise free output is

$$u_T(x) = \left(1 - \exp(-\pi^2/2)\right)\sin(\pi x/2), \quad x \in [0,1]. \quad (3.1.89)$$

The random noisy output data $u_T^\delta(x)$, $\|u_T - u_T^\delta\|_{L^2(0,l)} \leq \delta$, $\delta > 0$, is generated from (3.1.89) using the MATLAB random number generator function "rand $(\cdot)$". The TSVD regularization strategy defined by (3.1.87) is used in determination of the approximate solution $F_\alpha^{\delta,N} := \mathcal{R}_{\alpha(\delta)}^{N(\delta)} u_T^\delta$ from the noisy data $u_T^\delta(x)$.

Note that the smoothness of the above functions $G(t)$ and $u_T(x)$ are enough to ensure the fulfilment of the unique solvability condition (3.1.70) in Corollary 3.1.5.

For synthetic noise free output data (3.1.89) ($\delta = 0$) the TSVD formula (3.1.86) is used to obtain the approximate solution F^N. The accuracy error $E(n; F_\alpha^{\delta,n}; \delta) := \|F - F_\alpha^{\delta,n}\|_{L^2(0,l)}$ obtained for the values $N = 3 \div 4$ of the cutoff parameter is 1.7×10^{-6}. For noisy output data this error increase drastically, when $N > 3$, even for the small value $\delta = 0.01$ of the noise level.

For the synthetic noisy output data $u_T^\delta(x)$, with $\delta = 0.01$ and $\delta = 0.6$, the TSVD regularization strategy (3.1.87) is used to obtain the approximate solution $F_\alpha^{\delta,N}$. The dependence of the accuracy error $E(n; F_\alpha^{\delta,n}; \delta)$ on the parameter of regularization α and the cutoff parameter N are given in Tables 3.1 and 3.2. Bold-faced values of the accuracy error correspond to the optimal values of the cutoff parameter and agree with the estimate $N(\delta) = \mathcal{O}(\alpha(\delta)^{-1/8})$. For the optimal value $N = 3$ the parameter of regularization α should small to ensure the accuracy, but, at the same time, should

3.1 Inverse Source Problem for Heat Equation

Table 3.1 Dependence of the accuracy error on the parameter of regularization α and the cutoff parameter N: $\delta = 0.01$

$N\backslash\alpha$	0	0.1	0.01	0.001
3	2.5×10^{-2}	1.2×10^{-1}	**1.4×10^{-2}**	1.5×10^{-2}
5	1.5×10^{-1}	1.2×10^{-1}	1.5×10^{-2}	3.1×10^{-2}
10	3.9×10^{-1}	1.2×10^{-1}	1.5×10^{-2}	3.2×10^{-2}

Table 3.2 Dependence of the accuracy error on the parameter of regularization α and the cutoff parameter N: $\delta = 0.6$

$N\backslash\alpha$	0	0.1	0.01	0.001
3	1.5×10^{0}	**1.4×10^{-1}**	2.7×10^{-1}	2.8×10^{-1}
10	2.4×10^{1}	1.4×10^{-1}	3.4×10^{-1}	1.9×10^{-1}
20	7.9×10^{1}	1.5×10^{-1}	3.4×10^{-1}	2.0×10^{-1}

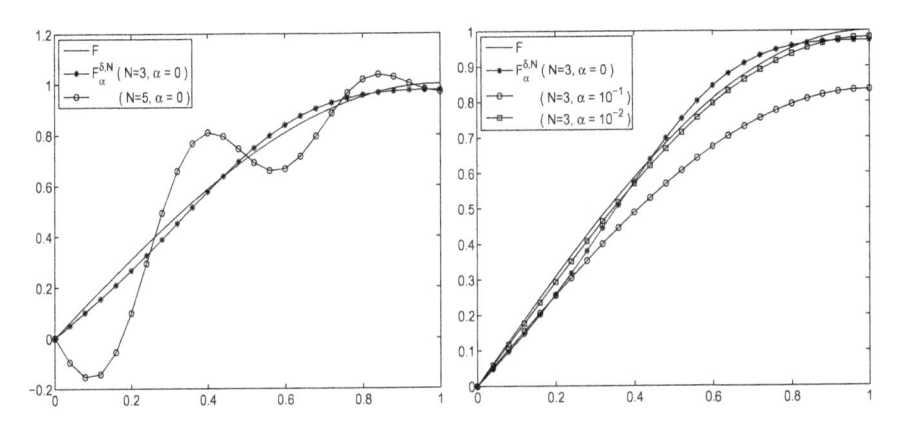

Fig. 3.1 Influence of cutoff (*left figure*) and regularization (*right figure*) parameters on reconstruction accuracy: $\delta = 0.01$

not be too small to ensure the stability. Both situations are is clearly illustrated in the third and fifth columns of the tables.

Although it would seem from the above tables that the difference between the errors for $N = 3$ and $N = 5$ are small, the deviation of the approximate solution F^N, corresponding to $N = 5$, from the exact solution is large enough, even at low-level noise $\delta = 0.01$, as the left Fig. 3.1 shows. The reconstructed from the noisy output data $u_T^{\delta}(x)$, with $\delta = 0.01$, approximate solutions F^N obtained for two values $\alpha_1 = 0.1$ and $\alpha_2 = 0.01$ of the parameter of regularization are plotted in the right Fig. 3.1. Better reconstruction obviously is obtained when $\alpha_2 = 0.01$, since in this case both conditions of the Theorem 3.1.5 are met approximately: $\alpha \ll 1$ and $\delta^2/\alpha \ll 1$. By increasing the noise level, impact of the first condition increases. Figure 3.2 illustrates the situation: for the higher noise level $\delta = 0.1$, better reconstruction is

Fig. 3.2 Influence of the parameter of regularization on accuracy of reconstruction: $\delta = 0.1$. Table reflects the accuracy error depending on cutoff and regularization parameters

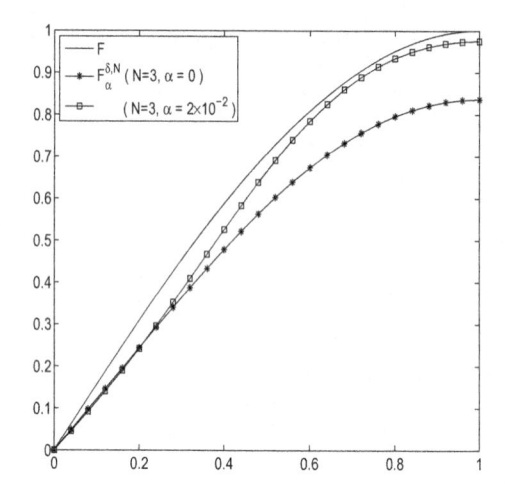

obtained when $\alpha_2 = 0.02$. Table here reflects the accuracy error depending on cutoff and regularization parameters. □

3.2 Inverse Source Problems for Wave Equation

In this section we formulate the spacewise-dependent source identification problem for $1D$ wave equation. Although the approach and methods given in the previous section can be used for this class of inverse source problems also (see [39]), we will demonstrate on a simple model that *the final data inverse source problem for wave equation has generic poor properties*.

Consider the problem of determining the unknown spatial load $F(x)$ in the hyperbolic problem

$$\begin{cases} u_{tt} = (\kappa(x)u_x)_x + F(x)G(t), & (x,t) \in \Omega_T := (0,l) \times (0,T); \\ u(x,0) = u_0(x), \quad u_t(x,0) = u_1(x), & x \in (0,l); \\ u(0,t) = 0, \quad u(l,t) = 0, & t \in (0,T), \end{cases} \quad (3.2.1)$$

from the *final state overdetermination*

$$u_T(x) := u(x,T), \quad x \in (0,l). \quad (3.2.2)$$

The one-dimensional Eq. (3.2.1) describes traveling-wave phenomena, and governs, for instance, the transverse displacements of an oscillating elastic string [99]. With Hooke's law, $u(x,t)$ is the displacement in the one-dimensional thin elastic string, $u(0,t)$ is a the deformation gradient, i.e. strain, at the left end $x = 0$ of a string, $\kappa > 0$ is the elastic modulus. The functions $F(x)$ and $G(t)$ represent a spatial

3.2 Inverse Source Problems for Wave Equation

and temporal load distributions along the one-dimensional elastic string. $u_0(x)$ and $u_1(x)$ are the initial strain and initial velocity, respectively.

We define the problem of determining the unknown spatial load $F(x)$, i.e. problem (3.2.1) and (3.2.2), as an *inverse source problem for wave equation with final state overdetermination*. Subsequently, for a given source $F \in L^2(0, l)$, the initial-boundary value problem (3.2.1) will be defined as the *direct problem*.

In practice, instead of the final state overdetermination (3.2.2) the *final velocity overdetermination*

$$v_T(x) := u_t(x, T), \quad x \in (0, l). \tag{3.2.3}$$

can be given as a *measured output*. In this case, (3.2.1) and (3.2.3) define the problem of determining the unknown spatial load $F(x)$, i.e. an *inverse source problem for wave equation with final velocity overdetermination*.

Consider the direct problem (3.2.1). In common from the point of view of general PDE theory, we will assume that

$$\begin{cases} 0 < \kappa_* \geq \kappa(x) \geq \kappa *^< +\infty, \quad \kappa \in C^1[0, l]; \\ u_0 \in \mathcal{V}(0, l), \ u_1 \in L^2(0, l), \ F \in L^2(0, l), \ G \in L^2(0, T). \end{cases} \tag{3.2.4}$$

where $\mathcal{V}(0, l) := \{v \in H^1(0, l) : v(0, t) = v(l, t) = 0\}$. It is known that (see [24], Chap. 7.2) under the conditions the weak solution $u \in L^2(0, T; \mathcal{V}(0, l))$, with $u_t \in L^2(0, T; L^2(0, l))$, $u_{tt} \in L^2(0, T; H^{-1}(0, l))$, of the initial-boundary value problem (3.2.1) exists, unique and satisfies the following estimate:

$$\max_{t \in [0,T]} \left(\|u\|_{\mathcal{V}(0,l)} + \|u_t\|_{L^2(0,l)} \right) + \|u_{tt}\|_{L^2(0,T;H^{-1}(0,l))} \leq$$
$$C_0 \left(\|F\|_{L^2(0,l)} \|G\|_{L^2(0,T)} + \|u_0\|_{\mathcal{V}(0,l)} + \|u_1\|_{L^2(0,l)} \right), \tag{3.2.5}$$

where the constant $C_0 > 0$ depends on l, $T > 0$ and the constants in (3.2.4).

In view of Theorem 3.1.1 this weak solution belongs to $L^\infty(0, T; \mathcal{V}(0, l))$, with $u_t \in L^\infty(0, T; L^2(0, l))$, $u_{tt} \in L^2(0, T; H^{-1}(0, l))$ (see also [24], Chap. 7.2, Theorem 5).

The following preliminary result will be useful in the sequel.

Lemma 3.2.1 *Let conditions (3.2.4) holds. Then the weak solution* $u \in L^2(0, T; \mathcal{V}(0, l))$ *of the initial-boundary value problem (3.2.1) satisfies the following energy identity:*

$$\int_0^l \left[u_t^2(x, t) + \kappa(x) u_x^2(x, t) \right] dx = 2 \int_0^l \int_0^t F(x) G(\tau) u_\tau(x, \tau) d\tau$$
$$+ \int_0^l \left[(u_1(x))^2 + \kappa(x) (u_0'(x))^2 \right] dx \tag{3.2.6}$$

for all $t \in [0, T]$.

Proof Multiplying both sides of Eq. (3.2.1) by $u_t(x, t)$, integrating on $[0, l]$ and then applying the integration by parts formula to the integral containing the term $(\kappa(x)u_x)_x$ we get:

$$\frac{1}{2}\frac{d}{dt}\int_0^l \left[u_t^2(x,t) + \kappa(x)u_x^2(x,t)\right]dx = \int_0^l F(x)G(t)u_t(x,t)dx,$$

for all $t \in [0, T]$. Integrating this identity on $[0, t]$ and using the non-homogeneous initial conditions we arrive at the required identity. □

Consider the problem (3.2.1) and (3.2.2) of determining the unknown spatial load $F(x)$ from final state overdetermination. Denote by $u := u(x, t; F)$ the unique weak solution of the direct problem (3.2.1), corresponding to a given source $F \in L^2(0, l)$. Introducing the *input-output operator* $\Phi : L^2(0, l) \mapsto L^2(0, l)$, defined as $(\Phi F)(x) := u(x, T; F)$, we can reformulate the inverse problem as the following operator equation

$$\Phi F = u_T, \quad F \in L^2(0, l), \quad u_T \in L^2(0, l). \tag{3.2.7}$$

Lemma 3.2.2 *Let conditions (3.2.4) holds. Assume that $u_0(x) = u_1(x) = 0$. Then the input-output operator*

$$\Phi : F \in L^2(0, l) \mapsto u(x, T; F) \in L^2(0, l),$$

corresponding to the source problem (3.2.1) and (3.2.2) is a linear compact operator.

The proof is almost exactly the same as that of Lemma 3.1.1 and is left as an exercise to the reader. □

Thus, the hyperbolic inverse source problem (3.2.1) and (3.2.2) is ill-posed. Hence the approach given in Sect. 3.3 can be used for a minimum of the regularized Tikhonov functional

$$J_\alpha(F) = \frac{1}{2}\int_0^l [(\Phi F)(x) - u_T(x)]^2 dx + \frac{1}{2}\alpha \|F\|_{L^2(0,l)}, \tag{3.2.8}$$

where $\alpha > 0$ is the parameter of regularization.

Consider now the problem of determining the unknown spatial load $F(x)$ from final velocity overdetermination, i.e. the inverse problem defined by (3.2.1) and (3.2.3). We define the *input-output operator* $\Psi : L^2(0, l) \mapsto L^2(0, l)$, $(\Psi F)(x) := u_t(x, T; F)$. In a similar way, we can prove that the problem

$$\Psi F = v_T, \quad F \in L^2(0, l), \quad v_T \in L^2(0, l) \tag{3.2.9}$$

3.2 Inverse Source Problems for Wave Equation

is also ill-posed. The regularized Tikhonov functional corresponding to the inverse problem defined by (3.2.1) and (3.2.3) is the functional

$$J_\alpha(F) = \frac{1}{2}\int_0^l [(\Psi F)(x) - v_T(x)]^2 dx + \frac{1}{2}\alpha \|F\|_{L^2(0,l)}. \qquad (3.2.10)$$

3.2.1 Non-uniqueness of a Solution

Consider first the inverse problem (3.2.1) and (3.2.2) assuming that the initial strain $u_0(x)$ and the initial velocity $u_1(x)$ are zero. To prove an existence of unique minimum of functional (3.2.8) we need to apply Theorem 2.5.1 in Sect. 2.5. For this aim we analyze some properties of the input-output operator $\Phi : L^2(0,l) \mapsto L^2(0,l)$ and the operator equation (3.2.7), adopting Lemma 3.1.4 to the considered case.

Lemma 3.2.3 *Let conditions (3.2.4) hold. Then the input-output operator* $\Phi : L^2(0,l) \mapsto L^2(0,l)$, $(\Phi F)(x) := u(x,T;F)$, *corresponding to the inverse problem (3.2.1) and (3.2.2), is self-adjoint. Furthermore,*

$$(\Phi \varphi_n)(x) = \kappa_n \varphi_n(x), \qquad (3.2.11)$$

that is, $\{\kappa_n, \varphi_n\}$ *is the eigensystem of the input-output operator* Φ,

$$\kappa_n = \frac{1}{\sqrt{\lambda_n}} \int_0^T \sin\left(\sqrt{\lambda_n}(T-t)\right) G(t) dt, \quad n = 1, 2, 3, \ldots \qquad (3.2.12)$$

and $\{\varphi_n\}_{n=1}^\infty$ *are orthonormal eigenvectors corresponding to eigenvalues* $\{\lambda_n\}_{n=1}^\infty$ *of the differential operator* $\ell : \mathcal{V}(0,l) \mapsto L^2(0,l)$, $\mathcal{V}(0,l) := \{v \in H^1(0,l) : v(0) = v(l) = 0\}$, *defined by*

$$\begin{cases} (\ell\varphi)(x) := -(k(x)\varphi'(x))' = \lambda\varphi(x), & x \in (0,l); \\ \varphi(0) = 0, \quad \varphi(l) = 0, \end{cases} \qquad (3.2.13)$$

Proof Let $\{\varphi_n\}_{n=1}^\infty$ be the orthonormal eigenvectors corresponding to the positive eigenvalues $\{\lambda_n\}_{n=1}^\infty$, $0 < \lambda_1 < \lambda_2 < \lambda_3 \ldots$ of the self-adjoint positive defined differential operator $\ell : \mathcal{V}(0,l) \mapsto L^2(0,l)$ defined by (3.2.13). Then we can make use of Fourier series expansion

$$u(x,t) = \sum_{n=1}^\infty u_n(t)\varphi_n(x) \qquad (3.2.14)$$

of the solution of the initial boundary value problem (3.2.1) via the orthonormal basis $\{\varphi_n\}_{n=1}^\infty$ in $L^2(0,l)$. Here

$$u_n(t) = \frac{F_n}{\sqrt{\lambda_n}} \int_0^t \sin\left(\sqrt{\lambda_n}(t-\tau)\right) G(\tau) d\tau, \quad t \in (0, T] \quad (3.2.15)$$

is the solution of the Cauchy problem

$$\begin{cases} u_n''(t) + \lambda_n u_n(t) = F_n G(t), & t \in (0, T), \\ u_n(0) = 0, \quad u_n'(0) = 0, \end{cases}$$

for each $n = 1, 2, 3, \ldots$ and $F_n := (F, \varphi_n)_{L^2(0,l)}$ is the Fourier coefficient of the function $F \in L^2(0, l)$.

Now we can use (3.2.14) and (3.2.15) to obtain the Fourier series expansion of the input-output operator Φ, defined as $(\Phi F)(x) := u(x, T; F)$:

$$(\Phi F)(x) = \sum_{n=1}^{\infty} (F, \varphi_n)_{L^2(0,l)} \kappa_n \varphi_n(x), \quad (3.2.16)$$

where κ_n is defined by (3.2.12).

To show that $\{\varphi_n\}_{n=1}^{\infty}$ are eigenvectors of the input-output operator Φ, corresponding to eigenvalues $\{\kappa_n\}_{n=1}^{\infty}$, we replace $F(x)$ by $\varphi_m(x)$ in (3.2.16):

$$(\Phi \varphi_m)(x) = \sum_{n=1}^{\infty} (\varphi_m, \varphi_n)_{L^2(0,l)} \kappa_n \varphi_n(x) = \kappa_m \varphi_m(x).$$

This implies (3.2.11).

The proof of the self-adjointness of the operator $\Phi : L^2(0, l) \mapsto L^2(0, l)$ is the same as in Lemma 3.1.4. □

The assertions of Lemma 3.2.3 can be proved for the problem of determining the unknown spatial load $F(x)$ from final velocity overdetermination.

Lemma 3.2.4 *Let conditions (3.2.4) hold. Then the input-output operator $\Psi : L^2(0, l) \mapsto L^2(0, l)$, $(\Psi F)(x) := u_t(x, T; F)$, corresponding to the inverse problem (3.2.1) and (3.2.3), is self-adjoint. Furthermore,*

$$(\Psi \varphi_n)(x) = \kappa_n \varphi_n(x),$$

that is, $\{\kappa_n, \varphi_n\}$ is the eigensystem of the input-output operator Φ, $\{\varphi_n\}_{n=1}^{\infty}$ are orthonormal eigenvectors corresponding to eigenvalues $\{\lambda_n\}_{n=1}^{\infty}$ of the differential operator $\ell : \mathcal{V}(0, l) \mapsto L^2(0, l)$, defined by (3.2.13), and

$$\kappa_n = \int_0^T \cos(\sqrt{\lambda_n}(T-t)) G(t) dt, \quad n = 1, 2, 3, \ldots. \quad (3.2.17)$$

3.2 Inverse Source Problems for Wave Equation

The Fourier series expansion of the input-output operator Ψ corresponding to the inverse problem (3.2.1) and (3.2.3) is

$$(\Psi F)(x) = \sum_{n=1}^{\infty} (F, \varphi_n)_{L^2(0,l)} \, \kappa_n \varphi_n(x), \qquad (3.2.18)$$

where κ_n is defined by (3.2.17).

The Fourier series expansions (3.2.16) and (3.2.18) of the input-output operators Φ and Ψ, corresponding to the above inverse problems, show that the necessary condition for bijectivity of these operators is the condition $\kappa_n \neq 0$. Remark that this is a main condition of Theorem 2.5.1 in Sect. 2.5. It follows from formulae (3.2.12) and (3.2.17) for the eigenvalues κ_n that even positivity of the time dependent source $G \in L^2(0, T)$ can not guarantee the condition $\kappa_n \neq 0$ for unique determination of the unknown source $F(x)$. To show this explicitly consider the following examples.

Example 3.2.1 Identification of a spatial load in constant coefficient wave equation from final state overdetermination

Consider the problem of determining the unknown spatial load $F(x)$ in

$$\begin{cases} u_{tt} = u_{xx} + F(x), & (x, t) \in \Omega_T := (0, 1) \times (0, T); \\ u(x, 0) = 0, \; u_t(x, 0) = 0, & x \in (0, 1); \\ u(0, t) = 0, \; u(l, t) = 0, & t \in (0, T), \end{cases} \qquad (3.2.19)$$

from the final state overdetermination (3.2.2).

For $G(t) \equiv 1$ formula (3.2.12) implies:

$$\kappa_n = \frac{1}{\lambda_n}[1 - \cos(\sqrt{\lambda_n} T)], \; \lambda_n = \pi^2 n^2, \; n = 1, 2, 3, \ldots . \qquad (3.2.20)$$

Formula (3.2.12) shows that $\kappa_n = 0$, if $nT = 2m$, where m is a natural number. Hence $\kappa_n \neq 0$ for all $n = 1, 2, 3, \ldots$ if and only if

$$T \neq \frac{2m}{n}, \text{ for all } n = 1, 2, 3, \ldots . \qquad (3.2.21)$$

Thus, *for unique determination of the unknown source $F(x)$ from final state overdetermination* (3.2.2), *the final time $T > 0$ must satisfy the condition* (3.2.21). □

Example 3.2.2 Identification of a spatial load in constant coefficient wave equation from final velocity overdetermination

Consider now the problem of determining the unknown spatial load $F(x)$ in (3.2.19) from the final velocity overdetermination (3.2.3). Formula (3.2.17) implies $(G(t) \equiv 1)$:

$$\kappa_n = \frac{1}{\sqrt{\lambda_n}} \sin(\sqrt{\lambda_n} T), \ \lambda_n = \pi^2 n^2, \ n = 1, 2, 3, \ldots. \quad (3.2.22)$$

Hence $\kappa_n = 0$, if $nT = m$, where $m = m(T)$ is a natural number. This means that the final time should satisfy the condition

$$T \neq \frac{m}{n}, \ for all \ n = 1, 2, 3, \ldots. \quad (3.2.23)$$

Thus, *for unique determination of the unknown source $F(x)$ from the final velocity overdetermination* (3.2.2) the final time $T > 0$ must satisfy condition (3.2.23). □

As a matter of fact, both conditions (3.2.21) and (3.2.23) are equivalent and mean that the value of $T > 0$ the final time cannot be a rational number. Evidently, in practice for arbitrary given final time $T > 0$ the fulfilment of this necessary condition is impossible. Thus, both final data overspecifications (3.2.2) and (3.2.3) for the wave equation are not feasible.

3.3 Backward Parabolic Problem

As a next application of the above approach we consider now the backward parabolic problem. Specifically, consider the problem of determining the unknown initial temperature $f(x) \in L^2(0, l)$ in

$$\begin{cases} u_t = (k(x)u_x)_x, & (x, t) \in \Omega_T := (0, l) \times (0, T], \\ u(x, 0) = f(x), & x \in (0, l), \\ u(0, t) = 0, \ u_x(l, t) = 0, & t \in (0, T), \end{cases} \quad (3.3.1)$$

from the measured temperature $u_T(x)$ at the final time $t = T$:

$$u_T(x) := u(x, T), \quad x \in (0, l). \quad (3.3.2)$$

This problem is defined as a *backward parabolic problem* (BPP).

We apply the above approach to the backward parabolic problem defined by (3.3.1) and (3.3.2). For this end, we denote by $u(x, t; f)$ be the weak solution of the forward problem (3.3.1), for a given initial data $f \in L^2(0, l)$, and introduce the *input-output operator* $\Phi : L^2(0, l) \mapsto L^2(0, l)$, defined by $(\Phi f)(x) := u(x, 0; f), x \in (0, l)$. Then the backward parabolic problem can be reformulated as the following operator equation:

$$\Phi f = u_T, \ f \in L^2(0, l). \quad (3.3.3)$$

Using Lemma 3.1.1, it can be easily proved that (3.3.3) is a linear compact operator. This, in particular, implies that backward parabolic problem (3.3.1) and (3.3.2)

3.3 Backward Parabolic Problem

is ill-posed. We may use now Tikhonov regularization introducing the regularized functional

$$J_\alpha(F) := J(F) + \frac{1}{2}\alpha \|f\|^2_{L^2(0,l)}, \quad (3.3.4)$$

where

$$J(F) = \frac{1}{2}\|u(\cdot, T; f) - u_T\|^2_{L^2(0,l)}, \quad f \in L^2(0, l) \quad (3.3.5)$$

and $\alpha > 0$ is the parameter of regularization.

The following lemma shows that the Fréchet gradient $J'(F)$ of Tikhonov functional (3.3.5) can be derived via the solution $\psi(x, t) := \psi(x, t; f)$ of the adjoint problem solution

$$\begin{cases} \psi_t = -(k(x)\psi_x)_x, & (x, t) \in \Omega_T; \\ \psi(x, T) = u(x, T; f) - u_T(x), & x \in (0, l); \\ \psi(0, t) = 0, \ \psi_x(l, t) = 0, & t \in [0, T). \end{cases} \quad (3.3.6)$$

Note that this is the same adjoint problem (3.1.19), corresponding to the parabolic inverse problem, with $u(x, T; F)$ replaced by $u(x, T; f)$.

Lemma 3.3.1 *Let conditions (3.1.3) hold. Denote by $u(x, t; f)$ the weak solution of the parabolic problem (3.3.1) corresponding to a given initial data $f \in L^2(0, l)$. Then Tikhonov functional (3.3.5) is Fréchet differentiable and for the Fréchet gradient $J'(f)$ the following formula holds:*

$$J'(f)(x) = \psi(x, 0; f), \quad \text{for a.e. } x \in (0, l), \quad (3.3.7)$$

where $\psi(x, t; f)$ is the weak solution of the adjoint problem (3.3.6).

Proof Let $f, f + \delta f \in L^2(0, l)$ be given initial data and $u(x, t; f), u(x, t; f + \delta f)$ be corresponding solutions of (3.3.1). Then $\delta u(x, t; f) := u(x, t; f + \delta f) - u(x, t; f)$ is the weak solution of the following parabolic problem

$$\begin{cases} \delta u_t = (k(x)\delta u_x)_x, & (x, t) \in \Omega_T, \\ \delta u(x, 0) = \delta f(x), & x \in (0, l), \\ \delta u(0, t) = 0, \ \delta u_x(l, t) = 0, & t \in (0, T]. \end{cases} \quad (3.3.8)$$

We calculate the increment $\delta J(f) := J(f + \delta f) - J(f)$ of functional (3.3.5):

$$\delta J(f) = \int_0^l [u(x, T; f) - u_T(x)]\delta u(x, T; f)dx + \frac{1}{2}\int_0^l [\delta u(x, T; f)]^2 dx \quad (3.3.9)$$

and then transform the first right hand side integral. Assuming that $\psi(x,t) := \psi(x,t;f)$ and $\delta u(x,t;f) := \delta u(x,t;f)$ are solutions of problems (3.3.6) and (3.3.8), accordingly, we have:

$$\int_0^l [u(x,T;f) - u_T(x)]\delta u(x,T;f)dx = \int_0^l \psi(x,T)\delta u(x,T)dx$$

$$= \int_0^l \left\{ \int_0^T (\psi(x,t)\delta u(x,t))_t \, dt \right\} dx + \int_0^l \psi(x,0)\delta f(x)dx$$

$$= \int_0^T \int_0^l [\psi_t \delta u + \psi \delta u_t] \, dx dt + \int_0^l \psi(x,0)\delta f(x)dx$$

$$= \int_0^T \int_0^l [-(k(x)\psi_x)_x \delta u + \psi(k(x)\delta u_x)_x] \, dx dt + \int_0^l \psi(x,0)\delta f(x)dx$$

$$= \int_0^T [-k(x)\psi_x \delta u + k(x)\delta u_x \psi]_{x=0}^{x=l} \, dt + \int_0^l \psi(x,0)\delta f(x)dx.$$

Taking into account here the initial/final and boundary conditions in (3.3.6) and (3.3.8) we obtain the following integral identity:

$$\int_0^l [u(x,T;f) - u_T(x)]\delta u(x,T;f)dx = \int_0^l \psi(x,0;f)\delta f(x)dx,$$

for all $f, \delta f \in L^2(0,l)$. With formula (3.3.9) this implies:

$$\delta J(F) = \int_0^l \psi(x,0;f)\delta f(x)dx + \frac{1}{2}\int_0^l [\delta u(x,T;f)]^2 dx. \quad (3.3.10)$$

The last right hand side integral is of the order $\mathcal{O}\left(\|\delta f\|_{L^2(0,l)}^2\right)$ by estimate (3.1.8). This completes the proof of lemma. $\square$

Although the Lipschitz continuity of the Fréchet gradient (3.3.7) can be obtained from the results given in Sect. 2.1, to show an explicit form of the Lipschitz constant, we will prove it.

Lemma 3.3.2 *Let conditions (3.1.3) hold. Then Fréchet gradient of Tikhonov functional (3.3.5) is Lipschitz continuous with the Lipschitz constant $L_3 = 1$:*

$$\|J'(f + \delta f) - J'(f)\|_{L^2(0,l)} \leq \|\delta f\|_{L^2(0,l)}, \quad \forall f, \delta f \in L^2(0,l). \quad (3.3.11)$$

Proof By definition,

$$\|J'(f + \delta f) - J'(f)\|_{L^2(0,l)} := \|\delta \psi(\cdot, 0; f)\|_{L^2(0,l)}, \quad (3.3.12)$$

where $\delta\psi(x,t;f) := \psi(x,t;f+\delta f) - \psi(x,t;f)$ is the weak solution of the problem:

3.3 Backward Parabolic Problem

$$\begin{cases} \delta\psi_t = -(k(x)\delta\psi_x)_x, & (x,t) \in \Omega_T; \\ \delta\psi(x,T) = \delta u(x,T;f), & x \in (0,l); \\ \delta\psi(0,t) = 0, \ \psi_x(l,t) = 0, & t \in [0,T). \end{cases} \qquad (3.3.13)$$

and $\delta u(x,t;f)$ is the weak solution of the auxiliary problem (3.3.8). We use the energy identity

$$\frac{1}{2}\frac{d}{dt}\int_0^l (\delta\psi(x,t))^2\,dx = \int_0^l k(x)\,(\delta\psi_x(x,t))^2\,dx$$

for the well-posed backward problem (3.3.13), integrate it on $[0,T]$ and use the final condition $\delta\psi(x,T) = \delta u(x,T;f)$, to deduce the estimate

$$\int_0^l (\delta\psi(x,0;f))^2\,dx \le \int_0^l (\delta u(x,T;f))^2\,dx.$$

Then we use the energy identity

$$\frac{1}{2}\frac{d}{dt}\int_0^l (\delta u(x,t))^2\,dx + \int_0^l k(x)\,(\delta u_x(x,t))^2\,dx = 0$$

for the auxiliary problem (3.3.8) to obtain the estimate

$$\int_0^l (\delta u(x,T;f))^2\,dx \le \int_0^l (\delta f(x))^2\,dx, \quad \forall f \in L^2(0,l).$$

These estimates imply $\|\delta\psi(\cdot,0;f)\|_{L^2(0,l)} \le \|\delta f\|_{L^2(0,l)}$. With (3.3.12) this yields the proof of the theorem. $\square$

We can prove analogues of Corollaries 3.1.1 and 3.1.2 for the backward parabolic problem (3.3.1) and (3.3.2) in the same way.

Lemma 3.1.4 can be adopted to the backward parabolic problem to derive main properties of the input-output operator.

Lemma 3.3.3 *Let conditions (3.1.3) hold. Then the input-output operator* $\Phi : L^2(0,l) \mapsto L^2(0,l)$, $(\Phi f)(x) := u(x,0;f)$, *corresponding to the backward problem defined by (3.3.1) and (3.3.2), is self-adjoint. Moreover,*

$$(\Phi\varphi_n)(x) = \kappa_n \varphi_n(x), \qquad (3.3.14)$$

that is, $\{\kappa_n, \varphi_n\}$ *is the eigensystem of the input-output operator* Φ, *where* $\{\varphi_n\}_{n=1}^\infty$ *are orthonormal eigenvectors of the differential operator* $\ell : \mathcal{V}(0,l) \mapsto L^2(0,l)$, *defined by (3.1.34) corresponding to the eigenvalues* $\{\lambda_n\}_{n=1}^\infty$, *and*

$$\sigma_n \equiv \kappa_n = \exp(-\lambda_n T), \quad n = 1,2,3,\ldots \qquad (3.3.15)$$

The proof follows from the proof of Lemma 3.1.4 and is left as an exercise to the reader.

Thus the singular system of the (compact) input-output operator Φ is $\{\kappa_n, \varphi_n, \varphi_n\}$. The eigenvalues λ_n of the differential operator $\ell : \mathcal{V}(0, l) \mapsto L^2(0, l)$, defined in (3.1.34), are of order $\lambda_n = \mathcal{O}(n^2)$ Then, as it follows from (3.3.15) that the singular values κ_n of the input-output operator Φ are of order $\mathcal{O}(\exp(-n^2))$. This means that the backward parabolic problem (3.3.1) and (3.3.2) is *severely ill-posed*.

Since $\{\varphi_n\}_{n=1}^{\infty}$ forms a complete orthonormal system in $L^2(0, l)$ and $\Phi : L^2(0, l) \mapsto L^2(0, l)$, $(\Phi f)(x) := u(x, 0; f)$ is a self-adjoint operator, we have $\mathcal{N}(\Phi) = \mathcal{N}(\Phi^*) = \{0\}$. Moreover, $\mathcal{D}(\Phi^\dagger) = \mathcal{R}(\Phi)$ is dense in $L^2(0, l)$. Hence, for $u_T \in L^2(0, l)$ the first condition $u_T \in \mathcal{N}(\Phi^*)^\perp$ in Picard's Theorem 2.4.1 of Sect. 2.4 holds. Then we can reformulate this theorem for the backward problem defined by (3.3.1) and (3.3.2).

Theorem 3.3.1 *Let conditions (3.1.3) hold. Assume that $u_T \in L^2(0, l)$ is a noise free measured output data defined by (3.3.2). Then the operator equation $\Phi f = u_T$, corresponding to backward problem (3.3.1)–(3.3.2), has a solution if and only if*

$$u_T \in L^2(0, l) \text{ and } \sum_{n=1}^{\infty} \exp(\lambda_n T) u_{T,n}^2 < \infty. \quad (3.3.16)$$

In this case

$$f(x) = \sum_{n=1}^{\infty} \exp(\lambda_n T) u_{T,n} \varphi_n(x), \quad x \in (0, l) \quad (3.3.17)$$

is the solution of the operator equation $\Phi F = u_T$, where $u_{T,n} := (u_T, \varphi_n)$ is the nth Fourier coefficient of the measured output data $u_T(x)$ and $\varphi_n(x)$ are the normalized eigenfunctions corresponding to the eigenvalues λ_n of the operator $\ell : \mathcal{V}(0, l) \mapsto L^2(0, l)$ defined by (3.1.34).

The proof of this theorem is the same as proof of Theorem 3.1.3. □

The following example show that neither the truncated SVD nor Tikhonov regularization can recover information about higher Fourier coefficients in (3.3.17).

Example 3.3.1 The truncated SVD and Tikhonov regularization for the backward parabolic problem.

Consider the simplest backward parabolic problem, i.e. the problem of determining the unknown initial temperature $f(x) \in L^2(0, l)$ in

$$\begin{cases} u_t = u_{xx}, & (x, t) \in \Omega_T := (0, \pi) \times (0, 1], \\ u(x, 0) = f(x), & x \in (0, \pi), \\ u(0, t) = 0, \ u(l, t) = 0, & t \in (0, 1), \\ u_T(x) := u(x, 1), & x \in (0, \pi), \end{cases} \quad (3.3.18)$$

3.3 Backward Parabolic Problem

assuming $k(x) \equiv 1$, $l = \pi$ and $T = 1$. In this case the eigenvalues λ_n and the normalized eigenfunctions $\varphi_n(x)$ of the operator $(\ell\varphi)(x) := -\varphi''(x)$, with the Dirichlet conditions $u(0, t) = u(l, t) = 0$, are

$$\lambda_n = n^2, \quad \varphi_n(x) = \sqrt{2/\pi} \sin nx, \quad n = 1, 2, 3, \ldots.$$

The singular values of the input-output operator Φ are:

$$\kappa_n = \exp(-n^2), \quad n = 1, 2, 3, \ldots.$$

Let $u_T^\delta \in L^2(0, 1)$ be a noisy data and we wish to reconstruct the initial data $f^\delta \in L^2(0, 1)$ by the truncated SVD. Then assuming $1/N$ a parameter of regularization we may consider the regularization strategy

$$\left(\mathcal{R}^{N(\delta)} u_T^\delta\right)(x) := \sum_{n=1}^{N(\delta)} \exp(n^2) \, u_{T,n}^\delta \varphi_n(x), \qquad (3.3.19)$$

where $u_{T,n}^\delta = (u_T^\delta, \phi_n)_{L^2(0,1)}$ is the nth Fourier coefficient of the noisy output data $u_T^\delta(x)$. Let us estimate the approximation error $\|\mathcal{R}^{N(\delta)} u_T^\delta - \Phi^\dagger u_T\|_{L^2(0,1)}$ as in Sect. 1.4. We have:

$$\|\mathcal{R}^{N(\delta)} u_T^\delta - \Phi^\dagger u_T\|_{L^2(0,1)}^2 \le 2 \left\| \sum_{n=1}^{N(\delta)} \exp(n^2) \, (u_{T,n}^\delta - u_{T,n}) \varphi_n(x) \right\|_{L^2(0,1)}^2$$

$$+ 2 \left\| \sum_{n=N(\delta)+1}^{\infty} \exp(n^2) \, u_{T,n} \varphi_n(x) \right\|_{L^2(0,1)}^2.$$

Hence

$$\|\mathcal{R}^{N(\delta)} u_T^\delta - \Phi^\dagger u_T\|_{L^2(0,1)}^2$$
$$\le 2 \exp(2N(\delta)^2)\delta^2 + 2 \sum_{n=N(\delta)+1}^{\infty} \exp(2n^2) u_{T,n}^2. \qquad (3.3.20)$$

The first term of the right hand side of (3.3.20), i.e. the data error increase dramatically by increasing $N(\delta)$, due to the term $\exp(2N(\delta)^2)$, if even the noise level $\delta > 0$ is small enough. This means that we may use only the first few Fourier coefficients of (3.3.19) to obtain an approximate solution by the truncated SVD. For example, if even $\delta = 10^{-2}$, we can use only the first two terms of this series, since for $N(\delta) = 2$ the data error is $2\exp(2N(\delta)^2)\delta^2 \approx 6000 \times 10^{-4} = 0.6$. By adding the next term will make this error equals to 1.31×10^4! Moreover, one needs to take into account also the contribution of the second term of the right hand side of (3.3.20), i.e. the approximation error. The term $\exp(2n^2)$ here shows that the approximation error also does dramatically increase, as N increases (except the case when the Fourier coefficients $u_{T,n}$ of the output decay faster than $\exp(-N(\delta)^2)$, which holds only for infinitely differentiable functions).

Let us apply now Tikhonov regularization to the inverse problem (3.3.18). We use estimate (3.3.21) in Theorem 3.1.5, adopting it to the considered problem. Then we get:

$$\|\mathcal{R}_{\alpha(\delta)}^{N(\delta)} u_T^\delta - f\|_{L^2(0,l)}^2 \leq$$

$$\frac{\delta^2}{\alpha} + \frac{4\alpha(\delta)}{(1+\sqrt{\alpha(\delta)})^2} \sum_{n=1}^{N(\delta)} \exp(2n^2) u_{T,n}^2 + 2 \sum_{N(\delta)+1}^{\infty} \exp(2n^2) u_{T,n}^2. \quad (3.3.21)$$

Following the selection (3.1.83) of the cutoff parameter $N(\delta)$ and taking into account that $\kappa_n = \exp(-n^2)$, we deduce: no matter how small the parameter of regularization $\alpha(\delta) > 0$ was selected, there exists such a natural number $N = N(\delta)$ that

$$\min_{1 \leq n \leq N(\delta)} \exp(-2n^2) = \exp(-2N(\delta)^2) \geq \sqrt{\alpha} > \exp(-2(N(\delta)+1)^2).$$

This implies that $N(\delta) = \mathcal{O}(\sqrt{\ln \alpha(\delta)}/2)$. Assume again that the noise level is very low: $\delta = 10^{-3}$. We choose the parameter of regularization as $\alpha(\delta) = 10^{-5}$ from the requirements (3.1.78) of Theorem 3.1.5. Then $\sqrt{\ln \alpha(\delta)}/2 \approx 1.7$, which means that only the first two Fourier coefficients can be used in the Nth partial sum

$$\left(\mathcal{R}_{\alpha(\delta)}^{N(\delta)} u_T^\delta\right)(x) := \sum_{n=1}^{N(\delta)} \frac{q(\alpha(\delta); \kappa_n)}{\kappa_n} u_{T,n}^\delta \varphi_n(x), \quad q(\alpha; \kappa) = \frac{\kappa^2}{\kappa^2 + \alpha}$$

for recovering the initial data. Remark that we did not take into account an influence of the last right hand side terms of (3.3.21) to the above error, although these terms need to be taken into account also. □

Remark, that Duhamel's Principle for heat equation illustrates relationship between the singular values corresponding to the inverse source problem and the backward problem for heat equation. Let $v(x, t; \tau)$, $(x, t) \in \Omega_\tau$, be the solution of the problem

$$\begin{cases} v_t(x, t; \tau) = (k(x) v_x(x, t; \tau))_x, & (x, t) \in \Omega_\tau; \\ v(x, \tau; \tau) = F(x) G(\tau), & x \in (0, l); \\ v(0, t; \tau) = 0, \quad v_x(l, t) = 0, & t \in (0, T] \end{cases} \quad (3.3.22)$$

corresponding to a fixed value of the parameter $\tau \in [0, T]$, where $\Omega_\tau := \{(x, t) \in \mathbb{R}^2 : x \in (0, l), t \in (\tau, T]\}$. Then, by Duhamel's Principle,

$$u(x, t) = \int_0^t v(x, t; \tau) d\tau, \quad \tau \in (0, t], \ t \in (0, T] \quad (3.3.23)$$

3.3 Backward Parabolic Problem

is the solution of the problem

$$\begin{cases} u_t = (k(x)u_x)_x + F(x)G(t), & (x,t) \in \Omega_T; \\ u(x,0) = 0, & x \in (0,l); \\ u(0,t) = 0, \ u_x(l,t) = 0, & t \in (0,T]. \end{cases} \quad (3.3.24)$$

For each fixed $\tau \in (0,T]$, the problem of determining the initial data $F(x)$ in (3.3.22) from the final output data $u_T(x) := u(x,T)$ is severely ill-posed. While the problem of determining the unknown space-wise dependent source $F \in L^2(0,l)$ in (3.3.24) from the same final data is moderately ill-posed. This change of degree of ill-posedness is due to the integration in (3.3.23). Specifically, formula (3.1.35) is obtained from formula (3.3.15) via this integration (3.3.23) taking into account the factor $G(t)$.

Finally, to compare the behavior of the singular values, we consider the inverse source problem (3.1.31) for heat equation (ISPH), the backward parabolic problem (BPP) defined by (3.3.1) and (3.3.2) and the inverse source problem (3.2.1) and (3.2.2) for wave equation (ISPW). Assuming, for simplicity, that $G(t) \equiv 1, k(x) = 1$, $\kappa(x) = 1, l = \pi, T = 1$, we obtain the following formulae for the singular values $\sigma_n \equiv \kappa_n$ of these problems:

$$\begin{cases} \sigma_n = [1 - \exp(-(n-1/2)^2)](n+1/2)^{-2} & \text{ISPH}; \\ \sigma_n = \exp(-(n-1/2)^2), \ n = 1,2,3,\ldots & \text{BPP}; \\ \sigma_n = [1 - \cos n]n^{-2} & \text{ISPW}. \end{cases}$$

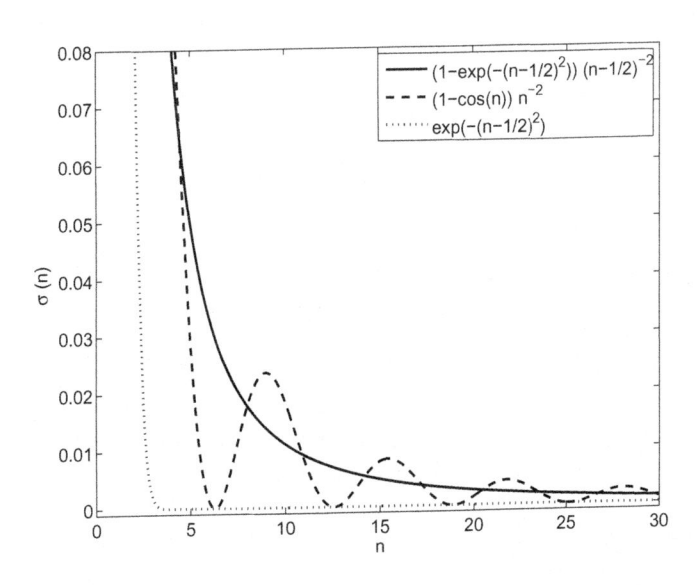

Fig. 3.3 Behavior of the singular values

The behavior of the singular values σ_n, depending on $n = \overline{1,30}$, is shown in Fig. 3.3. This figure shows n^{-2}-decay and $\exp(n^{-2})$-decay characters of the singular values corresponding to ISPH and BPP, respectively. The behavior of the singular values corresponding to ISPW is as an oscillating function of n that also decays, as $n \to \infty$.

3.4 Computational Issues in Inverse Source Problems

Any numerical method for inverse problems related to PDEs requires, first of all, construction of an optimal computational mesh for solving corresponding direct problem. This mesh needs to be fine enough in order to obtain an accurate numerical solution of the direct problem, i.e. to minimize the *computational noise level* in synthetic output data, on one hand. On the other hand, an implementation of any iterative (gradient) method for an inverse problem requires solving the direct and the corresponding adjoint problems at each iteration step. Hence, a too fine mesh for numerical solving of these problems is sometimes unnecessary and increases the computational cost. Taking into account that the weak solution of the direct problem (3.1.1) is in an appropriate Sobolev space, the most appropriate method for the discretization and then solving of these problems, is the Finite Element Method (FEM). This method with continuous piecewise quadratic basic functions, which is most appropriate for the weak solution, is used here in all computational experiments. The second important point is that, the approach given in Sect. 3.3 allows to derive an explicit gradient formula. Hence, having explicit gradient formula for the above inverse problem with final data, it is reasonable to use in numerical solving gradient type algorithms, in particular, the Conjugate Gradient (CG) Algorithm. Moreover, as we explain below, the Lipschitz continuity of the gradient of the Tikhonov functional $J(F)$ implies monotonicity of the numerical sequence $\{J(F^{(n)})\}$, where $F^{(n)}$ is the nth iteration of the CG algorithm. As a result, the CG algorithm is an optimal algorithm in numerical solving of this class of inverse problems.

The last important point in numerical solving of ill-posed problems is that employing the same mathematical model to generate a synthetic output data, as well as to invert it, may lead to trivial inversion. This situation is defined in [19] as an *"inverse crime"*. As it is stated there, to avoid this trivial inversion "it is crucial that the synthetic data be obtained by a forward solver which has no connection to the inverse solver". To avoid an inverse crime, in all computational results below coarser mesh is used for generation of a synthetic output data, and a finer mesh is used in the inversion algorithm. All numerical analysis in this section will be given for the inverse problem (3.1.31) related to the heat equation.

3.4 Computational Issues in Inverse Source Problems

3.4.1 Galerkin FEM for Numerical Solution of Forward Problems

In this subsection we will consider this direct problem governed by the following initial-boundary value problem:

$$\begin{cases} u_t = (k(x)u_x)_x + F(x)G(t), & (x,t) \in \Omega_T := (0,l) \times (0,T]; \\ u(x,0) = u_0(x), & x \in (0,l); \\ u(0,t) = 0, \ u_x(1,t) = 0, & t \in (0,T]. \end{cases} \quad (3.4.1)$$

First we will introduce semi-discrete analogue of this problem, using the continuous piecewise quadratic basic functions. Then applying time discretization, we will derive fully discrete analogue of the direct problem (3.1.1).

Let $\omega_h := \{x_i \in (0,l] : x_i = ih, \ i = \overline{1, N_x}\}$ be a uniform space mesh with the mesh parameter $h = 1/N_x > 0$. The quadratic polynomials

$$\begin{cases} \xi_i(x) = \begin{cases} 2(x - x_{i-1/2})(x - x_{i-1})/h^2, & x \in [x_{i-1}, x_i), \\ 2(x - x_{i+1/2})(x - x_{i+1})/h^2, & x \in [x_i, x_{i+1}), \\ 0, & x \notin [x_{i-1}, x_{i+1}], \end{cases} \\ \qquad\qquad\qquad\qquad\qquad\qquad\qquad\qquad\qquad i = \overline{1, N_x - 1}; \\ \xi_N(x) = \begin{cases} 0, & x \in [0, x_{N_x-1}), \\ 2(x - x_{N-1/2})(x - x_{N_x-1})/h^2, & x \in [x_{N_x-1}, x_{N_x}]; \end{cases} \\ \eta_i(x) = \begin{cases} -4(x - x_{i-1})(x - x_i)/h^2, & x \in [x_{i-1}, x_i], \\ 0, & x \notin [x_{i-1}, x_i], \ i = \overline{1, N_x}, \end{cases} \end{cases}$$

with the compact supports are defined as $1D$ continuous piecewise quadratic basic functions in FEM (Fig. 3.4). Here the midpoints $x_{i-1/2} := (i - 1/2)h$, $i = \overline{1, N_x}$ are also defined as auxiliary nodes, in addition to the nodal points $x_i \in \omega_h$ of the uniform space mesh ω_h.

Then the function

$$u_h(x,t) = \sum_{i=1}^{N} c_i(t)\xi_i(x) + \sum_{i=0}^{N_x} d_i(t)\eta_i(x)$$

is a piecewise-quadratic approximation of $u(x,t)$ in the finite-dimensional subspace $\mathcal{V}_h \subset \mathcal{H}^1(0,l)$ spanned by the set of basis functions $\{\xi_i(x)\} \cup \{\eta_i(x)\}$, where $c_i(t) := u_h(x_i, t)$ and $d_i(t) := u_h(x_{i-1/2}, t)$. For convenience we treat here and below $t \in (0,T]$ as a parameter and the functions $u(t)$ and $u_h(t)$ as mappings $u : [0,T] \mapsto \mathcal{V}$ and $u_h : [0,T] \mapsto \mathcal{V}_h$, defined as $u(t)(x) := u(x,t)$ and $u_h(t)(x) := u_h(x,t)$, $x \in (0,l)$. For the interpolation with piecewise quadratic basic functions the following estimates hold:

$$\|u - u_h\|_{\mathcal{V}(0,l)} \le Ch^2 |u|_{H^3(0,l)}, \ C > 0,$$

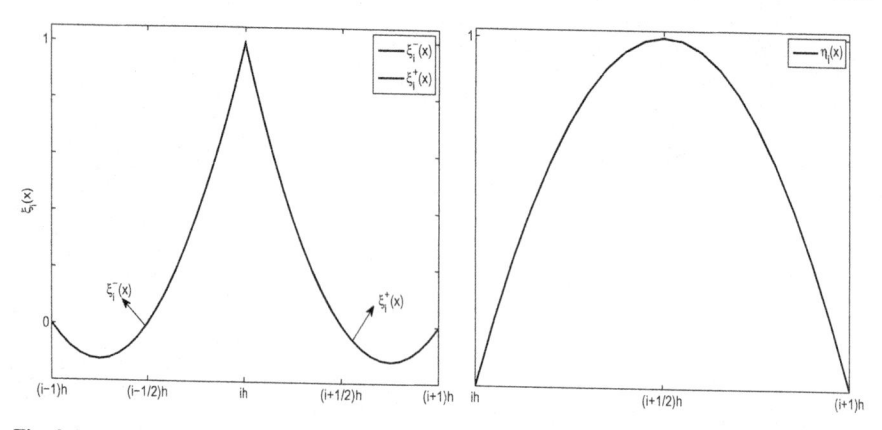

Fig. 3.4 The piecewise quadratic basis functions

which means the second-order accuracy in space, in the norm of $H^1(0,l)$ [101]. Here $|\cdot|_{H^3(0,l)}$ is the semi-norm of $u \in H^3(0,l)$ and $\mathcal{V} := \{v \in H^1(0,l) : v(0,t) = 0\}$.

We will apply Galerkin FEM for discretization of the direct and adjoint problems (3.4.1) and (3.1.19). Find $u_h(t) \in \mathcal{S}_N$ such that

$$\begin{cases} \langle \partial_t u_h, \xi_i \rangle + a(u_h, \xi_i) = G(t)\langle F(x), \xi_i \rangle, & t \in (0,T]; \\ \langle u_h(0), \xi_i \rangle := \langle u_0, \xi_i \rangle, & i = \overline{1, N_x}, \\ \\ \langle \partial_t u_h, \eta_i \rangle + a(u_h, \eta_i) = G(t)\langle F(x), \eta_i \rangle, & t \in (0,T]; \\ \langle u_h(0), \eta_i \rangle := \langle u_0, \eta_i \rangle, & i = \overline{0, N_x}, \end{cases} \quad (3.4.2)$$

where

$$\begin{cases} \langle \partial_t u_h, \xi_i \rangle := \int_{x_i}^{x_{i+1}} u_t(x,t) \xi_i(x) dx, \\ a(u_h, \xi_i) := \int_{x_i}^{x_{i+1}} k(x) u'_h(x) \xi'_i(x) dx, \\ \langle F(x), \xi_i \rangle := \int_{x_i}^{x_{i+1}} F(x) \xi_i(x) dx, \ i = \overline{0, N_x}; \\ \langle \partial_t u_h, \eta_i \rangle := \int_{x_i}^{x_{i+1}} u_t(x,t) \eta_i(x) dx, \\ a(u_h, \eta_i) := \int_{x_i}^{x_{i+1}} k(x) u'_h(x) \eta'_i(x) dx, \\ \langle F(x), \eta_i \rangle := \int_{x_i}^{x_{i+1}} F(x) \eta_i(x) dx, \ i = \overline{0, N_x}. \end{cases} \quad (3.4.3)$$

Introducing the uniform time mesh $\omega_\tau := \{t_j \in (0,T] : t_j = j\tau, \ j = \overline{1, N_t}\}$, with mesh parameter $\tau = T/N_t$, by standard Galerkin FEM procedure, we then obtain from the semi-discrete Galerkin discretization (3.4.2)–(3.4.3) the following full discretization by using Crank-Nicolson method:

3.4 Computational Issues in Inverse Source Problems

$$\begin{cases} \left\langle \dfrac{u_h^j + u_h^{j+1}}{\tau}, \xi_i \right\rangle + \dfrac{a(u_h^j, \xi_i) + a(u_h^{j+1}, \xi_i)}{2} = \dfrac{G(t_j) + G(t_{j+1})}{2} \langle F(x), \xi_i \rangle; \\ \langle u_h(0), \xi_i \rangle := \langle u_0, \xi_i \rangle, \quad t_j \in \omega_\tau, \; x_i \in \omega_h. \\ \left\langle \dfrac{u_h^j + u_h^{j+1}}{\tau}, \eta_i \right\rangle + \dfrac{a(u_h^j, \eta_i) + a(u_h^{j+1}, \eta_i)}{2} = \dfrac{G(t_j) + G(t_{j+1})}{2} \langle F(x), \eta_i \rangle; \\ \langle u_h(0), \eta_i \rangle := \langle u_0, \eta_i \rangle, \quad t_j \in \omega_\tau, \; x_i \in \omega_h. \end{cases} \quad (3.4.4)$$

The FEM scheme (3.4.4) is used in subsequent computational experiments for solving the forward and backward parabolic problems.

3.4.2 The Conjugate Gradient Algorithm

Above derived explicit gradient formulae for Tikhonov functionals allow use of the classical version of the Conjugate Gradient Algorithm for the numerical reconstruction of an unknown source. We define the *convergence error* (or *discrepancy*) $e(n; F; \delta)$ and the *accuracy error* $E(n; F; \delta)$:

$$\begin{cases} e(n; F^{(n)}; \delta) := \|\Phi F^{(n)} - u_T^\delta\|_{L^2(0,T)}, \\ E(n; F^{(n)}; \delta) := \|F - F^{(n)}\|_{L^2(0,l)}, \end{cases} \quad (3.4.5)$$

where $\Phi : L^2(0, l) \mapsto L^2(0, l), (\Phi F)(x) := u(x, T; F)$ is the input-output operator. Here and below $u_T^\delta(x)$ is the noisy data:

$$\|u_T - u_T^\delta\|_{L^2(0,l)} \leq \delta, \quad \|u_T^\delta\|_{L^2(0,l)} > \delta, \; \delta > 0. \quad (3.4.6)$$

The Conjugate Gradient Algorithm discussed below is applied to the Tikhonov functional

$$J(F) = \frac{1}{2} \|\Phi F - u_T^\delta\|_{L^2(0,l)}^2, \quad (3.4.7)$$

corresponding to the normal equation $\Phi^* \Phi F = \Phi^* u_T^\delta$, although the same technique remains hold for the regularized form

$$J_\alpha(F) = J(F) + \frac{1}{2}\alpha \|F\|_{L^2(0,l)}^2 \quad (3.4.8)$$

of the Tikhonov functional corresponding to the regularized form of normal equation $(\Phi^* \Phi + \alpha I) F_\alpha = \Phi^* u_T^\delta$.

The *iterative Conjugate Gradient Algorithm* (subsequently, *CG-algorithm*) for the functional (3.4.7) consists of the following basic steps. All norms and scalar products below are in $L^2(0, l)$ and will be omitted.

Step 1. For $n = 0$ choose the initial iteration $F^{(0)}(x)$.

Step 2. Compute the initial descent direction

$$p^{(0)}(x) := J'(F^{(0)})(x). \tag{3.4.9}$$

Step 3. Find the descent direction parameter

$$\beta_n = \frac{\left(J'(F^{(n)}), p^{(n)}\right)}{\|\Phi p^{(n)}\|^2}. \tag{3.4.10}$$

Step 4. Find next iteration

$$F^{(n+1)}(x) = F^{(n)}(x) - \beta_n p^{(n)}(x) \tag{3.4.11}$$

and compute the convergence error $e(n; F^{(n)}; \delta)$.

Step 5. If the *stopping condition*

$$e(n; F^{(n)}; \delta) \leq \tau_M \delta < e(n; F^{(n-1)}; \delta), \ \tau_M > 1, \ \delta > 0 \tag{3.4.12}$$

holds, then go to Step 7.

Step 6. Set $n := n+1$ and compute

$$\begin{cases} p^{(n)}(x) := J'(F^{(n)})(x) + \gamma_n p^{(n-1)}(x), \\ \gamma_n = \frac{\|J'(F^{(n)})\|^2}{\|J'(F^{(n-1)})\|^2} \end{cases} \tag{3.4.13}$$

and go to Step 3.

Step 7. Stop the iteration process.

This version of the CG-algorithm is usually called in literature as CGNE, i.e. the CG-algorithm applied to the Normal Equation [23]. We will follow here the version of the CG-algorithm given in [54], adopting the results given here to the considered class of inverse problems.

First of all we make some important remarks concerning formulae (3.4.9) and (3.4.10). It follows from gradient formula (3.1.29) the Fréchet derivative of Tikhonov functional that

$$J'(F^{(n)}) = \Phi^*\left(\Phi F^{(n)} - u_T\right), \ F^{(n)} \in L^2(0, l). \tag{3.4.14}$$

Hence, if the CG-algorithm starts with the *initial iteration* $F^{(0)}(x) \equiv 0$, then

$$J'_\alpha(F^{(0)}) = -\Phi^* u_T. \tag{3.4.15}$$

3.4 Computational Issues in Inverse Source Problems

With formula (3.4.9) this implies that

$$p^{(0)} = -\Phi^* u_T. \tag{3.4.16}$$

This initial descent direction is used in some versions of the CG-algorithm.

Taking into account above gradient formula (3.4.14), we can rewrite formula (3.4.10) as follows:

$$\frac{\left(J'(F^{(n)}), p^{(n)}\right)}{\|\Phi p^{(n)}\|^2} = \frac{(\Phi F^{(n)} - u_T, \Phi p^{(n)})}{\|\Phi p^{(n)}\|^2}, \tag{3.4.17}$$

Lemma 3.4.1 *Formula (3.4.10) for the descent direction parameter β_n is equivalent to the following formula:*

$$\beta_n = \frac{\|J'(F^{(n)})\|^2}{\|\Phi p^{(n)}\|^2}. \tag{3.4.18}$$

Proof By (3.4.11) and (3.4.14) we have:

$$J'(F^{(n+1)}) = \Phi^*\left(\Phi(F^{(n)} - \beta_n p^{(n)}) - u_T\right) = \Phi^*\left(\Phi F^{(n)} - u_T\right) - \beta_n \Phi^* \Phi p^{(n)}.$$

Hence

$$J'(F^{(n+1)}) = J'(F^{(n)}) - \beta_n \Phi^* \Phi p^{(n)}. \tag{3.4.19}$$

Using this formula we show the orthogonality

$$\left(p^{(n)}, J'\left(F^{(n+1)}\right)\right) = 0. \tag{3.4.20}$$

Indeed,

$$\begin{aligned}\left(p^{(n)}, J'(F^{(n+1)})\right) &= \left(p^{(n)}, J'(F^{(n)})\right) - \beta_n \left(p^{(n)}, \Phi^* \Phi p^{(n)}\right) \\ &= \left(p^{(n)}, J'(F^{(n)})\right) - \beta_n \left(\Phi p^{(n)}, \Phi p^{(n)}\right) = 0,\end{aligned}$$

by formula (3.4.10).

Substituting now formulae (3.4.11) and (3.4.13) in (3.4.10) and using the the orthogonality (3.4.20), we deduce:

$$\beta_n = \frac{\left(J'(F^{(n)}), J'(F^{(n)})\right) + \gamma_n p^{(n-1)}\right)}{\|\Phi p^{(n)}\|^2} = \frac{\|J'(F^{(n)})\|^2}{\|\Phi p^{(n)}\|^2}.$$

This implies (3.4.18). $\square$

Note that formula (3.4.18) is more convenient in computations.

Some orthogonality properties of the parameters of the CG-algorithm are summarized in the following lemma.

Lemma 3.4.2 *Let conditions* (3.1.3) *and* (3.1.69) *hold. Denote by* $\{J'(F^{(n)})\} \subset L^2(0, l)$ *and* $\{p^{(n)}\} \subset L^2(0, l)$, $n = 0, 1, 2, \ldots$ *the sequences of gradients and descent directions obtained by CG-algorithm. Then the gradients are orthogonal and the descent directions are* Φ-*conjugate, that is,*

$$\begin{aligned} \left(J'(F^{(n)}), J'(F^{(k)})\right) &= 0, \\ \left(\Phi p^{(n)}, \Phi p^{(k)}\right) &= 0, \quad \text{for all } n \neq k, \end{aligned} \quad (3.4.21)$$

where $\Phi : L^2(0, l) \mapsto L^2(0, l)$ *is the input-output mapping corresponding to the inverse source problem* (3.1.1) *and* (3.1.2).

Proof To prove both assertions of (3.4.21) we use simultaneous induction with respect to n. Let $n = 1$. For $k = 0$ we have:

$$\left(J'(F^{(0)}), J'(F^{(1)})\right) := \left(J'(F^{(0)}), J'(F^{(0)})\right) - \beta_0 \left(J'(F^{(0)}), \Phi^* \Phi p^{(0)}\right)$$

$$= \|J'(F^{(0)})\|^2 - \frac{\|J'(F^{(0)})\|^2}{\|\Phi p^{(0)}\|^2} \left(\Phi J'(F^{(0)}), \Phi p^{(0)}\right) = 0,$$

due to (3.4.9) and (3.4.10). Further, using this orthogonality we deduce:

$$\left(\Phi p^{(0)}, \Phi p^{(1)}\right) = \left(\Phi^* \Phi p^{(0)}, p^{(1)}\right) = \frac{1}{\beta_0} \left(J'(F^{(0)}) - J'(F^{(1)}), p^{(1)}\right)$$

$$= \frac{1}{\beta_0} \left(J'(F^{(1)}) + \frac{\|J'(F^{(1)})\|^2}{\|J'(F^{(0)})\|^2} J'(F^{(0)}), J'(F^{(0)}) - J'(F^{(1)})\right) = 0,$$

by (3.4.13) and (3.4.19).

Assume now that the assertions

$$\begin{aligned} \left(J'(F^{(n)}), J'(F^{(k)})\right) &= 0, \\ \left(\Phi p^{(n)}, \Phi p^{(k)}\right) &= 0, \quad \text{for all } k = \overline{1, n-1} \end{aligned} \quad (3.4.22)$$

hold. We need to prove that

$$\begin{aligned} \left(J'(F^{(n+1)}), J'(F^{(k)})\right) &= 0, \\ \left(\Phi p^{(n+1)}, \Phi p^{(k)}\right) &= 0, \quad \text{for all } k = \overline{1, n}. \end{aligned} \quad (3.4.23)$$

First we prove these assertions for $k = \overline{1, n-1}$. To prove the first assertion of (3.4.23), we use (3.4.19), then the first assertion of (3.4.22), and finally formula $J'(F^{(n)})(x) = p^{(n)}(x) - \gamma_n p^{(n-1)}(x)$ obtained from (3.4.13). Then we get:

3.4 Computational Issues in Inverse Source Problems

$$\begin{aligned}\left(J'(F^{(n+1)}), J'(F^{(k)})\right) &= \left(J'(F^{(n)}) - \beta_n \Phi^*\Phi p^{(n)}, J'(F^{(k)})\right) \\ &= -\beta_n \left(\Phi^*\Phi p^{(n)}, J'(F^{(k)})\right) \\ &= -\beta_n \left(\Phi^*\Phi p^{(n)}, p^{(k)} - \gamma_k p^{(k-1)}\right) \\ &= -\beta_n \left(\Phi p^{(n)}, \Phi p^{(k)} - \gamma_k \Phi p^{(k-1)}\right),\end{aligned} \quad (3.4.24)$$

for all $k = \overline{1, n-1}$. By (3.4.22) the right hand side scalar product is zero.

We prove the second assertion of (3.4.23) for $k = \overline{1, n-1}$, in the similar way by using assertions (3.4.22) with formulae (3.4.13) and (3.4.19). We have:

$$\begin{aligned}\left(\Phi p^{(n+1)}, \Phi p^{(k)}\right) &= \left(\Phi J'(F^{(n+1)}) + \gamma_n \Phi p^{(n)}, \Phi p^{(k)}\right) \\ &= \left(\Phi J'(F^{(n+1)}), \Phi p^{(k)}\right) = \left(J'(F^{(n+1)}), \Phi^*\Phi p^{(k)}\right) \\ &= \tfrac{1}{\beta_n} \left(J'(F^{(n+1)}), J'(F^{(k)}) - J'(F^{(k+1)})\right) = 0.\end{aligned}$$

To complete the induction we need to prove assertions (3.4.23) for $k = n$. For the first of them we use the same technique as in (3.4.24):

$$\begin{aligned}\left(J'(F^{(n+1)}), J'(F^{(n)})\right) &= \left(J'(F^{(n)}) - \beta_n \Phi^*\Phi p^{(n)}, J'(F^{(n)})\right) \\ &= \|J'(F^{(n)})\|^2 - \beta_n \left(\Phi^*\Phi p^{(n)}, J'(F^{(n)})\right) \\ &= \|J'(F^{(n)})\|^2 - \beta_n \left(\Phi p^{(n)}, \Phi J'(F^{(n)})\right) \\ &= \|J'(F^{(n)})\|^2 - \beta_n \left(\Phi p^{(n)}, \Phi(p^{(n)} - \gamma_n p^{(n-1)})\right) \\ &= \|J'(F^{(n)})\|^2 - \tfrac{\|J'(F^{(n)})\|^2}{\|\Phi p^{(n)}\|^2} \left(\Phi p^{(n)}, \Phi p^{(n)}\right) = 0.\end{aligned}$$

The second assertion of (3.4.23)

$$\left(\Phi p^{(n+1)}, \Phi p^{(n)}\right) = 0.$$

can be proved in the same way. $\square$

The detailed results related to application of the CG-algorithm to ill-posed problems are treated in great detail in books [23, 34, 52] and references given therein, we will use only some of these results, concerning to the considered here class of inverse problems. The most important convergence result is given in the convergence theorem (Theorem 7.9 [23]). We give this theorem slightly modifying it.

Theorem 3.4.1 *Let $F^{(n)}$ be iterates defined by the CG-algorithm with the stopping rule (3.4.12).*
(a) If $u_T^\delta \in \mathcal{D}(\Phi^\dagger)$, then the iterates $F^{(n)}$ converge to $\Phi^\dagger u_T$, as $n \to \infty$.
(b) If $u_T \notin \mathcal{D}(\Phi^\dagger)$, then $\|F^{(n)}\| \to \infty$, as $n \to \infty$.

This theorem implies, in particular, that the iteration must be terminated appropriately when dealing with perturbed (or noisy) data $u_T^\delta \notin \mathcal{D}(\Phi^\dagger)$, due to numerical instabilities. Remark that here and below $u_T^\delta \in L^2(0,l)$ is the noisy data with the noise level $\delta > 0$, by definition (3.4.6), that is, $\|u_T^\delta - u_T\|_{L^2(0,l)} \leq \delta$.

The *termination index* $n = n(\delta, u_T^\delta)$ in the stopping condition (3.4.12) of the CG-algorithm is defined according to Morozov's discrepancy principle given in Sect. 2.6. If $u_T^\delta \in \mathcal{D}(\Phi^\dagger)$ and the CG-algorithm is stopped according to discrepancy principle, then this algorithm guarantees a finite termination index $n(\delta, u_T^\delta)$.

3.4.3 Convergence of Gradient Algorithms for Functionals with Lipschitz Continuous Fréchet Gradient

Let us return to the inverse source problem discussed in Sect. 3.3. Lemma 3.1.3 asserts the Lipschitz continuity $J \in C^{1,1}$ of Fréchet gradient of Tikhonov functional $J(F)$ corresponding to the final data inverse problem for heat equation. Here we will show that an important advantage of gradient methods comes when dealing with this class of functionals.

We start with a simple gradient-type iteration algorithm, usually used in solving ill-posed problems and defined also as *Landweber iteration algorithm*. Consider the functional $J : H \mapsto \mathbb{R}_+$ defined on a real Hilbert space H. Denote by $\{v^{(n)}\} \subset U$ the iterations defined as follows:

$$v^{(n+1)} = v^{(n)} - \omega_n J'(v^{(n)}), \quad n = 0, 1, 2, \ldots, \quad (3.4.25)$$

where $v^{(0)} \in U$ is an initial iteration and $\omega_n > 0$ is a relaxation parameter, defined by the minimum problem:

$$f_n(\omega_n) := \inf_{\omega \geq 0} f_n(\omega), \quad f_n(\omega) := J\left(v^{(n)} - \omega J'(v^{(n)})\right), \quad (3.4.26)$$

for each $n = 0, 1, 2, \ldots$. We assume here and below, without loss of generality, that $J'(v^{(n)}) \neq 0$, for all $n = 0, 1, 2, \ldots$.

Lemma 3.4.3 *Let $U \subset H$ be a convex set of a real Hilbert space, $J : U \mapsto \mathbb{R}_+$ be a functional defined on $U \subset H$. Assume that the functional $J(u)$ has Lipschitz continuous Fréchet gradient, i.e. $J \in C^{1,1}$, that is, for all $u_1, u_2 \in U$*

$$|J'(u) - J'(v)| \leq L\|u - v\|_H, \quad (3.4.27)$$

where $L > 0$ is the Lipschitz constant. Then for all $u_1, u_2 \in U$ the following inequality hods:

$$|J(u_1) - J(u_2) - (J'(u_2), u_1 - u_2)| \leq \frac{1}{2} L \|u_1 - u_2\|_H^2. \quad (3.4.28)$$

3.4 Computational Issues in Inverse Source Problems

Proof Assuming $u_1 = u + h$, $u_2 = u$ in the increment formula

$$J(u+h) - J(u) = \int_0^1 \left(J'(u+\theta h)), h\right) d\theta,$$

we get

$$J(u_1) - J(u_2) = \int_0^1 \left(J'(u_2 + \theta(u_1 - u_2)), u_1 - u_2\right) d\theta.$$

Using the Lipschitz condition (3.4.27) we conclude that

$$|J(u_1) - J(u_2) - \left(J'(u_2), u_1 - u_2\right)|$$

$$= \int_0^1 \left(J'(u_2 + \theta(u_1 - u_2)) - J'(u_2), u_1 - u_2\right) d\theta$$

$$\leq L \|u_1 - u_2\|_H^2 \int_0^1 \theta d\theta = \frac{1}{2} L \|u_1 - u_2\|_H^2.$$

This completes the proof. □

Lemma 3.4.4 *Let conditions of Lemma 3.4.3 hold. Assume that $J_* := \inf_U J(u) > -\infty$. Denote by $\{v^{(n)}\} \subset U$ the a sequence of iterations defined by the gradient algorithm (3.4.25) and (3.4.26). Then $\{J(v^{(n)})\} \subset \mathbb{R}_+$ is a monotone decreasing sequence and*

$$\lim_{n \to \infty} \|J'(v^{(n)})\|_H = 0, \qquad (3.4.29)$$

Proof Use inequality (3.4.28), taking here $v = v^{(n)}$ and $u = v^{(n)} - \omega J'(v^{(n)})$, $\omega > 0$. Then we get:

$$J\left(v^{(n)} - \omega_n J'(v^{(n)})\right) - J(v^{(n)}) + \omega_n \|J'(v^{(n)})\|_H^2 \leq \frac{L\omega_n^2}{2} \|J'(v^{(n)})\|_H^2.$$

Due to (3.4.25), $J(v^{(n+1)}) = J\left(v^{(n)} - \omega_n J'(v^{(n)})\right)$. Using this in the above inequality we obtain:

$$J(v^{(n)}) - J(v^{(n+1)}) \geq \omega_n (1 - L\omega_n/2) \|J'(v^{(n)})\|_H^2, \quad \forall \omega \geq 0.$$

The function $g(\omega_n) = \omega_n (1 - L\omega_n/2)$, $\omega_n > 0$ reaches its minimum value $g_* := g(\tilde{\omega}_n) = 1/(2L)$ at $\tilde{\omega}_n = 1/L$. Hence

$$J(v^{(n)}) - J(v^{(n+1)}) \geq \frac{1}{2L} \|J'(v^{(n)})\|_H^2, \quad \forall v^{(n)}, v^{(n+1)} \in U. \qquad (3.4.30)$$

The right hand side is positive, since $J'(v^{(n)}) \neq 0$, which means that the sequence $\{J(v^{(n)})\} \subset \mathbb{R}_+$ is monotone decreasing. Since this sequence is bounded below, this assertion also implies convergence of the numerical sequence $\{J(v^{(n)})\}$. Then passing to the limit in the above inequality, we obtain the second assertion (3.4.29) of the lemma. □

Lemma 3.4.4 shows that in the case of Lipschitz continuity of the gradient $J'(v)$ the relaxation parameter $\omega > 0$ can be estimated via the Lipschitz constant $L > 0$.

Let us apply this lemma to the Tikhonov functional given by (3.4.7) and find the value of the relaxation parameter $\omega_n > 0$ at nth iteration via the gradient $J'(F^{(n)})$. Using (3.4.26) we conclude that

$$f(\omega) := J\left(F^{(n)} - \omega_n J'(F^{(n)})\right) = \frac{1}{2} \left\| \Phi F^{(n)} - \omega_n \Phi J'(F^{(n)}) - u_T^\delta \right\|^2$$
$$= \frac{1}{2} \left\| \Phi F^{(n)} - u_T^\delta \right\|^2 - \omega_n \left(\Phi F^{(n)} - u_T^\delta, \Phi J'(F^{(n)}) \right) + \frac{1}{2} \omega_n^2 \left\| \Phi J'(F^{(n)}) \right\|^2.$$

Since the relaxation parameter is defined as a solution of the minimum problem (3.4.26), from the condition $f'(\omega_n) = 0$ we obtain:

$$\omega_n := \frac{\left(\Phi F^{(n)} - u_T^\delta, \Phi J'(F^{(n)}) \right)}{\left\| \Phi J'(F^{(n)}) \right\|^2} = \frac{\left(\Phi^*(\Phi F^{(n)} - u_T^\delta), J'(F^{(n)}) \right)}{\left\| \Phi J'(F^{(n)}) \right\|^2}.$$

But it follows from the gradient formula (3.4.14) that $J'(F^{(n)}) = \Phi^*\left(\Phi F^{(n)} - u_T\right)$. Hence we arrive at the formula

$$\omega_n = \frac{\left\| J'(F^{(n)}) \right\|^2}{\left\| \Phi J'(F^{(n)}) \right\|^2}. \tag{3.4.31}$$

This formula with (3.4.18) illustrate the similarity and dissimilarity between the descent direction parameter β_n in the CG-algorithm and the relaxation parameter ω_n in the Landweber iteration algorithm. These parameter coincide only for $n = 0$, i.e. $\beta_0 = \omega_0$, due to formula (3.4.9). As a consequence, the first iterations obtained by these algorithms are also the same. Comparison of the numerical results obtained by these algorithms will be given below.

It turns out that, for a functional with Lipschitz continuous Fréchet gradient the rate of convergence of the Landweber iteration algorithm can also be estimated.

Lemma 3.4.5 *Let, in addition to conditions of Lemma 3.4.3, $J : H \mapsto \mathbb{R}_+$ with $J_* := \inf_U J(u) > -\infty$ is a continuous convex functional. Assume that the set $\mathcal{M}(v^{(0)}) := \{u \in H : J(u) \leq J(v^{(0)})\}$, where $v^{(0)}$ is an initial iteration, is bounded. Then the a sequence of iterations $\{v^{(n)}\} \subset U$ defined by the gradient algorithm (3.4.25) and (3.4.26) is a minimizing sequence, i.e.*

$$\lim_{n \to \infty} J(v^{(n)}) = J_*.$$

3.4 Computational Issues in Inverse Source Problems

Moreover, for the rate of convergence of the sequence $\{J(v^{(n)})\}$ following estimate holds:

$$0 \leq J(v^{(n)}) - J_* \leq 2L\,d\,n^{-1}, \quad n = 1, 2, 3, \ldots, \qquad (3.4.32)$$

where $L > 0$ is the Lipschitz constant and $d := \sup \|u - v\|_H$, $u, v \in \mathcal{M}(v^{(0)})$ is the diameter of the set $\mathcal{M}(v^{(0)})$.

Proof It is known that the minimum problem

$$J_* = \inf_{\mathcal{M}(u^{(0)})} J(u)$$

for a continuous convex functional in a bounded closed and convex set has a solution. Hence each minimizing sequence $\{v^{(n)}\} \subset \mathcal{M}(v^{(0)})$ weakly converges to an element $u_* \in U_*$ of the solution set $U_* := \{u \in \mathcal{M}(v^{(0)}) : J(u) = J_*\}$, that is $J(v^{(n)}) \to J(u_*)$, as $n \to \infty$.

To prove the rate of the convergence, we introduce the numerical sequence $\{a_n\}$ defined as

$$a_n := J(v^{(n)}) - J(u_*), \quad n = 1, 2, 3, \ldots. \qquad (3.4.33)$$

For a functional with Lipschitz continuous Fréchet gradient convexity is equivalent to the following inequality:

$$J(v^{(n)}) - J(u_*) \leq \left(J'(v^{(n)}), v^{(n)} - u_*\right).$$

Applying to the right hand the Cauchy-Schwartz inequality we conclude that

$$a_n \leq \|J'(v^{(n)})\| \|v^{(n)} - u_*\| \leq d \|J'(v^{(n)})\|.$$

With estimate (3.4.28) and (3.4.33) this implies:

$$a_n^2 \leq 2L\,d^2 \left[J(v^{(n)}) - J(v^{(n+1)})\right] = 2L\,d^2[a_n - a_{n+1}].$$

Thus the numerical sequence $\{a_n\}$ defined by (3.4.33) has the following properties:

$$a_n^2 > 0, \quad a_n - a_{n+1} \geq \frac{1}{2L\,d^2} a_n^2, \quad n = 1, 2, 3, \ldots. \qquad (3.4.34)$$

Evidently $\{a_n\}$ is a monotone decreasing sequence with $a_n/a_{n+1} > 1$. Using these properties we prove now that $a_n = \mathcal{O}(n^{-1})$. Indeed, for all $k = \overline{0, n-1}$ we have:

$$\frac{1}{a_{k+1}} - \frac{1}{a_k} = \frac{a_k - a_{k+1}}{a_k a_{k+1}} \geq \frac{1}{2L\,d^2} \frac{a_k^2}{a_k a_{k+1}}$$

$$\geq \frac{1}{2L\,d^2} \frac{a_k}{a_{k+1}} > \frac{1}{2L\,d^2}, \quad n = 1, 2, 3, \ldots .$$

Summing up these inequalities and using inequalities (3.4.34) for each summand we deduce that

$$\frac{1}{a_{n+1}} - \frac{1}{a_1} := \sum_{k=1}^{n} \left(\frac{1}{a_{k+1}} - \frac{1}{a_k} \right) \geq \frac{n}{2L\,d^2}, \quad n = 1, 2, 3, \ldots .$$

This implies that $a_n = \mathcal{O}(n^{-1})$. Taking into account (3.4.33) we arrive at the estimate of the rate of the convergence (3.4.32). □

3.4.4 Numerical Examples

In this subsection we present some numerical examples to show the performance analysis of the CG-algorithm. Consider the inverse problem of determining an unknown source term $F(x)$ in

$$\begin{cases} u_t = (k(x)u_x)_x + F(x)G(t), & (x,t) \in \Omega_T := (0,l) \times (0,T]; \\ u(x,0) = 0, & x \in [0,l]; \\ u(0,t) = 0, \quad u(l,t) = 0, & t \in (0,T), \end{cases} \quad (3.4.35)$$

from the final data

$$u_T(x) := u(x,T), \quad x \in [0,l]. \tag{3.4.36}$$

In all examples below the fine mesh with the mesh parameters $N_x = 201$ and $N_t = 801$ is used in the FEM scheme (3.4.4) to generate the noise free synthetic output data $u_{T,h}$. Remark that the computational noise level defined as $\delta_c := \|u_T - u_{T,h}\|_{L_h^2(0,1)} / \|u_T\|_{L_h^2(0,1)}$, where u_T and $u_{T,h}$ outputs obtained from the exact and numerical solutions of the direct problem, on this fine mesh is estimated as 10^{-8}. With this accuracy the synthetic output data $u_{T,h}$ generated on the fine mesh will be assumed as a noise free.

The coarser mesh with the mesh parameters $N_x = 100$ and $N_t = 201$ is used in numerical solution of inverse problems. The noisy output data $u_{T,h}^\delta$, with $\|u_{T,h} - u_{T,h}^\delta\|_{L_h^2(0,1)} = \delta$, is generated by employing the "randn" function in MATLAB, that is,

$$u_{T,h}^\delta(x) = u_{T,h}(x) + \gamma \|u_{T,h}\|_{L_h^2(0,1)} \texttt{randn}(N), \tag{3.4.37}$$

3.4 Computational Issues in Inverse Source Problems

where $\gamma > 0$ is the *noise level*. The parameter $\tau_M > 1$ in the stopping condition (3.4.12) is taken as $\tau_M = 1.1$.

Example 3.4.1 The performance analysis of CG-algorithm (3.4.25): reconstruction of a smooth function with one critical point

To generate the noise free synthetic output data $u_{T,h}$ the simplest concave function

$$F(x) = 10x(1-x) \qquad (3.4.38)$$

with $k(x) = 1 + x^2$, $G(t) = \exp(-t)$, $t \in [0, 1]$, is used as an input data in the direct (3.4.35). The function $u_{T,h}$, obtained from the numerical solution of the direct problem, is assumed to be the noise free output data. Then, using (3.4.37), the noisy output data $u^\delta_{T,h}$ is generated. For the value $\gamma = 5\%$ of the noise level, the parameter $\delta > 0$ is obtained as $\delta := \|u_{T,h} - u^\delta_{T,h}\|_{L^2_h(0,1)} = 4.6 \times 10^{-3}$.

The CG-algorithm is employed with and without regularization. The left Fig. 3.5 displays the reconstructions of the unknown source $F(x)$ from the noisy data $u^\delta_{T,h}$, with the noise level $\gamma = 5\%$. The function $F(x)$, defined by (3.4.38), is plotted by the solid line. This figure also shows the iteration numbers (n) corresponding to the values of the parameter of regularization $\alpha > 0$ are given

Table 3.3 shows that the value $\alpha = 10^{-6}$ of the parameter of regularization defined from the both conditions $\alpha \ll 1$ and $\delta^2/\alpha < 1$ is an optimal one. Remark that the reconstructions obtained with the optimal value $\alpha = 10^{-6}$ of the parameter of

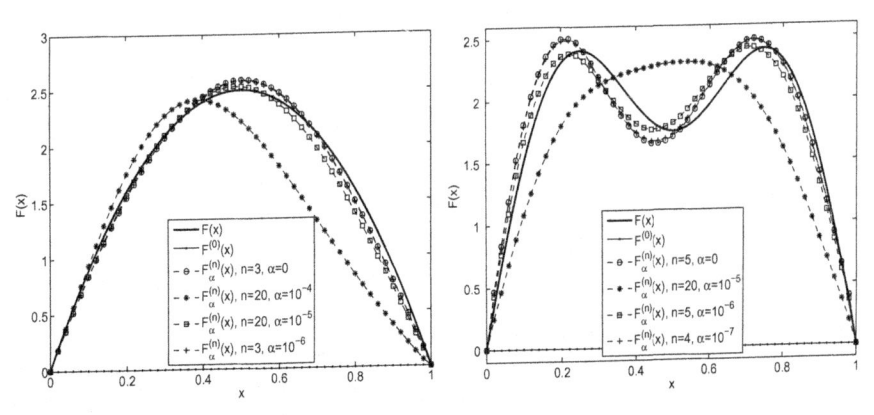

Fig. 3.5 Reconstructions of the smooth function with one critical point: $\gamma = 5\%$ (*left figure*), and the smooth function with three critical points: $\gamma = 3\%$ (*right figure*)

Table 3.3 Errors depending on the parameter of regularization: $\gamma = 5\%$

α	0	1.0×10^{-4}	1.0×10^{-5}	1.0×10^{-6}
$e(n; \alpha; \delta)$	4.7×10^{-3}	7.3×10^{-3}	4.7×10^{-3}	4.7×10^{-3}
$E(n; \alpha; \delta)$	0.2030	0.3342	0.2019	0.2028

regularization and without regularization $\alpha = 0$ are very close, as the left Fig. 3.5 shows. This illustrates the regularizing property of the CG-algorithm. □

Example 3.4.2 The performance analysis of CG-algorithm: reconstruction of a function with three critical points

In this example, the function

$$F(x) = 5(x - x^2) + \sin(3\pi x), \quad x \in [0, 1] \tag{3.4.39}$$

with the input data $k(x)$ and $G(t)$ from Example 3.4.1 is used to generate the noise free synthetic output data $u_{T,h}$. Then, assuming the function $u_{T,h}$, obtained from the numerical solution of the direct problem, to be the noise free output data, the noisy output data $u_{T,h}^\delta$ is generated for the value $\gamma = 3\%$ of the noise level in (3.4.37). This level corresponds, the value $\delta := \|u_{T,h} - u_{T,h}^\delta\|_{L_h^2(0,1)} = 4.3 \times 10^{-4}$ of the parameter $\delta > 0$.

Table 3.4 illustrates the convergence error $e(n; F; \delta)$ and the accuracy error $E(n; F; \delta)$, computed by formulae (3.4.5). The right Fig. 3.5 presents the reconstructed sources for the different values of the parameter of regularization. □

Example 3.4.3 The performance analysis of CG-algorithm: mixed boundary conditions in the direct problem

In both above examples we assumed in the direct problem (3.4.35) the homogeneous Dirichlet conditions $u(0, t) = u(l, t) = 0$. In this example we assume that the mixed boundary conditions $u(0, t) = u_x(l, t) = 0$ are given in the direct problem, instead of the homogeneous Dirichlet conditions.

The noise free synthetic output data $u_{T,h}$ here is generated from the following input data

$$F(x) = \sin(\pi x) + \sqrt{x}; , \quad x \in [0, 1],$$
$$k(x) = 1 + x^2,$$
$$G(t) = \exp(-t), \quad t \in [0, 1],$$

Table 3.4 Errors depending on the parameter of regularization: $\gamma = 3\%$

α	0	1.0×10^{-5}	1.0×10^{-6}	1.0×10^{-7}
$e(n; \alpha; \delta)$	4.5×10^{-4}	1.6×10^{-3}	4.9×10^{-4}	4.5×10^{-4}
$E(n; \alpha; \delta)$	0.1443	0.4331	0.1472	0.1423

Table 3.5 Errors depending on the parameter of regularization: $\gamma = 3\%$

α	0	1.0×10^{-3}	1.0×10^{-4}	1.0×10^{-5}	1.0×10^{-6}
$e(n; \alpha; \delta)$	8.0×10^{-3}	1.5×10^{-2}	8.2×10^{-3}	8.0×10^{-3}	8.0×10^{-3}
$E(n; \alpha; \delta)$	0.1165	0.2754	0.1306	0.1060	0.1152

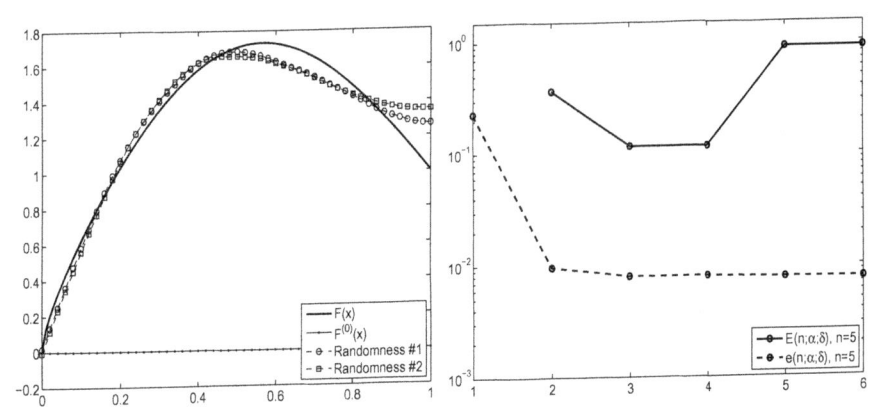

3.4 Computational Issues in Inverse Source Problems

Fig. 3.6 Influence of randomness: $\alpha = 10^{-6}$ (*left figure*); and the termination index $n(\delta) = 3$ shows the beginning of the iteration from convergence to divergence (*right figure*) ($\gamma = 3\%$)

assuming in the direct problem (3.4.35) the mixed boundary conditions $u(0, t) = u_x(l, t) = 0$ instead of the homogeneous Dirichlet conditions. The noisy output data $u_{T,h}^\delta$ is generated for the noise level $\gamma = 5\%$, which corresponds to $\delta := \|u_{T,h} - u_{T,h}^\delta\|_{L_h^2(0,1)} = 8.0 \times 10^{-3}$. The convergence error and the accuracy error are give in Table 3.5 illustrates the convergence error $e(n; F; \delta)$ and the accuracy error $E(n; F; \delta)$, computed by formulae (3.4.5).

From a computational viewpoint the change of the Dirichlet conditions in the direct problem by the mixed boundary conditions means that the order of approximation of FE-scheme at $x = l$ will be less than the order approximation at the interior mesh points. Natural effect of this case is seen from the reconstructed sources in the left Fig. 3.6 near the point $x = 1$. This figure also shows the degree of influence of the randomness to the reconstructions.

The right Fig. 3.6 explains that the termination index $n(\delta) = 3$ shows the beginning of the iteration from convergence to divergence. □

The above computational results demonstrate that CG-algorithm is effective, robust against a middle level noise on data ($\gamma = 3 \div 5\%$), and provides satisfactory reconstructions. The accuracy in the recovery of the spacewise dependent source decreases as the noise level increases. With an appropriately chosen parameter of regularization, the number of iterations of the CG-algorithm to reach the condition (3.4.12) with $\tau_M = 1.1$ was about $3 \div 5$.

Part II
Inverse Problems for Differential Equations

Chapter 4
Inverse Problems for Hyperbolic Equations

In the first part of this chapter we study two inverse source problems related to the second order hyperbolic equations $u_{tt} - u_{xx} = \rho(x,t)g(t)$ and $u_{tt} - u_{xx} = \rho(x,t)\varphi(x)$ for the quarter plane $\mathbb{R}_+^2 = \{(x,t)|\, x > 0, t > 0\}$, with Dirichlet type measured output data $f(t) := u(x,t)|_{x=0}$. The time-dependent source $g(t)$ and the spacewise-dependent source $\varphi(x)$ are assumed to be unknown in these inverse problems. Next, we study more complex problem, namely the problem of recovering the potential $q(x)$ in the string equation $u_{tt} - u_{xx} - q(x)u = 0$ from the Neumann type measured output data $f(t) := u_x(x,t)|_{x=0}$. We prove the uniqueness of the solution and then derive the global stability estimate. In the final part of the chapter, inverse coefficient problems for layered media is studied as an application.

4.1 Inverse Source Problems

We begin with the simplest linear inverse source problems for wave equation. In the first case, we assume that the time dependent source term $g(t)$ in the wave equation $u_{tt} - u_{xx} = \rho(x,t)g(t)$, $(x,t) \in \mathbb{R}_+^2$, is unknown and needs to be identified from boundary information. In the second case, based on the same boundary information, the inverse source problem of determining the unknown spacewise dependent source term $\varphi(x)$ in the wave equation $u_{tt} - u_{xx} = \rho(x,t)\varphi(x)$ in studied. In both cases, the reflection method [24] is used to derive solution of the inverse problems via the solution of the Volterra equation of the second kind.

4.1.1 Recovering a Time Dependent Function

Consider the initial-boundary value problem

$$\begin{cases} \left(\dfrac{\partial^2}{\partial t^2} - \dfrac{\partial^2}{\partial x^2}\right) u(x,t) = p(x,t)g(t), \ (x,t) \in \mathbb{R}^2_+, \\ u|_{t=0} = 0, \quad u_t|_{t=0} = 0, \qquad \qquad x > 0, \\ u_x|_{x=0} = 0 \qquad \qquad \qquad \qquad \quad t > 0 \end{cases} \quad (4.1.1)$$

for the inhomogeneous wave equation on $\mathbb{R}^2_+ = \{(x,t) | \, x > 0, t > 0\}$. If the right-hand side in the wave Eq. (4.1.1) is given, then the initial-boundary value problem (4.1.1) is a well-posed problem of mathematical physics. Let us assume that the right-hand side is known only partially, namely, the function $p(x,t)$ is known, but the time dependent function $g(t)$ is unknown.

Consider the following *inverse source problem*: find the time dependent function $g(t)$ for a given function $p(x,t)$ and the given trace

$$f(t) := u(x,t)|_{x=0}, \quad t \geq 0, \quad (4.1.2)$$

defined on the half-axis $x = 0, t \geq 0$, of the solution $u(x,t)$ of problem (4.1.1). The function $f(t)$ in (4.1.2) is called *Dirichlet type measured output*.

The problem (4.1.1) and (4.1.2) is called *an inverse source problem of determining an unknown time dependent source in the wave equation from Dirichlet boundary data*. Accordingly, for a given admissible function $g(t)$, the initial-boundary value problem (4.1.1) is defined as the *direct problem*.

We assume that the functions $p(x,t)$ and $g(t)$ satisfy the following conditions:

$$p \in C(\overline{\mathbb{R}^2_+}), \ p_x \in C(\overline{\mathbb{R}^2_+}), \ g \in C[0,\infty). \quad (4.1.3)$$

The last condition in (4.1.3) suggests that we will look for a solution of the inverse problem in the space $C[0, \infty)$.

Now we use the *reflection method* extending the functions $u(x,t)$ and $p(x,t)$ to the domain $\mathbb{R}^2_- = \{(x,t) | \, x < 0, t > 0\}$ as even, with respect to x, functions, that is, $u(x,t) = u(-x,t)$ and $p(x,t) = p(-x,t)$, for $x < 0$. Then the condition $u_x|_{x=0} = 0$ is automatically satisfied. Moreover, the extended function $u(x,t)$ solves the Cauchy problem

$$\begin{cases} \left(\dfrac{\partial^2}{\partial t^2} - \dfrac{\partial^2}{\partial x^2}\right) u(x,t) = p(x,t)g(t), \ x \in \mathbb{R}, \ t > 0, \\ u|_{t=0} = 0, \quad u_t|_{t=0} = 0, \qquad \qquad t > 0. \end{cases} \quad (4.1.4)$$

The unique solution of this problem is given by d'Alembert's formula

4.1 Inverse Source Problems

$$u(x,t) = \frac{1}{2} \iint_{\Delta(x,t)} \rho(\xi,\tau)g(\tau)\,d\xi d\tau,$$

where $\Delta(x,t) = \{(\xi,\tau) \in \mathbb{R}^2 \mid 0 \leq \tau \leq t - |x - \xi|\}$ is the characteristic triangle with the vertex at the point (x,t). Taking into account the additional condition (4.1.2) we get:

$$f(t) = \frac{1}{2} \int_{\Delta(0,t)} \rho(\xi,\tau)g(\tau)\,d\xi d\tau = \int_{\Delta_+(0,t)} \rho(\xi,\tau)g(\tau)\,d\xi d\tau, \quad t \geq 0,$$

where $\Delta_+(0,t) = \{(\xi,\tau) \in \mathbb{R}^2 \mid 0 \leq \tau \leq t,\ 0 \leq \xi \leq t - \tau\}$. This implies the following equation with respect to the unknown function $g(t)$:

$$\int_0^t g(\tau) \int_0^{t-\tau} \rho(\xi,\tau)\,d\xi d\tau = f(t), \quad t \geq 0. \tag{4.1.5}$$

Thus the inverse problem (4.1.1) and (4.1.2) is reformulated as the integral Eq. (4.1.5).

We deduce from Eq. (4.1.5) that under conditions (4.1.3) the function $f(t)$ satisfies the following conditions:

$$f \in C^2[0,\infty), \quad f(0) = f'(0) = 0. \tag{4.1.6}$$

Indeed, differentiating (4.1.5), we get:

$$\begin{aligned} f'(t) &= \int_0^t g(\tau)\rho(t-\tau,\tau)\,d\tau, \\ f''(t) &= g(t)\rho(0,t) + \int_0^t g(\tau)\rho_x(t-\tau,\tau)\,d\tau,\ t \geq 0, \end{aligned} \tag{4.1.7}$$

which means, fulfilment of conditions (4.1.6). The conditions $f(0) = f'(0) = 0$ can be treated as *consistency conditions for the output data* $f(t)$.

The above derived conditions (4.1.6) for the function $f(t)$ are the necessary conditions for solvability of the inverse problem (4.1.1) and (4.1.2). We prove that under the additional condition

$$\rho(0,t) \neq 0, \text{ for all } t \in [0,T], \tag{4.1.8}$$

these conditions are also sufficient conditions for the unique solvability of the inverse problem on the closed interval $[0, T]$.

Theorem 4.1.1 *Let the source function $\rho(x,t)$ satisfies conditions (4.1.3) and (4.1.8) in $D_T := \{(x,t) \in \mathbb{R}^2_+ \mid 0 \leq t \leq T - x\},\ T > 0$. Then the inverse problem (4.1.1) and (4.1.2) is uniquely solvable in $C[0,T]$ if and only if the function $f(t)$ satisfies conditions $f(0) = f'(0) = 0,\ f(t) \in C^2[0,T]$.*

Proof We only need to prove that the conditions on the function $f(t)$ allows to find uniquely the function $g(t)$. Divide the both sides of the second equation in (4.1.7) by $\rho(0, t)$. Then we obtain the following *Volterra equation* of the second kind:

$$g(t) + \int_0^t K(t - \tau, \tau)g(\tau)d\tau = F(t), \quad t \in [0, T], \tag{4.1.9}$$

with the kernel $K(x, t)$ and the right hand side $F(t)$, depending on input and output data:

$$K(x, t) = \rho_x(x, t)/\rho(0, t), \quad F(t) = f''(t)/\rho(0, t), \quad t \in [0, T]. \tag{4.1.10}$$

By conditions (4.1.3) and (4.1.6) these functions are continuous, that is, $K \in C(D_T)$ and $F \in C[0, T]$. It is well known, that the Volterra equation of the second kind with such kernel and right-hand side has the unique solution $g \in C[0, T]$ [100]. This completes the proof of the theorem. $\square$

Remark that Eq. (4.1.9) can easily be solved, for example, by Picard's method of successive approximations:

$$g_0(t) = F(t),$$
$$g_n(t) + \int_0^t K(t - \tau, \tau)g_{n-1}(\tau)d\tau = F(t), \quad n = 1, 2, \ldots,$$

which uniformly converges in $C[0, T]$.

Remark 4.1.1 As formula (4.1.10) shows, the right hand side $F(t)$ of the integral Eq. (4.1.9) contains the second derivative $f''(t)$ of the output data. This implies that the inverse problem (4.1.1) and (4.1.2) is ill-posed. This linear ill-posedness arises from the fact that the measured output data $f(t)$ have to be differentiated twice. If, for example, $f^\delta(t) = f(t) + \delta \sin(t/\delta)$ is a noisy data, then $\|f^\delta - f\|_{C[0,T]} \to 0$, as $\delta \to 0$, while $\|F^\delta - F\|_{C[0,T]} \to \infty$.

4.1.2 Recovering a Spacewise Dependent Function

Consider the inverse source problem of identifying the *unknown spacewise dependent source* $\varphi(x)$ in

$$\begin{cases} \left(\dfrac{\partial^2}{\partial t^2} - \dfrac{\partial^2}{\partial x^2}\right)u(x, t) = \rho(x, t)\varphi(x), & (x, t) \in \mathbb{R}_+^2, \\ u|_{t=0} = 0, \quad u_t|_{t=0} = 0, & x > 0, \\ u_x|_{x=0} = 0, & t > 0 \end{cases} \tag{4.1.11}$$

from the trace of the function $u(x, t)$ given by (4.1.2). The function $\rho(x, t)$ is assumed to be known.

4.1 Inverse Source Problems

Let us use again the reflection method, with the same even extensions $u(x,t) = u(-x,t)$ and $\rho(x,t) = \rho(-x,t)$ to the domain $\mathbb{R}_-^2 = \{(x,t)| x < 0, t > 0\}$. As noted above, the condition $u_x|_{x=0} = 0$ is automatically satisfied. Then we can derive the solution $u(x,t)$ of the direct problem (4.1.11) via d'Alembert's formula:

$$u(x,t) = \frac{1}{2} \iint_{\Delta(x,t)} \rho(\xi,\tau)\varphi(\xi)\, d\xi d\tau.$$

With the additional condition (4.1.2) this implies:

$$f(t) = \frac{1}{2} \int_{\Delta(0,t)} \rho(\xi,\tau)\varphi(\xi)\, d\xi d\tau = \int_{\Delta_+(0,t)} \rho(\xi,\tau)\varphi(\xi)\, d\xi d\tau.$$

Hence the unknown spacewise dependent source $\varphi(x)$ satisfies the integral equation:

$$\int_0^t \varphi(\xi) \int_0^{t-\xi} \rho(\xi,\tau)\, d\tau d\xi = f(t), \quad t \geq 0. \tag{4.1.12}$$

As in the previous problem, we conclude again that the function $f(t)$ satisfies conditions (4.1.6). Indeed, differentiating (4.1.12), one get

$$\begin{aligned} f'(t) &= \int_0^t \varphi(\xi)\rho(\xi, t-\xi)\, d\xi, \\ f''(t) &= \varphi(t)\rho(t,0) + \int_0^t \varphi(\tau)\rho_t(\xi, t-\xi)\, d\xi, \quad t \geq 0. \end{aligned} \tag{4.1.13}$$

Thus, we arrive at almost the same necessary conditions (4.1.6) for a solvability of the inverse problem.

Now we return to the inverse problem defined by (4.1.11) and (4.1.2). Assuming that $\rho(t,0) \neq 0$, for all $t > 0$, we can formulate the necessary and sufficient conditions for the unique solvability of this inverse problem, similar to Theorem 4.1.1.

Theorem 4.1.2 *Let the source function $\rho(x,t)$ satisfies the following conditions:*

$$\rho \in C(D_T), \quad \rho_t \in C(D_T), \quad \rho(t,0) \neq 0, \quad t \in [0,T], \ T > 0.$$

Then the inverse problem defined by (4.1.11) and (4.1.2) is uniquely solvable in $C[0,T]$ if and only if the function $f(t)$ belongs to $C^2[0,T]$ and satisfies to the conditions $f(0) = f'(0) = 0$.

Proof Dividing the both sides of the second equation in (4.1.13) by $\rho(t,0)$ we obtain the Volterra equation of the second kind

$$\varphi(t) + \int_0^t K_2(\xi, t-\xi)\varphi(\xi)\, d\xi = F_2(t), \quad t \in [0,T], \tag{4.1.14}$$

which has the unique solution $\varphi \in C[0,T]$, where $K_2(x,t) = \rho_t(x,t)/\rho(t,0)$, $F_2(t) = f''(t)/\rho(t,0)$. Evidently, $K_1 \in C(D_T)$, $F_1 \in C[0,T]$. Then we conclude that Eq. (4.1.14) has the unique solution $\varphi \in C[0,T]$. □

Note that the solution of the Eq. (4.1.14) can be found by the same method of successive approximations.

Remark 4.1.2 If function $\rho(x,t) = 0$ for $0 < t < x$, then the solution to problem (4.1.11) satisfies the equality $u(x,t) = 0$ for $0 < t < x$. In this case Eq. (4.1.12) takes the form

$$f(t) = \int_0^{t/2} \varphi(\xi) \int_\xi^{t-\xi} \rho(\xi,\tau)\,d\tau d\xi, \quad t \geq 0. \tag{4.1.15}$$

The derivatives of $f(t)$ up to the second order are defined then by the formulae:

$$\begin{aligned} f'(t) &= \int_0^{t/2} \varphi(\xi)\rho(\xi, t-\xi)\,d\xi, \\ f''(t) &= \tfrac{1}{2}\varphi\left(\tfrac{t}{2}\right)\rho\left(\tfrac{t}{2},\tfrac{t}{2}\right) + \int_0^{t/2} \varphi(\tau)\rho_t(\xi, t-\xi)\,d\xi, \quad t \geq 0. \end{aligned} \tag{4.1.16}$$

These considerations lead to the following result.

Theorem 4.1.3 Let $\rho(x,t) = 0$ for $0 < t < x \leq T-t$ and $\rho \in C(D_T')$, $\rho_t \in C(D_T')$, $D_T' = \{(x,t) \in \mathbb{R}^2 \mid 0 \leq x \leq t \leq T-x\}$, for some positive T and $\rho(t,t) \neq 0$ for $t \in [0, T/2]$. Then the inverse problem defined by (4.1.11) and (4.1.2) is uniquely solvable in the space $C[0, T/2]$ if and only if function $f(t)$ belong to $C^2[0, T]$ and satisfy to the consistency conditions $f(0) = f'(0) = 0$.

Remark 4.1.3 The statement of the above theorems remain to hold if we replace the wave operator in Eqs. (4.1.1) and (4.1.11) by more general operator $\partial^2/\partial t^2 - \partial^2/\partial x^2 + q(x)$, where $q(x)$, $x \geq 0$, is a given continuous function for $x \geq 0$.

4.2 Problem of Recovering the Potential for the String Equation

In this section we consider the inverse problem of *recovering the potential* $q(x)$, $x \in \mathbb{R}_+ := \{x \geq 0\}$ in

$$\begin{cases} L_q u := \left(\frac{\partial^2}{\partial t^2} - \frac{\partial^2}{\partial x^2} - q(x)\right) u = 0, & (x,t) \in \Omega, \\ u|_{x=0} = h(t), \quad u|_{t=0} = 0, \quad u_t|_{t=0} = 0, \end{cases} \tag{4.2.1}$$

from the output data

$$f(t) := u_x|_{x=0}, \quad t \in (0, T], \tag{4.2.2}$$

4.2 Problem of Recovering the Potential for the String Equation

given as the trace of the derivative u_x of the solution $u(x, t)$ to problem (4.2.2) at $x = 0$. Here $\Omega = \mathbb{R}_+ \times [0, T]$. It is assumed that

$$h \in C^2[0, T], \quad h(0) \neq 0. \tag{4.2.3}$$

The problem (4.2.1) and (4.2.2) of recovering the potential in the string equation can be treated as a simplest *inverse coefficient problem* for the differential operator $L_q = \partial^2/\partial t^2 - \partial^2/\partial x^2 - q$, with Neumann data $f(t)$. The hyperbolic problem (4.2.1) will be defined as a *direct problem*.

Assuming $q \in C[0, \infty)$, we demonstrate that the potential $q(x)$ can be determined uniquely and *in a stable way* from the output data (4.2.2), for $x \in [0, T/2]$. Moreover, we will prove that the solution of the inverse problem exists for small $T > 0$, if the function $f(t)$ satisfies some necessary conditions.

4.2.1 Some Properties of the Direct Problem

The necessary conditions for unique solvability of the inverse problem, obtained in the next subsection, follow from properties of the solution of the direct problem (4.2.1). The lemma below shows that the solution $u(x, t)$ of the hyperbolic problem (4.2.1) can be represented as a product of a smooth function and the Heaviside step function.

Lemma 4.2.1 *Let the boundary data $h(t)$ satisfies conditions (4.2.3) and $q \in C[0, T/2]$, $T > 0$. Then the solution of the direct problem (4.2.1) exists in the triangle $\Delta_T := \{(x, t) \in \mathbb{R}^2 \mid 0 \leq t \leq T - x, x \geq 0\}$. Moreover, it represented as a product*

$$u(x, t) = \overline{u}(x, t)\theta_0(t - x). \tag{4.2.4}$$

of a smooth function $\overline{u} \in C^2(\Omega_T)$, $\Omega_T := \{(x, t) \in \mathbb{R}^2 \mid 0 \leq x \leq t \leq T - x\}$, and the Heaviside step function

$$\theta_0(t) = \begin{cases} 1, & t \geq 0, \\ 0, & t < 0. \end{cases} \tag{4.2.5}$$

Proof The function $u(x, t)$ is identically zero, for all $0 < t < x, t \leq T - x$, since in this domain it solves the Cauchy problem for the homogeneous equation $L_q u = 0$ with homogeneous initial data. This implies the presentation (4.2.4).

We introduce now the function $u_0(x, t) = h(t - x)\theta_0(t - x)$, where $h(t)$ is the Dirichlet input data in the direct problem (4.2.1). We represent the function $u(x, t)$ in the form:

$$u(x, t) = u_0(x, t) + v(x, t). \tag{4.2.6}$$

Then $v(x, t)$ solves the problem

$$\begin{cases} L_q v = q(x) u_0(x, t), & 0 < x < t \leq T - x, \\ v|_{x=0} = 0, \quad v|_{t=x} = 0. \end{cases} \quad (4.2.7)$$

Let us explain the appearance of the condition $v|_{t=x} = 0$ in (4.2.7), on the characteristic line $t = x$. For this aim we extend the functions $v(x, t)$, $u_0(x, t)$ to the domain $x < 0$ as an odd functions with respect to x, that is, $v(x, t) = -v(-x, t)$, $u_0(x, t) = -u_0(-x, t)$. We also extend the coefficient $q(x)$ to the domain $x < 0$, as an even function, i.e., $q(x) = q(-x)$. Then we may use d'Alembert's formula for the solution of the hyperbolic problem (4.2.7) to get

$$v(x, t) = \frac{1}{2} \int_{\Delta(x,t)} q(x)(v(\xi, \tau) + u_0(\xi, \tau)) \, d\xi d\tau$$

$$= \frac{1}{2} \int_{\Diamond(x,t)} q(x)(v(\xi, \tau) + u_0(\xi, \tau)) \, d\xi d\tau, \quad t \geq |x|.$$

where $\Diamond(x, t) = \{(\xi, \tau) \in \mathbb{R}^2 \mid |\xi| \leq \tau \leq t - |x - \xi|\}$. The area of the domain $\Diamond(x, t)$ tends to zero, as $t \to |x| + 0$. Therefore $v(x, |x| + 0) = 0$, which implies the fulfilment of the second condition $v|_{t=x} = 0$ in (4.2.7).

Introduce now the functions

$$v_1(x, t) := \frac{\partial v}{\partial t} + \frac{\partial v}{\partial x}, \quad v_2(x, t) := \frac{\partial v}{\partial t} - \frac{\partial v}{\partial x}. \quad (4.2.8)$$

Then the following relations hold:

$$L_q v = \frac{\partial v_1}{\partial t} - \frac{\partial v_1}{\partial x} - qv = \frac{\partial v_2}{\partial t} + \frac{\partial v_2}{\partial x} - qv, \quad \frac{\partial v}{\partial t} = \frac{v_1 + v_2}{2}. \quad (4.2.9)$$

Note that

$$v_1|_{t=x} = 0, \quad v_2|_{x=0} = -v_1(0, t). \quad (4.2.10)$$

Now we are going to obtain integral relationships between the above introduced functions v_1, v_2, v. Integrating the equation $L_q v = q u_0$ on the plane (ξ, τ) along the line $\xi + \tau = x + t$, then using relations (4.2.9) and the first condition of (4.2.10) we find:

$$v_1(x, t) =$$
$$\int_{(t+x)/2}^{t} q(x + t - \tau)(u_0(x + t - \tau, \tau) + v(x + t - \tau, \tau)) \, d\tau. \quad (4.2.11)$$

Similarly, integrating the equation $L_q v = q u_0$ on the plane (ξ, τ) along the line $\xi - \tau = x - t$, using (4.2.9) and the second condition of (4.2.10) we obtain

4.2 Problem of Recovering the Potential for the String Equation

$$v_2(x, t) =$$
$$-v_1(0, t - x) + \int_{t-x}^{t} q(x - t + \tau)(u_0(x - t + \tau, \tau) + v(x - t + \tau, \tau)) d\tau.$$

Use (4.2.11) in the first right hand side term of this relation equation. Then we get:

$$v_2(x, t) =$$
$$- \int_{(t-x)/2}^{t-x} q(t - x - \tau)(u_0(t - x - \tau, \tau) + v(t - x - \tau, \tau)) d\tau \quad (4.2.12)$$
$$+ \int_{t-x}^{t} q(x - t + \tau)(u_0(x - t + \tau, \tau) + v(x - t + \tau, \tau)) d\tau.$$

Finally, integrate the second relation in (4.2.9) along the line $\xi = x$ to find

$$v(x, t) = \frac{1}{2} \int_{x}^{t} (v_1(x, \tau) + v_2(x, \tau)) d\tau. \quad (4.2.13)$$

The Eqs. (4.2.11)–(4.2.13) form the system of integral equations with respect to the functions v_1, v_2, v defined in the domain $\Omega_T := \{(x, t) | 0 \leq x \leq t \leq T - x\}$. We prove that this system has a unique solution in Ω_T. For this aim, we represent the functions $v_p(x, t)$, $p = 1, 2$ and $v(x, t)$ in the form

$$v_p(x, t) = \sum_{n=1}^{\infty} v_p^n(x, t), \ p = 1, 2, \quad v(x, t) = \sum_{n=1}^{\infty} v^n(x, t), \quad (4.2.14)$$

and look for the series solution of this system. The first terms ($n = 1$) of these series are given by the formulae:

$$v_1^1(x, t) = \int_{(t+x)/2}^{t} q(x + t - \tau) u_0(x + t - \tau, \tau) d\tau,$$
$$v_2^1(x, t) = - \int_{(t-x)/2}^{t-x} q(t - x - \tau) u_0(t - x - \tau, \tau) d\tau \quad (4.2.15)$$
$$+ \int_{t-x}^{t} q(x - t + \tau) u_0(x - t + \tau, \tau) d\tau,$$
$$v^1(x, t) = \frac{1}{2} \int_{x}^{t} (v_1^1(x, \tau) + v_2^1(x, \tau)) d\tau.$$

For $n \geq 2$ other terms of series (4.2.14) are defined recursively as follows:

$$v_1^n(x, t) = \int_{(t+x)/2}^{t} q(x + t - \tau) v^{n-1}(x + t - \tau, \tau) d\tau,$$
$$v_2^n(x, t) = - \int_{(t-x)/2}^{t-x} q(t - x - \tau) v^{n-1}(t - x - \tau, \tau) d\tau \quad (4.2.16)$$
$$+ \int_{t-x}^{t} q(x - t + \tau) v^{n-1}(x - t + \tau, \tau) d\tau,$$
$$v^n(x, t) = \frac{1}{2} \int_{x}^{t} (v_1^n(x, \tau) + v_2^n(x, \tau)) d\tau.$$

Obviously $v_1^n(x,t)$, $v_2^n(x,t)$, $v^n(x,t)$ are continuous functions in Ω_T together with the first derivatives with respect to x and t, for all $n \geq 1$. We use now Mathematical Induction to prove the uniform convergence of the series (4.2.14) in Ω_T.

Denote by

$$h_0 = \|h\|_{C[0,T]}, \quad q_0 = \|q\|_{C[0,T/2]}$$

the norms. For $n=1$ we can easily derive the estimates:

$$|v_1^1(x,t)| \leq h_0 q_0 \frac{(t-x)}{2},$$

$$|v_2^1(x,t)| \leq h_0 q_0 \frac{(t+x)}{2} \leq \frac{h_0 q_0 T}{2},$$

$$|v^1(x,t)| \leq \frac{h_0 q_0 T (t-x)}{2}, \quad (x,t) \in \Omega_T,$$

by using (4.2.15).

Assume that for $n = k \geq 1$ the estimates

$$|v_1^k(x,t)| \leq \frac{\gamma_k}{2^{k+1}T} \frac{(t-x)^k}{k!}, \quad |v_2^k(x,t)| \leq \frac{\gamma_k}{2^{k+1}} \frac{(t-x)^{k-1}}{(k-1)!},$$
$$|v^k(x,t)| \leq \frac{\gamma_k}{2^{k+1}} \frac{(t-x)^k}{k!}, \quad \gamma_k = h_0 q_0^k T^k. \tag{4.2.17}$$

hold. Using these estimates in (4.2.16), for $n = k+1$ we find:

$$|v_1^{k+1}(x,t)| \leq \frac{q_0 \gamma_k}{2^{k+1} k!} \int_{(t+x)/2}^{t} (2\tau - x - t)^k \, d\tau = \frac{q_0 \gamma_k}{2^{k+2}} \frac{(t-x)^{k+1}}{(k+1)!},$$

$$|v_2^{k+1}(x,t)| \leq \frac{q_0 \gamma_k}{2^{k+1} k!} \left[\int_{(t-x)/2}^{t-x} (2\tau - t + x)^k \, d\tau + \int_{t-x}^{t} (t-x)^k \, d\tau \right]$$
$$= \frac{q_0 \gamma_k (t-x)^k}{2^{k+1} k!} \left(\frac{t-x}{2(k+1)} + x \right) \leq \frac{\gamma_{k+1}}{2^{k+2}} \frac{(t-x)^k}{k!},$$

$$|v^{k+1}(x,t)| \leq \frac{\gamma_{k+1}}{2^{k+2}} \frac{(t-x)^{k+1}}{(k+1)!}, \quad \gamma_k = h_0 q_0^k T^k.$$

Therefore estimates (4.2.17) hold for all $n \geq 1$.

Further, due to $t - x \leq T$ in $\Omega_T := \{(x,t) \in \mathbb{R}^2 \mid 0 \leq x \leq t \leq T - x\}$, all series in (4.2.14) converge uniformly and, as a result, their sums are continuous functions in Ω_T. Furthermore, for these functions the following estimates hold:

4.2 Problem of Recovering the Potential for the String Equation

$$|v_1(x,t)| \le \frac{h_0}{2T}\left(\exp\left(\frac{1}{2}q_0 T(t-x)\right) - 1\right)$$
$$\le \frac{h_0}{2T}\left(\exp\left(\frac{1}{2}q_0 T^2\right) - 1\right) := v_{10},$$

$$|v_2(x,t)| \le \frac{h_0 q_0 T}{2}\exp\left(\frac{1}{2}q_0 T(t-x)\right)$$
$$\le \frac{h_0 q_0 T}{2}\exp\left(\frac{1}{2}q_0 T^2\right) := v_{20}, \qquad (4.2.18)$$

$$|v(x,t)| \le \frac{h_0}{2}\left(\exp\left(\frac{1}{2}q_0 T(t-x)\right) - 1\right)$$
$$\le \frac{h_0}{2}\left(\exp\left(\frac{1}{2}q_0 T^2\right) - 1\right) := v_0,$$

for all $(x,t) \in \Omega_T$. Note that $v \in C^1(\Omega_T)$, by the relations $v_t = (v_1 + v_2)/2$ and $v_x = (v_1 - v_2)/2$. It follows from (4.2.11) and (4.2.12) that the functions $v_1(x,t)$ and $v_2(x,t)$ are continuously differentiable with respect to x and t in Ω_T. To verify this, one needs to rewrite Eqs. (4.2.11) and (4.2.12) in the following form:

$$v_1(x,t) = \int_x^{(t+x)/2} q(\xi)(u_0(\xi, x+t-\xi) + v(\xi, x+t-\xi))\,d\xi, \qquad (4.2.19)$$

$$v_2(x,t) = -\int_0^{(t-x)/2} q(\xi)(u_0(\xi, t-x-\xi) + v(\xi, t-x-\xi))\,d\xi \qquad (4.2.20)$$
$$+ \int_0^x q(\xi)(u_0(\xi, t-x+\xi) + v(\xi, t-x+\xi))\,d\xi.$$

Since v_1 and v_2 belong to $C^1(\Omega_T)$, the function $v(x,t)$ is twice continuously differentiable in Ω_T. Then, it follows from (4.2.6) that the function $\bar{u}(x,t)$ is also twice continuously differentiable in Ω_T. This completes the proof of the lemma. □

4.2.2 Existence of the Local Solution to the Inverse Problem

First, we establish a relationship between the input $h(t)$ and the output $f(t)$ of the inverse problem (4.2.1) and (4.2.2) and the functions $v_1(x,t)$, $v_2(x,t)$ introduced in (4.2.8). It follows from (4.2.6) that

$$\bar{u}_x(x,t) = -h'(t-x) + v_x(x,t), \quad 0 \le x \le t,$$

where $h'(t)$ means the derivative of $h(t)$ with respect to its argument. Using formulae (4.2.8) we find that $v_x(x,t) = (v_1(x,t) - v_2(x,t))/2$. Then the additional condition (4.2.2) leads to the relation

$$f(t) := \bar{u}_x(0,t) = -h'(t) + (v_1(0,t) - v_2(0,t))/2, \quad t \in (0,T]. \qquad (4.2.21)$$

It follows from the Lemma 4.2.1 that $f \in C^1(0, T]$. Moreover, since $v_1(0, 0) = v_2(0, 0) = 0$, we conclude from (4.2.21) that

$$f(+0) = -h'(0). \qquad (4.2.22)$$

Hence, the conditions (4.2.22) and $f \in C^1(0, T]$ are the necessary conditions for solvability of the inverse problem in the class of continuous functions $q(x)$ on the segment $[0, T/2]$.

Let us show now that these conditions are also sufficient for a local unique solvability of the inverse problem.

Theorem 4.2.1 *Let $f \in C^1(0, T]$, $T > 0$, and $f(+0) = -h'(0)$. Then there exist a positive number $T_0 \leq T$ and unique function $q(x) \in C[0, T_0/2]$ such that the solution to problem (4.2.1) satisfies relation (4.2.2) for $t \leq T_0$.*

Proof We derive first the integral relation which follows from Eqs. (4.2.19), (4.2.20) and (4.2.21). Taking into account that $u_0(x, t) = h(t - x)$ for $t \geq x \geq 0$ we obtain:

$$\int_0^{t/2} q(\xi)(h(t - 2\xi) + v(\xi, t - \xi)) d\xi = \hat{f}(t), \qquad (4.2.23)$$

where

$$\hat{f}(t) := f(t) + h'(t) \in C^1[0, T]. \qquad (4.2.24)$$

Taking now the derivative with respect to t of both sides of (4.2.23) and using $v(x, x) = 0$, we arrive at the equations:

$$\frac{1}{2} q(t/2) h(0) + \int_0^{t/2} q(\xi)(h'(t - 2\xi) + v_t(\xi, t - \xi)) d\xi = \hat{f}'(t), \quad t \in [0, T],$$

or

$$q(x) + \frac{1}{h(0)} \int_0^x q(\xi) \left(2h'(2x - 2\xi) + v_1(\xi, 2x - \xi) + v_2(\xi, 2x - \xi)\right) d\xi$$
$$= F(x), \ x \in [0, T/2], \ (4.2.25)$$

with the right hand side

$$F(x) = \frac{2\hat{f}'(2x)}{h(0)} = \frac{2(f'(2x) + h''(2x))}{h(0)}. \qquad (4.2.26)$$

Equations (4.2.11)–(4.2.13) and (4.2.25) form a system of integral equations with respect to unknown functions $q(x)$ and $v_1(x, t), v_2(x, t), v(x, t)$.

In order to simplify further notations, it is convenient to transform, first, the Eqs. (4.2.11) and (4.2.12), then to rewrite the system of integral equations in terms

4.2 Problem of Recovering the Potential for the String Equation

of an operator equation. For this aim we use (4.2.21) and (4.2.24) to deduce that both functions v_1 and v_2 are known at $x = 0$:

$$v_1(0, t) = \hat{f}(t), \quad v_2(0, t) = -\hat{f}(t), \quad t \in [0, T]. \tag{4.2.27}$$

Using these conditions we integrate the equation $L_q v = q u_0$ on the plane (ξ, τ) along the line $\xi + \tau = x + t$ from point $(x, t) \in \Omega_T$ till the axis $\xi = 0$ to obtain the equation

$$v_1(x, t) = \hat{f}(t + x) - \int_0^x q(\xi)(u_0(\xi, x + t - \xi) + v(\xi, x + t - \xi))\, d\xi. \tag{4.2.28}$$

Similarly, integrating the equation $L_q v = q u_0$ on the plane (ξ, τ) along the line $\xi - \tau = x - t$ from point $(x, t) \in \Omega_T$ till the axis $\xi = 0$, we obtain the second equation

$$\begin{aligned}v_2(x, t) &= -\hat{f}(t - x) \\ &+ \int_0^x q(\xi)(u_0(\xi, t - x + \xi) + v(\xi, t - x + \xi))\, d\xi.\end{aligned} \tag{4.2.29}$$

Now we rewrite the system formed, respectively, by Eqs. (4.2.25), (4.2.28), (4.2.29) and (4.2.13), in the form of the operator equation

$$\begin{aligned}\varphi &= A\varphi, \\ \varphi &= (\varphi_1, \varphi_2, \varphi_3, \varphi_4) := (q(x), v_1(x, t), v_2(x, t), v(x, t)),\end{aligned} \tag{4.2.30}$$

where

$$(A\varphi)_1(x) = F(x) - \tfrac{1}{h(0)} \int_0^x \varphi_1(\xi)(2h'(2x - 2\xi) + \varphi_2(\xi, 2x - \xi) \\ + \varphi_3(\xi, 2x - \xi))\, d\xi, \quad x \in [0, T/2],$$

$$(A\varphi)_2(x, t) = \hat{f}(t + x) - \int_0^x \varphi_1(\xi)(h(x + t - 2\xi) \\ + \varphi_4(\xi, x + t - \xi))\, d\xi, \tag{4.2.31}$$

$$(A\varphi)_3(x, t) = -\hat{f}(t - x) + \int_0^x \varphi_1(\xi)(h(t - x) + \varphi_4(\xi, t - x + \xi))\, d\xi,$$

$$(A\varphi)_4(x, t) = \tfrac{1}{2}\int_x^t (\varphi_2(x, \tau) + \varphi_3(x, \tau))\, d\tau, \quad (x, t) \in \Omega_T.$$

Let $\varphi^0 = (\varphi_{10}, \varphi_{20}, \varphi_{30}, \varphi_{40})$ be a vector with components

$$\varphi_{10} = F(x), \quad \varphi_{20} = \hat{f}(t + x), \quad \varphi_{30} = -\hat{f}(t - x), \quad \varphi_{40} = 0.$$

Denote by $\mathbf{C}(\Omega_T)$ the space of continuous vector functions, with the norm

$$\|\varphi\|_{C(\Omega_T)} = \max_{k=\overline{1,4}} \|\varphi_k\|_{C(\Omega_T)}.$$

Since $\varphi_0 \in \mathbf{C}(\Omega_T)$, all vector functions defined by (4.2.31) are evidently elements of $\mathbf{C}(\Omega_T)$. We introduce in this Banach space the closed ball

$$\mathbf{B}_T := \{\varphi \in \mathbf{C}(\Omega_T) : \|\varphi - \varphi^0\|_{C(\Omega_T)} \leq \|\varphi^0\|_{C(\Omega_T)}\}. \quad (4.2.32)$$

of radius $\|\varphi^0\|_{C(\Omega_T)} > 0$ centered at $\varphi^0 \in \mathbf{C}(\Omega_T)$. Evidently,

$$\|\varphi^0\|_{C(\Omega_T)} \leq a_0(T) := \max(\|F\|_{C[0,T/2]}; \|\hat{f}\|_{C[0,T]}), \quad (4.2.33)$$

where $\hat{f}(t)$ and $F(x)$ are defined by (4.2.24) and (4.2.26), respectively.

Hereafter we assume that $F(x)$ and $\hat{f}(t)$ are given fixed functions. Then their norms $\|F\|_{C[0,T/2]}$, $\|\hat{f}\|_{C[0,T]}$ depend on T only. Taking it into account we have used in (4.2.33) the notation $a_0(T)$ for the maximum of these two norms. The similar notations for some values we shall use and later on in order indicate on a dependence of these values on T.

Now we are going to prove that the operator A, defined by (4.2.30) and (4.2.31) is a contraction on the Banach space $\mathbf{B}_T$, if the final time $T > 0$ is small enough. Recall that an operator is named contracting one on $\mathbf{B}_T$, if the following two conditions hold:
(c1) $A\varphi \in \mathbf{B}_T$, for all $\varphi \in \mathbf{B}_T$;
(c2) for all $\varphi^1, \varphi^2 \in \mathbf{B}_T$, the condition

$$\|A\varphi^1 - A\varphi^2\|_{C(\Omega_T)} \leq \rho \|\varphi^1 - \varphi^2\|_{C(\Omega_T)}$$

holds with some $\rho \in (0, 1)$.

We verify the first condition (c1). Let $\varphi \in \mathbf{B}_T$. Then

$$\|\varphi\|_{C(\Omega_T)} \leq \|\varphi - \varphi^0\|_{C(\Omega_T)} + \|\varphi^0\|_{C(\Omega_T)} \leq 2\|\varphi^0\|_{C[0,T]}) \leq 2a_0(T),$$

by (4.2.33). Using this in (4.2.31) we estimate the norms $|(A\varphi)_k - \varphi_{k0}|$, $k = \overline{1,4}$ as follows:

$$|(A\varphi)_1 - \varphi_{10}| \leq \frac{1}{|h(0)|} \int_0^x |\varphi_1(\xi)| (2 |h'(2x - 2\xi)|$$
$$+ |\varphi_2(\xi, 2x - \xi)| + |\varphi_3(\xi, 2x - \xi)|) \, d\xi$$
$$\leq \frac{T}{|h(0)|} (\|h'\|_{C[0,T]} + 2a_0(T)) \|\varphi\| := a_1(T) \|\varphi\|,$$

$$|(A\varphi)_2 - \varphi_{20}| \leq \int_0^x |\varphi_1(\xi)| (|h(x + t - 2\xi)| + |\varphi_4(\xi, x + t - \xi)|) \, d\xi$$
$$\leq \frac{T}{2} (\|h\|_{C[0,T]} + 2a_0(T)) \|\varphi\| := a_2(T) \|\varphi\|,$$

4.2 Problem of Recovering the Potential for the String Equation

$$|(A\varphi)_3 - \varphi_{30}| \leq \int_0^x |\varphi_1(\xi)|(|h(t-x)| + |\varphi_4(\xi, t-x+\xi)|)\,d\xi$$
$$\leq \frac{T}{2}(\|h\|_{C[0,T]} + 2a_0(T))\|\varphi\| := a_3(T)\|\varphi\|,$$

$$|(A\varphi)_4 - \varphi_{40}| \leq \frac{1}{2}\int_x^t (|\varphi_2(x,\tau)| + |\varphi_3(x,\tau)|)\,d\tau \leq T\|\varphi\| := a_4(T)\|\varphi\|.$$

Therefore $A\varphi \in \mathbf{B}_T$, if the following condition holds:

$$\max_{k=1,2,3,4} a_k(T) \leq 1. \tag{4.2.34}$$

We verify the second condition **(c2)**. Let $\varphi^k := (\varphi_1^k, \varphi_2^k, \varphi_3^k, \varphi_4^k)$ and $\varphi^k \in \mathbf{B}_T$, $k = 1, 2$. Then one has

$$|(A\varphi^1 - A\varphi^2)_1| \leq \frac{1}{|h(0)|}\int_0^x \Big(|\varphi_1^1(\xi) - \varphi_1^2(\xi)|(2|h'(2x-2\xi)|$$
$$+ |\varphi_2^1(\xi, 2x-\xi)| + |\varphi_3^1(\xi, 2x-\xi)|) + |\varphi_1^2(\xi)|(|\varphi_2^1(\xi, 2x-\xi) - \varphi_2^2(\xi, 2x-\xi)|$$
$$+ |\varphi_3^1(\xi, 2x-\xi) - \varphi_3^2(\xi, 2x-\xi)|)\Big)\,d\xi$$
$$\leq \frac{T}{|h(0)|}(\|h'\|_{C[0,T]} + 4a_0(T))\|\varphi^1 - \varphi^2\| := b_1(T)\|\varphi^1 - \varphi^2\|,$$

Similarly,

$$|(A\varphi^1 - A\varphi^2)_2| \leq \int_0^x \Big(|\varphi_1^1(\xi) - \varphi_1^2(\xi)|(|h(x+t-2\xi)| + |\varphi_4^1(\xi, x+t-\xi)|)$$
$$+ |\varphi_1^2(\xi)||\varphi_4^1(\xi, x+t-\xi) - \varphi_4^2(\xi, x+t-\xi)|\Big)\,d\xi$$
$$\leq \frac{T}{2}(\|h\|_{C[0,T]} + 4a_0(T))\|\varphi^1 - \varphi^2\| := b_2(T)\|\varphi^1 - \varphi^2\|,$$

$$|(A\varphi^1 - A\varphi^2)_3| \leq \int_0^x \Big(|\varphi_1^1(\xi) - \varphi_1^2(\xi)|(|h(t-x)| + |\varphi_4^1(\xi, t-x-\xi)|)$$
$$+ |\varphi_1^2(\xi)||\varphi_4^1(\xi, t-x-\xi) - \varphi_4^2(\xi, t-x-\xi)|\Big)\,d\xi$$
$$\leq \frac{T}{2}(\|h\|_{C[0,T]} + 4a_0(T))\|\varphi^1 - \varphi^2\| := b_3(T)\|\varphi^1 - \varphi^2\|,$$

$$|(A\varphi^1 - A\varphi^2)_4| \leq \frac{1}{2}\int_x^t (|\varphi_2^1(x,\tau) - \varphi_2^2(x,\tau)| + |\varphi_3^1(x,\tau) - \varphi_3^2(x,\tau)|)\,d\tau$$
$$\leq T\|\varphi^1 - \varphi^2\| := b_4(T)\|\varphi^1 - \varphi^2\|.$$

Hence, $\|A\varphi^1 - A\varphi^2\| \leq \rho\|\varphi^1 - \varphi^2\|$ with $\rho < 1$, if T satisfies the conditions

$$\max_{k=1,2,3,4} b_k(T) \leq \rho < 1. \qquad (4.2.35)$$

Thus, if the final time $T > 0$ is chosen so (small) that both conditions (4.2.34) and (4.2.35) hold, then the operator A is contracting on $\mathbf{B}_T$. Then, according to Banach Contraction Mapping Principle, there exists a unique solution of the operator Eq. (4.2.30) in $\mathbf{B}_T$. This completes the proof of the theorem. □

4.2.3 Global Stability and Uniqueness

Now we state a stability estimate and a uniqueness theorem when the final time T is an arbitrary fixed positive number.

Denote by $Q(q_0)$ the set of functions $q \in C[0, T/2]$, satisfying the inequality $|q(x)| \leq q_0$, for $x \in [0, T/2]$, with the positive constant q_0. Let $\mathcal{H}(h_0, h_1, d)$ be the set of functions $h(t)$ satisfying for some fix positive constants h_0, h_1 and d the conditions:

(1) $h \in C^2[0, T]$,
(2) $\|h(t)\|_{C[0,T]} \leq h_0$, $\|h'(t)\|_{C[0,T]} \leq h_1$,
(3) $|h(0)| \geq d > 0$.

Evidently, for $q \in Q(q_0)$ and $h \in \mathcal{H}(h_0, h_1, d)$, estimates (4.2.18) remain true in Ω_T for the solution of the system of integral Eqs. (4.2.11)–(4.2.13).

Theorem 4.2.2 *Let $q_k \in Q(q_0)$ be a solution of the inverse problem (4.2.1), (4.2.2) corresponding to the data $h_k \in \mathcal{H}(h_0, h_1, d)$, $f_k \in C^1[0, T]$, for each $k = 1, 2$. Then there exists a positive number $C = C(q_0, h_0, h_1, d, T)$ such that the following stability estimate holds:*

$$\|q_1 - q_2\|_{C[0,T/2]} \leq C\left(\|f_1 - f_2\|_{C^1[0,T]} + \|h_1 - h_2\|_{C^2[0,T]}\right). \qquad (4.2.36)$$

Proof To prove this theorem we use Lemma 4.2.1. For this aim, similar to representations (4.2.4) and (4.2.6), we represent the solution of the direct problem (4.2.1) corresponding to $q_k(x) \in Q(q_0)$ and $h_k(t) \in \mathcal{H}(h_0, h_1, d), k = 1, 2$, in the following form:

$$u_k(x, t) = [h_k(t - x) + v^k(x, t)]\theta_0(t - x), \quad k = 1, 2. \qquad (4.2.37)$$

Then the functions $v_1^k = v_t^k + v_x^k$, $v_2^k = v_t^k - v_x^k$, v^k, q_k, $k = 1, 2$, satisfy in $\Omega_T := \{(x, t) | 0 \leq x \leq t \leq T - x\}$ the integral equations:

4.2 Problem of Recovering the Potential for the String Equation

$$v_1^k(x,t) = f_k(t+x) + h'_k(t+x)$$
$$- \int_0^x q_k(\xi)(h_k(x+t-2\xi) + v^k(\xi, x+t-\xi))\,d\xi,$$
$$v_2^k(x,t) = -f_k(t-x) - h'_k(t-x)$$
$$+ \int_0^x q_k(\xi)(h_k(t-x) + v^k(\xi, t-x+\xi))\,d\xi,$$
$$v^k(x,t) = \tfrac{1}{2}\int_x^t (v_1^k(x,\tau) + v_2^k(x,\tau))\,d\tau,$$
$$q_k(x)h_k(0) + \int_0^x q_k(\xi)(2h'_k(2x-2\xi) + v_1^k(\xi, 2x-\xi)$$
$$+ v_2^k(\xi, 2x-\xi))\,d\xi = 2(f'_k(2x) + h''_k(2x)).$$

Introduce the differences:

$$\tilde{v}_1 = v_1^1 - v_1^2, \quad \tilde{v}_2 = v_2^1 - v_2^2, \quad \tilde{v} = v_1 - v_2,$$
$$\tilde{q} = q_1 - q_2, \quad \tilde{f} = f_1 - f_2, \quad \tilde{h} = h_1 - h_2.$$

Then the above equations imply that the following relations hold in Ω_T:

$$\tilde{v}_1(x,t) = \tilde{f}(t+x) + \tilde{h}'(t+x)$$
$$- \int_0^x \Big(\tilde{q}(\xi)(h_1(x+t-2\xi) + v^1(\xi, x+t-\xi))$$
$$+ q_2(\xi)(\tilde{h}(x+t-2\xi) + \tilde{v}(\xi, x+t-\xi))\Big)\,d\xi,$$
$$\tilde{v}_2(x,t) = -\tilde{f}(t-x) - \tilde{h}'(t-x)$$
$$+ \int_0^x \Big(\tilde{q}(\xi)(h_1(t-x) + v^1(\xi, t-x+\xi))$$
$$+ q_2(\xi)(\tilde{h}(t-x) + \tilde{v}(\xi, t-x+\xi))\Big)\,d\xi,$$
$$\tilde{v}(x,t) = \tfrac{1}{2}\int_x^t (\tilde{v}_1(x,\tau) + \tilde{v}_2(x,\tau))\,d\tau,$$
$$\tilde{q}(x)h_1(0) + q_2(x)\tilde{h}(0) + \int_0^x \Big(\tilde{q}(\xi)(2h'_1(2x-2\xi)$$
$$+ v_1^1(\xi, 2x-\xi) + v_2^1(\xi, 2x-\xi)) + q_2(\xi)(2\tilde{h}'(2x-2\xi)$$
$$+ \tilde{v}_1(\xi, 2x-\xi) + \tilde{v}_2(\xi, 2x-\xi))\Big)\,d\xi = 2(\tilde{f}'(2x) + \tilde{h}''(2x)).$$

Let

$$\psi(x) = \max\Big(|\tilde{q}(x)|;\ \max_{x\le t\le T-x}|\tilde{v}_1(x,t)|;\ \max_{x\le t\le T-x}|\tilde{v}_2(x,t)|;\ \max_{x\le t\le T-x}|\tilde{v}(x,t)|\Big).$$

Then using estimates (4.2.18) for v_1^1, v_2^1, v^1 we conclude:

$$|\tilde{v}_1(x,t)| \leq \|\tilde{f}\|_{C[0,T]} + \|\tilde{h}\|_{C^1[0,T]}$$
$$+ \int_0^x \left(\psi(\xi)(h_0+v_0) + q_0(\|\tilde{h}\|_{C[0,T]} + \psi(\xi))\right) d\xi,$$
$$|\tilde{v}_2(x,t)| \leq \|\tilde{f}\|_{C[0,T]} + \|\tilde{h}\|_{C^1[0,T]}$$
$$+ \int_0^x \left(\psi(\xi)(h_0+v_0) + q_0(\|\tilde{h}\|_{C[0,T]} + \psi(\xi))\right) d\xi,$$
$$|\tilde{v}(x,t)| \leq \tfrac{T}{2}(\max_{x\leq\tau\leq t}|\tilde{v}_1(x,\tau)| + \max_{x\leq\tau\leq t}|\tilde{v}_2(x,\tau)|)$$
$$\leq T\Big(\|\tilde{f}\|_{C[0,T]} + \|\tilde{h}\|_{C^1[0,T]}$$
$$+ \int_0^x \left(\psi(\xi)(h_0+v_0) + q_0(\|\tilde{h}\|_{C[0,T]} + \psi(\xi))\right) d\xi\Big),$$
$$|\tilde{q}(x)| \leq \tfrac{1}{d}\Big[q_0|\tilde{h}(0)| + \int_0^x \Big(\psi(\xi)(2h_1 + v_{01} + v_{02})$$
$$+ q_0\big(2\|\tilde{h}\|_{C^1[0,T]} + \psi(\xi)\big)\Big)d\xi + 2(\|\tilde{f}\|_{C^1[0,T]} + \|\tilde{h}\|_{C^2[0,T]})\Big].$$

It follows from these estimates that $\psi(x)$ satisfies the following integral inequality:

$$\psi(x) \leq C_1 \left(\|\tilde{f}\|_{C^1[0,T]} + \|\tilde{h}\|_{C^2[0,T]}\right) + C_2 \int_0^x \psi(\xi)\, d\xi,$$

for all $x \in [0, T/2]$, with the constants

$$C_1 = \max(1 + q_0 T/2;\ (2+2q_0)/d)\max(1;\ T),$$
$$C_2 = \max((h_0 + v_0 + q_0)\max(1;\ T);\ (2h_1 + v_{01} + v_{02} + q_0)/d).$$

Using Gronwall's inequality, one gets the estimate

$$|\psi(x)| \leq C_1\left(\|\tilde{f}\|_{C^1[0,T]} + \|\tilde{h}\|_{C^2[0,T]}\right)\exp(C_2 x),\ x \in [0, T/2]\ . \quad (4.2.38)$$

The required stability estimate (4.2.36) follows from (4.2.38) with $C = C_1 \exp(C_2 T/2)$. This completes the proof. □

We derive now one of the most important application of this theorem. The result below shows that the uniqueness, which valid for any $T > 0$, of the solution of inverse problem (4.2.1) and (4.2.2) is a consequence of the above proved stability theorem.

Theorem 4.2.3 *Let the input $h(t)$ in the direct problem (4.2.1) satisfies conditions (4.2.3), that is, $h \in C^2[0, T]$ and $h(0) \neq 0$. Assume that $q_1, q_2 \in C[0, T/2]$ are two arbitrary solutions of the inverse problem (4.2.1) and (4.2.2). Then $q_1(x) = q_2(x)$ for all $x \in [0, T/2]$.*

4.3 Inverse Coefficient Problems for Layered Media

Consider the following initial-boundary value problem for the acoustic equation:

$$\begin{cases} u_{tt} - \rho^{-1}\operatorname{div}_x(\rho c^2 \nabla_x u) = 0, & (x,t) \in \mathbb{R}^m_+ \times \mathbb{R}, \\ u|_{t=0} = 0, \quad u_t|_{t=0} = 0, \quad u|_{x_1=0} = g(x',t), \quad t \in \mathbb{R}, \end{cases} \quad (4.3.1)$$

where $x = (x_1, x_2 \ldots, x_m) := (x_1, x')$ and $\mathbb{R}^m_+ := \{x \in \mathbb{R}^m \mid x_1 > 0\}$.

The coefficients $c = c(x) > 0$ and $\rho = \rho(x) > 0$ in the acoustic Eq. (4.3.1) are the sound speed and the density of a medium, correspondingly. We assume here that the medium is layered, that is, the coefficients depend only on the variable $x_1 \in \mathbb{R}$: $c = c(x_1)$ and $\rho = \rho(x_1)$.

For a given admissible coefficients $c = c(x_1)$ and $\rho = \rho(x_1)$ the problem of finding a function $u(x,t)$ satisfying initial-boundary value problem (4.3.1), which is defined as a *direct problem*, is a well-posed problem.

The *inverse coefficient problem* here consists of the determination of the unknown coefficients $c(x_1)$ and $\rho(x_1)$ based on the following information on the solution to problem (4.3.1) on the hyperplane $x_1 = 0$ for all $x' \in \mathbb{R}^{m-1}$ and $t \in (0, T]$, $T > 0$:

$$F(x',t) := \rho(0) c^2(0) \left. \frac{\partial u}{\partial x_1} \right|_{x_1=0}, \quad x' \in \mathbb{R}^{m-1}, \quad t \in (0,T]. \quad (4.3.2)$$

Physically, the condition (4.3.2) means that the acoustic pressure is given at $x_1 = 0$.

Let us show that if the acoustic rigidity $\rho(x_1)c(x_1)$ is known at $x_1 = 0$, i.e. if $\rho(0)c(0)$ is given, then the inverse problem defined by (4.3.1) and (4.3.2) can be reduced to the previously considered one.

Indeed, consider the Fourier transform $\tilde{\varphi}(\lambda')$ of a function $\varphi(x')$, $x' \in \mathbb{R}^{m-1}$, with respect to x', i.e.,

$$\tilde{\varphi}(\lambda') = \int_{\mathbb{R}^{m-1}} \varphi(x') \exp\{i(x' \cdot \lambda')\} dx'.$$

Here $\lambda' := (\lambda_2, \ldots, \lambda_m)$ is the parameter of the transform and the symbol $(x' \cdot \lambda')$ means the scalar product of vectors x' and λ'. Applying the Fourier transform to the direct problem (4.3.1) as well as to the additional condition (4.3.2), we arrive at the following *transformed inverse problem*:

$$\begin{cases} \tilde{u}_{tt} - \rho^{-1}(\rho c^2 \tilde{u}_{x_1})_{x_1} + c^2 |\lambda'|^2 \tilde{u} = 0, & (x_1,t) \in \mathbb{R}_+ \times \mathbb{R}, \\ \tilde{u}|_{t=0} = 0, \quad \tilde{u}_t|_{t<0} = 0, \quad \tilde{u}|_{x_1=0} = \tilde{g}(\lambda',t), \end{cases} \quad (4.3.3)$$

$$\rho(0)c^2(0) \left. \frac{\partial \tilde{u}}{\partial x_1} \right|_{x_1=0} = \tilde{F}(\lambda',t), \quad t \in (0,T], \quad (4.3.4)$$

where $\tilde{u}$ and $\tilde{F}$ are the Fourier images of the functions u and F, respectively.

Now we use the transformation

$$z = \int_0^{x_1} \frac{d\xi}{c(\xi)} \qquad (4.3.5)$$

to introduce the new independent variable z, instead of x_1. Then $x_1 = h(z)$ is a monotone increasing function of z. Denote by

$$c(h(z)) := \hat{c}(z), \quad \rho((h(z)) := \hat{\rho}(z), \quad S(z) := \sqrt{[\rho(0)c(0)]/\hat{\rho}(z)\hat{c}(z)}$$

and introduce the new function $v(z, t, \lambda')$:

$$S(z)v(z, t, \lambda') := \tilde{u}(h(z), \lambda', t).$$

Notice that

$$\rho c^2 \tilde{u}_{x_1} = \hat{\rho}(z)\,\hat{c}(z)\tilde{u}_z(h(z), t)|_{z=z(x_1)},$$

$$\rho^{-1}(\rho c^2 \tilde{u}_{x_1})_{x_1} = (\hat{\rho}(z)\,\hat{c}(z))^{-1}(\hat{\rho}(z)\,\hat{c}(z)\tilde{u}_z)_z|_{z=z(x_1)}.$$

In terms of the function $v(z, t, \lambda')$ the transformed inverse problem (4.3.3) and (4.3.4) becomes

$$\begin{cases} v_{tt} - v_{zz} - q(z, \lambda')v = 0 & (z, t) \in \mathbb{R}_+ \times \mathbb{R}, \\ v|_{t=0} = 0, \quad v_t|_{t=0} = 0, \quad v|_{z=0} = \tilde{g}(\lambda', t), \end{cases} \qquad (4.3.6)$$

$$\rho(0)c(0)\,(v_z + Hv)_{z=0} = \tilde{F}(\lambda', t), \quad t \in (0, T]. \qquad (4.3.7)$$

where $q(z, \lambda') := q_0(z) + |\lambda'|^2 q_1(z)$, $H := S'(0)$, and

$$q_0(z) = (\ln S(z))'' - \left[(\ln S(z))'\right]^2, \quad q_1(z) = -\hat{c}^2(z).$$

By the boundary condition $v|_{z=0} = \tilde{g}(\lambda', t)$ in (4.3.6), the trace $v|_{z=0}$ is a known function. Using this in the additional condition (4.3.7) we transform it as follows:

$$v_z|_{z=0} = \hat{F}(\lambda', t), \quad t \in (0, T], \qquad (4.3.8)$$

where $\hat{F}(\lambda', t) = \tilde{F}(\lambda', t)/(\rho(0)c(0)) - H\tilde{g}(\lambda', t)$.

Let us fix the parameter λ' and assume that

$$c(x_1) \in C^2[0, \infty), \quad \rho(x_1) \in C^2[0, \infty), \quad \tilde{g}(\lambda', t) \in C^2[0, \infty)), \quad \tilde{g}(\lambda', 0) \neq 0.$$

Then $q(z, \lambda') \in C[0, \infty)$ and we obtain the inverse problem which is very close to the inverse problem of recovering the potential $q(x)$, considered in Sect. 4.2. It differs from the previous one only by the presence of unknown parameter H here. But this problem can be overcome using the property

4.3 Inverse Coefficient Problems for Layered Media

$$\hat{F}(\lambda', +0) = -\tilde{g}'(\lambda', 0). \tag{4.3.9}$$

of the function $\hat{F}(\lambda', t) \in C^1[0, \infty)$. Here $\tilde{g}'$ means the derivative of $\tilde{g}$ with respect to t. Relation (4.3.9) is the complete analogue of formula (4.2.22). From formula (4.3.9) one finds

$$H = \frac{\tilde{F}(\lambda', +0) + \rho(0)c(0)\tilde{g}'(\lambda', 0)}{\rho(0)c(0)\tilde{g}(\lambda', 0)}.$$

Hence, H becomes known. This implies that the transformed inverse problem, defined by (4.3.6) and (4.3.8), is quite similar to the inverse problem (4.2.1) and (4.2.2), under the assumption that the parameter λ' is fixed. Therefore for each fixed λ' function $q(\lambda', z)$ can be determined uniquely and by a stable way, on the interval $z \in [0, T/2]$, by the given function $\hat{F}(\lambda', t), t \in [0, T]$.

Further, then taking two values of λ' with different $|\lambda'|$, one can find the coefficients $q_0(z)$ and $q_1(z)$. The second coefficient determines the speed of the sound as function of z, i.e., $\hat{c}(z)$ and relation (4.3.5) allows to find the dependence between x_1 and z:

$$x_1 = h(z) := \int_0^z \hat{c}(z') \, dz', \quad 0 \le z \le T/2.$$

Thus, function $c(x_1)$ becomes known on the interval $[0, h(T/2)]$. Coefficient $q_0(z)$ determines the function $S(z)$. Indeed, $\ln S(z)$ satisfies to the ordinary differential equation

$$(\ln S(z))'' - \left[(\ln S(z))' \right]^2 = q_0(z), \quad 0 \le z \le T/2,$$

and the given initial Cauchy data: $S(0) = 1$, $S'(0) = H$. Therefore $S(z)$ is uniquely defined by $q_0(z)$ for $0 \le z \le T/2$. Using $S(z)$, one finds the density $\hat{\rho}(z)$, since $\rho(0)c(0)$ is known (by the assumption made above) and then $\rho(x_1)$ for $0 \le x_1 \le h(T/2)$.

Note that $\rho(0)c(0)$ can not be found in this problem because if one assumes that the medium is homogeneous, i.e. if ρ and c are constants, then ρ vanishes in relation (4.3.1).

Thus, for layered media some inverse problems with an output data given on the whole boundary of the half-space can be reduced to inverse problem for the string on the half-axis. The presence of the Fourier transform parameter allows to find several unknown coefficients or some combinations of them.

Remark 4.3.1 In the case when all components and the medium parameters depend on one space variable only, the same problem with $m = 3$ occurs in one-dimensional inverse problem related to elasticity equations.

For results related to multidimensional inverse problems for hyperbolic equations and some methods of solving them we refer [10–12, 14, 45, 46, 55, 59, 60, 85, 86].

Chapter 5
One-Dimensional Inverse Problems for Electrodynamic Equations

5.1 Formulation of Inverse Electrodynamic Problems

In this chapter we consider some inverse problems for the electrodynamics Maxwell equations

$$\nabla \times H = \varepsilon E_t + \sigma E + j, \quad \nabla \times E = -\mu H_t. \qquad (5.1.1)$$

Here $H = (H_1, H_2, H_3)$ and $E = (E_1, E_2, E_3)$ are vectors of electric and magnetic strengths, $\varepsilon > 0$, $\mu > 0$ and σ are the permittivity, permeability and conductivity coefficients, respectively, which define electro-dynamical parameters of a medium. The function $j = j(x, t)$ is the external current source which generates the electromagnetic waves. We consider the simplest physical model of a medium, assuming that the space $\mathbb{R}^3$ is divided in the two half-spaces $\mathbb{R}^3_- =: \{x \in \mathbb{R}^3 \mid x_3 < 0\}$ and $\mathbb{R}^3_+ =: \{x \in \mathbb{R}^3 \mid x_3 > 0\}$. Namely, in $\mathbb{R}^3_-$ the electro-dynamical parameters of the medium are assumed to be known and are constants $\varepsilon = \varepsilon^- > 0$, $\mu = \mu^- > 0$ and $\sigma = 0$.

The inverse coefficient problem consists in the determination of unknown parameters in $\mathbb{R}^3_+$, as functions of x, from observations of the electromagnetic field on the interface $S =: \{x \in \mathbb{R}^3 \mid x_3 = 0\}$. Generally speaking, the interface S is the discontinuity boundary for these parameters. For this reason we need to introduce special notations for limiting values of the parameters ε, μ on S taking from $\mathbb{R}^3_+$, namely, denote by $\varepsilon^+ = \varepsilon|_{x_3=0^+}$, $\mu^+ = \mu|_{x_3=0^+}$. Assume that the electromagnetic field vanishes until the moment $t = 0$, i.e.,

$$(E, H)_{t<0} \equiv 0, \quad j|_{t<0} \equiv 0, \qquad (5.1.2)$$

and it is generated by a source $j = j^0(x_1, x_2, t)\delta(x_3)$, $j^0 = (j_1^0, j_2^0, 0)$, located at the interface S and $j^0 = 0$ for $t < 0$. This assumption implies that on the interface S the following *transmission conditions* hold:

$$[H_1]_S = j_2(x_1, x_2, t), \quad [H_2]_S = -j_1(x_1, x_2, t). \qquad (5.1.3)$$

Here $[H_k]_S = H_k^+ - H_k^-$ and $H_k^+ = H_k|_{x_3=+0}$, $H_k^- = H_k|_{x_3=-0}$. These notations will also be used for other functions below. The tangential electric components of the vector E are continuous across S, i.e.,

$$[E_k]_S = 0, \quad k = 1, 2. \qquad (5.1.4)$$

One needs to add to Eq. (5.1.1) also the usual scalar equation of the electrodynamics

$$\nabla \cdot (\mu H) = 0.$$

However, this equation is a simple consequence of relations (5.1.1) and (5.1.2), which implies that the scalar equation does not play an independent role. Therefore, Eqs. (5.1.1), (5.1.2), (5.1.3) and (5.1.4) completely define the well-posed *direct problem*.

Let us formulate now the inverse problem in the time domain. For the determination of the unknown parameters ε, μ and σ in $\mathbb{R}_+^3$ the following information is usually used. The tangential components of E are given on S, i.e.,

$$E_k|_S = f_k(x_1, x_2, t), \quad t > 0, \quad k = 1, 2. \qquad (5.1.5)$$

Remark that in applications instead of $E_k|_S$ can be given $H_k^+|_S, k = 1, 2$.

The *inverse coefficient problem* here is to find the unknown parameters ε, μ and σ such that the solution to the direct problem (5.1.1), (5.1.2), (5.1.3) and (5.1.4) satisfies the additional conditions (5.1.5).

5.2 The Direct Problem: Existence and Uniqueness of a Solution

We consider the above defined inverse problem assuming that ε, μ and σ are functions of one variable x_3 only and j_k, $k = 1, 2$, do not depend on variables x_1, x_2, i.e., $j_k = j_k(t)$. Then the solution to the direct problem (5.1.1), (5.1.2), (5.1.3) and (5.1.4) does not depend on x_1 and x_2 and the problem can be decomposed into two independent sub problems for the functions (E_2, H_1, H_3) and (E_1, E_3, H_2). Consider the subsystem for (E_2, H_1, H_3). It has the form

$$\frac{\partial H_1}{\partial x_3} = \varepsilon \frac{\partial E_2}{\partial t} + \sigma E_2, \quad \frac{\partial E_2}{\partial x_3} = \mu \frac{\partial H_1}{\partial t}, \quad \frac{\partial \tilde{H}_3}{\partial t} = 0. \qquad (5.2.1)$$

For this subsystem we have the zero initial data

5.2 The Direct Problem: Existence and Uniqueness of a Solution

$$(E_2, H_1, H_3)_{t<0} \equiv 0, \tag{5.2.2}$$

and the following condition on the interface

$$[H_1]_{x_3=0} = j_2(t) := j(t), \quad [E_2]_{x_3=0} = 0, \quad t \geq 0. \tag{5.2.3}$$

Hereafter we suppose that $j(0) \neq 0$. It follows from (5.2.1) and (5.2.2) that $H_3 \equiv 0$. The information for determining the coefficients is of the form

$$E_2|_{x_3=0} = f_2(t) := f(t), \quad t > 0. \tag{5.2.4}$$

Eliminating H_1, one can find the following second-order equation for E_2:

$$\varepsilon \frac{\partial^2 E_2}{\partial t^2} + \sigma \frac{\partial E_2}{\partial t} - \frac{\partial}{\partial x_3}\left(\frac{1}{\mu}\frac{\partial E_2}{\partial x_3}\right) = 0 \tag{5.2.5}$$

with the initial zero data and the following conditions on the interface

$$[E_2]_{x_3=0} = 0, \quad \left[\frac{1}{\mu}\frac{\partial E_2}{\partial x_3}\right]_{x_3=0} = \frac{dj(t)}{dt}. \tag{5.2.6}$$

Thus, the inverse problem can be reduced to the problem for Eq. (5.2.5) with zero initial Cauchy data, the conditions (5.2.6) at the interface $x_3 = 0$ and the additional condition (5.2.4). One possible and standard way is use of these relations for analysis of direct and inverse problems. However, we prefer here another way which is based on use of the first order Eq. (5.2.1).

Let us transform Eq. (5.2.1) to a more convenient form. For this aim, introduce the new independent variable

$$z = \int_0^{x_3} \sqrt{\varepsilon(\xi)\mu(\xi)}\,d\xi, \tag{5.2.7}$$

which corresponds the travel time of a electromagnetic signal propagating with the speed $c = 1/\sqrt{\varepsilon\mu}$ along the axis x_3 from the origin to point x_3. Let $z = z(x_3)$ and $x_3 = h(z)$ be the adjoint inverse function $x_3 \equiv h(z(x_3))$. Denote by

$$E_2|_{x_3=h(z)} = \hat{E}_2(z,t), \quad H_1|_{x_3=h(z)} = \hat{H}_1(z,t),$$
$$\varepsilon(h(z)) = \hat{\varepsilon}(z), \mu(h(z)) = \hat{\mu}(z), \sigma(h(z)) = \hat{\sigma}(z). \tag{5.2.8}$$

Eq. (5.2.7) implies that

$$1 = h'(z)\sqrt{\hat{\varepsilon}(z)\hat{\mu}(z)}. \tag{5.2.9}$$

148   5 One-Dimensional Inverse Problems for Electrodynamic Equations

Hence,

$$h(z) = \int_0^z \frac{d\zeta}{\sqrt{\hat{\varepsilon}(\zeta)\hat{\mu}(\zeta)}}. \qquad (5.2.10)$$

This means that given $\hat{\varepsilon}(z)$, $\hat{\mu}(z)$ one can find the correspondence between x_3 and z of the form $x_3 = h(z)$. Eq. (5.2.1) in the new notations take the form

$$\sqrt{\hat{\varepsilon}\hat{\mu}}\,\frac{\partial \hat{H}_1}{\partial z} = \hat{\varepsilon}\,\frac{\partial \hat{E}_2}{\partial t} + \hat{\sigma}\,\hat{E}_2, \quad \sqrt{\hat{\varepsilon}\hat{\mu}}\,\frac{\partial \hat{E}_2}{\partial z} = \hat{\mu}\,\frac{\partial \hat{H}_1}{\partial t}, \qquad (5.2.11)$$

Introduce the new functions, called *Riemannian invariants*, by the formulae

$$\sqrt{\hat{\varepsilon}}\,\hat{E}_2 + \sqrt{\hat{\mu}}\,\hat{H}_1 = u_1, \quad \sqrt{\hat{\varepsilon}}\,\hat{E}_2 - \sqrt{\hat{\mu}}\,\hat{H}_1 = u_2. \qquad (5.2.12)$$

Then

$$\hat{E}_2 = \frac{u_1 + u_2}{2\sqrt{\hat{\varepsilon}}}, \quad \hat{H}_1 = \frac{u_1 - u_2}{2\sqrt{\hat{\mu}}}. \qquad (5.2.13)$$

Let us rewrite the Eq. (5.2.11) for the functions u_1, u_2. Assume that $\varepsilon(x_3)$, $\mu(x_3)$ are $C^1[0, \infty)$ functions and $\sigma(x_3) \in C[0, \infty)$. Dividing the first Eq. (5.2.11) by $\sqrt{\hat{\varepsilon}}$ and the second one by $\sqrt{\hat{\mu}}$, we obtain

$$\frac{\partial(\sqrt{\hat{\mu}}\hat{H}_1)}{\partial z} - \frac{\hat{\mu}'}{2\hat{\mu}}(\sqrt{\hat{\mu}}\hat{H}_1) = \frac{\partial(\sqrt{\hat{\varepsilon}}\hat{E}_2)}{\partial t} + \frac{\hat{\sigma}}{\hat{\varepsilon}}(\sqrt{\hat{\varepsilon}}\,\hat{E}_2),$$

$$\frac{\partial(\sqrt{\hat{\varepsilon}}\hat{E}_2)}{\partial z} - \frac{\hat{\varepsilon}'}{2\hat{\varepsilon}}(\sqrt{\hat{\varepsilon}}\hat{E}_2) = \frac{\partial(\sqrt{\hat{\mu}}\hat{H}_1)}{\partial t},$$

Now adding these equations and then subtracting the second equation from the first one and using (5.2.13), we transform the equations into the following ones:

$$\left(\frac{\partial}{\partial t} - \frac{\partial}{\partial z}\right)u_1 + \frac{1}{4}\left(\frac{\hat{\varepsilon}'}{\hat{\varepsilon}} + \frac{\hat{\mu}'}{\hat{\mu}} + \frac{2\hat{\sigma}}{\hat{\varepsilon}}\right)u_1 + \frac{1}{4}\left(\frac{\hat{\varepsilon}'}{\hat{\varepsilon}} - \frac{\hat{\mu}'}{\hat{\mu}} + \frac{2\hat{\sigma}}{\hat{\varepsilon}}\right)u_2 = 0,$$

$$\left(\frac{\partial}{\partial t} + \frac{\partial}{\partial z}\right)u_2 + \frac{1}{4}\left(\frac{\hat{\mu}'}{\hat{\mu}} - \frac{\hat{\varepsilon}'}{\hat{\varepsilon}} + \frac{2\hat{\sigma}}{\hat{\varepsilon}}\right)u_1 + \frac{1}{4}\left(\frac{2\hat{\sigma}}{\hat{\varepsilon}} - \frac{\hat{\varepsilon}'}{\hat{\varepsilon}} - \frac{\hat{\mu}'}{\hat{\mu}}\right)u_2 = 0.$$

We represent these equations together with initial data as follows

$$\left(I_2\frac{\partial}{\partial t} + K\frac{\partial}{\partial z} + A\right)U = 0, \quad U|_{t<0} \equiv 0, \qquad (5.2.14)$$

where the matrices I_2, K, A and the vector-column U are given by the formulae

5.2 The Direct Problem: Existence and Uniqueness of a Solution

$$U = \begin{pmatrix} u_1 \\ u_2 \end{pmatrix}, \quad I_2 = \begin{pmatrix} 1 & 0 \\ 0 & 1 \end{pmatrix}, \quad K = \begin{pmatrix} -1 & 0 \\ 0 & 1 \end{pmatrix},$$
$$A = \begin{pmatrix} q_1 + q_3 & q_2 + q_3 \\ q_3 - q_2 & q_3 - q_1 \end{pmatrix}, \tag{5.2.15}$$

and the coefficients q_1, q_2, q_3 are determined via $\hat{\varepsilon}, \hat{\mu}, \hat{\sigma}$ as follows:

$$q_1(z) = \frac{1}{4}\left(\frac{\hat{\varepsilon}'(z)}{\hat{\varepsilon}(z)} + \frac{\hat{\mu}'(z)}{\hat{\mu}(z)}\right), \quad q_2(z) = \frac{1}{4}\left(\frac{\hat{\varepsilon}'(z)}{\hat{\varepsilon}(z)} - \frac{\hat{\mu}'(z)}{\hat{\mu}(z)}\right),$$
$$q_3(z) = \frac{\hat{\sigma}(z)}{2\hat{\varepsilon}(z)}. \tag{5.2.16}$$

Then the interface (transmission) conditions take the form

$$\left[\frac{u_1 - u_2}{\sqrt{\hat{\mu}}}\right]_{z=0} = 2j(t), \quad \left[\frac{u_1 + u_2}{\sqrt{\hat{\varepsilon}}}\right]_{z=0} = 0, \quad t > 0,$$

or

$$\frac{u_1^+ - u_2^+}{\sqrt{\hat{\mu}^+}} - \frac{u_1^- - u_2^-}{\sqrt{\hat{\mu}^-}} = 2j(t), \quad \frac{u_1^+ + u_2^+}{\sqrt{\hat{\varepsilon}^+}} - \frac{u_1^- + u_2^-}{\sqrt{\hat{\varepsilon}^-}} = 0, \quad t > 0.$$

By the reasons that will be clearly below, we represent these conditions in the form

$$u_1^- = r_{11}u_1^+ + r_{12}u_2^- - r_{11}\sqrt{\hat{\mu}^+}\, j(t),$$
$$u_2^+ = -r_{12}u_1^+ + r_{22}u_2^- - r_{22}\sqrt{\hat{\mu}^-}\, j(t), \quad t > 0, \tag{5.2.17}$$

where r_{ij} given by the formulae

$$r_{11} = \frac{2\sqrt{\varepsilon^-\mu^-}}{\sqrt{\varepsilon^-\mu^+} + \sqrt{\varepsilon^+\mu^-}}, \quad r_{12} = \frac{\sqrt{\varepsilon^-\mu^+} - \sqrt{\varepsilon^+\mu^-}}{\sqrt{\varepsilon^-\mu^+} + \sqrt{\varepsilon^+\mu^-}},$$
$$r_{22} = \frac{2\sqrt{\varepsilon^+\mu^+}}{\sqrt{\varepsilon^-\mu^+} + \sqrt{\varepsilon^+\mu^-}}. \tag{5.2.18}$$

The information for the inverse problem is now given by the formula

$$u_1^+ + u_2^+ = 2\sqrt{\varepsilon^+}\, f(t), \quad t > 0. \tag{5.2.19}$$

Consider the direct problem defined by (5.2.14), (5.2.15), (5.2.16) and (5.2.17). Since the initial data are zero, one has

$$u_1(z,t) = u_2(z,t) \equiv 0, \quad 0 < t < |z|. \tag{5.2.20}$$

For arbitrary $T > 0$ introduce the notations $\Omega_T^- = \{(z,t) | 0 < -z \le t \le T + z\}$ and $\Omega_T^+ = \{(z,t) | 0 < z \le t \le T - z\}$. In the domain Ω_T^- we have Eq. (5.2.14) with the matrix $A = 0$ and the boundary conditions (5.2.17). The second Eq. (5.2.14) has the form

$$\frac{\partial u_2}{\partial t} + \frac{\partial u_2}{\partial z} = 0, \quad u_2|_{t=0} = 0, \quad z < 0.$$

This implies that $u_2(z,t) \equiv 0$ for all $(z,t) \in \Omega_T^-$. Hence $u_2^- = 0$. Then the second boundary condition (5.2.17) takes the form

$$u_2^+ = -r_{12} u_1^+ - r_{22}\sqrt{\mu^-}\, j(t), \quad t > 0. \tag{5.2.21}$$

It means that we can consider the direct problem in the domain Ω_T^+ with the boundary condition (5.2.21) and find $u_1(z,t)$ and $u_2(z,t)$ and then calculate $u_1^-(0,t)$ and $u_1(z,t)$ in Ω_T^- using the Eq. (5.2.14) and the first boundary condition (5.2.17). The latter is given by the formula

$$u_1(z,t) = r_{11} u_1^+(0, t+z) - r_{11}\sqrt{\mu^+}\, j(t+z), \quad (z,t) \in \Omega_T^-. \tag{5.2.22}$$

Write down integral equations for u_1 and u_2 in the domain Ω_T^+. Integrating the first component of the Eq. (5.2.14) on the plane (ζ, τ) along the characteristic line $\zeta + \tau = z + t$ from the point $((z+t)/2, (z+t)/2)$ till the point (z,t) and using that $u_1((z+t)/2, (z+t)/2) = 0$, we obtain

$$u_1(z,t) + \int_{(z+t)/2}^{t} (A(z+t-\tau) U(z+t-\tau, \tau))_1 \, d\tau = 0,$$
$$(z,t) \in \Omega_T^+. \tag{5.2.23}$$

Now we can rewrite the condition (5.2.21) in the form

$$u_2^+(0,t) = r_{12} \int_{t/2}^{t} (A(t-\tau) U(t-\tau, \tau))_1 \, d\tau - r_{22}\sqrt{\mu^-}\, j(t). \tag{5.2.24}$$

Integrating the second component of (5.2.14) along the characteristic line $\zeta - \tau = z - t$ from the point $(0, t-z)$ till the point (z,t) and using the boundary condition (5.2.24), we get

$$\begin{array}{l} u_2(z,t) + \int_{t-z}^{t} (A(z-t+\tau) U(z-t+\tau, \tau))_2 \, d\tau \\ -r_{12} \int_{(t-z)/2}^{t-z} (A(t-z-\tau) U(t-z-\tau, \tau))_1 \, d\tau \\ +r_{22}\sqrt{\mu^-}\, j(t-z) = 0, \quad (z,t) \in \Omega_T^+. \end{array} \tag{5.2.25}$$

5.2 The Direct Problem: Existence and Uniqueness of a Solution

Relations (5.2.23) and (5.2.25) form the closed system of integral equations in domain Ω_T^+. The properties of the solution to this system are given by the following existence and uniqueness theorem.

Theorem 5.2.1 *Let* $\hat{\varepsilon}(z) \in C^1[0, T/2]$, $\hat{\mu} \in C^1[0, T/2]$, $\hat{\sigma}(z) \in C[0, T/2]$ *and* $j(t) \in C^1[0, T]$, $T > 0$. *Assume, in addition, that the following bounds hold:*

$$\max_{z \in (0, T/2]} \left(\frac{1}{\hat{\varepsilon}(z)}, \frac{1}{\hat{\mu}(z)} \right) \leq q_0,$$
$$r_{22} \sqrt{\mu^-} \, \|j\|_{C(0,T]} \leq j_0,$$
$$\max(\|\hat{\varepsilon}'\|_{C(0,T/2]}, \|\hat{\mu}'\|_{C(0,T/2]}, \|\hat{\sigma}\|_{C(0,T/2]}) \leq q_{01},$$

with some constants q_0, q_{01} *and* j_0. *Then there exist a unique continuously differentiable solution to Eqs. (5.2.23) and (5.2.25) in the closed domain* $\overline{\Omega_T^+}$ *and this solution satisfies the conditions*

$$u_1(z, z) = 0,$$
$$u_2(z, z) = -r_{22}\sqrt{\mu^-} \, j(0) \exp\left(\int_0^z (q_1(\zeta) - q_3(\zeta)) \, d\zeta \right), \quad z \in [0, T/2], \quad (5.2.26)$$
$$\max(\|u_1\|_{C(\Omega_T^+)}, \|u_2\|_{C(\Omega_T^+)}) \leq C j_0,$$

where $C = C(q_0, q_{01}, T)$.

Proof To prove the existence of a continuous solution in $\overline{\Omega_T^+}$ one can use the method of successive approximation. For this goal, functions u_1 and u_2 are represented in the form of series

$$u_k(z, t) = \sum_{n=0}^{\infty} u_k^n(z, t), \quad k = 1, 2, \quad (5.2.27)$$

where u_1^n and u_2^n are defined by the formulae

$$u_1^0(z, t) = 0, \quad u_2^0(z, t) = -r_{22}\sqrt{\mu^-} \, j(t - z), \quad U^n = (u_1^n, u_2^n)^T,$$
$$u_1^n(z, t) = -\int_{(z+t)/2}^t (A(z + t - \tau) U^{n-1}(z + t - \tau, \tau))_1 \, d\tau, \quad n \geq 1,$$
$$u_2^n(z, t) = -\int_{t-z}^t (A(z + t - \tau) U^{n-1}(z + t - \tau, \tau))_2 \, d\tau$$
$$+ r_{12} \int_{(t-z)/2}^{t-z} (A(t - z - \tau) U^{n-1}(t - z - \tau, \tau))_1 \, d\tau, \quad n \geq 1.$$

Functions $u_1^n(z, t)$, $u_2^n(z, t)$ are continuous in $\overline{\Omega_T^+}$. Introduce

$$v^n(t) = \begin{cases} \max\limits_{k=1,2} \max\limits_{0 \leq z \leq t} |u_k^n(z, t)|, & t \in [0, T/2], \\ \max\limits_{k=1,2} \max\limits_{0 \leq z \leq T-t} |u_k^n(z, t)|, & t \in [T/2, T], \end{cases}$$

for $n = 0, 1, 2, \ldots$. Note that $|r_{12}| < 1$. Then one has

$$v^0(t) \le j_0,$$
$$|u_1^n(z,t)| \le 2q_0 q_{01} \int_{(z+t)/2}^t v^{n-1}(\tau)\,d\tau \le 2q_0 q_1 \int_0^t v^{n-1}(\tau)\,d\tau,$$
$$|u_2^n(z,t)| \le 2q_0 q_{01} \left(\int_{t-z}^t v^{n-1}(\tau)\,d\tau + \int_{(t-z)/2}^{t-z} v^{n-1}(\tau)\,d\tau \right)$$
$$\le 2q_0 q_{01} \int_0^t v^{n-1}(\tau)\,d\tau, \quad n \ge 1.$$

Hence,

$$v^n(t) \le 2q_0 q_{01} \int_0^t v^{n-1}(\tau)\,d\tau, \quad t \in [0,T], \quad n \ge 1. \tag{5.2.28}$$

From (5.2.28) we deduce the estimates

$$v^n(t) \le j_0 (2q_0 q_{01})^n \frac{t^n}{n!} \le j_0 (2q_0 q_{01})^n \frac{T^n}{n!}, \quad t \in [0,T], \quad n = 0,1,2,\ldots. \tag{5.2.29}$$

Therefore the series (5.2.27) uniformly convergence in $\overline{\Omega_T^+}$ and their sums are continuous functions in $\overline{\Omega_T^+}$. This means that there exists the solution to the integral Eqs. (5.2.23) and (5.2.25) and the solution satisfies the inequalities

$$|u_k(z,t)| \le j_0 \sum_{n=0}^{\infty} (2q_0 q_{01})^n \frac{T^n}{n!} \le j_0 \exp(2q_0 q_{01} T), \quad k = 1,2, \tag{5.2.30}$$

for all $(z,t) \in \overline{\Omega_T^+}$. Moreover, substituting in (5.2.23) and (5.2.25) $t = z$, we obtain $u_1(z,z) = 0$ and the integral relation for the function $\psi(z) = u_2(z,z)$ in the form

$$\psi(z) + \int_0^z (q_3(\zeta) - q_1(\zeta))\psi(\zeta)\,d\zeta = -r_{22}\sqrt{\mu^-}\,j(0).$$

The latter equation is equivalent to the Cauchy problem for the ordinary differential equation

$$\psi'(z) + (q_3(z) - q_1(z))\psi(z) = 0, \quad \psi(0) = -r_{22}\sqrt{\mu^-}\,j(0).$$

The solution of this problem is given by the formula

$$\psi(z) = \psi(0) \exp\left(\int_0^z (q_1(\zeta) - q_3(\zeta))\,d\zeta \right),$$

which coincides with the second formula (5.2.26).

To prove the uniqueness of the found solution, we suppose that there exist two solutions of the integral Eqs. (5.2.23) and (5.2.25), namely, u_1, u_2 and $\bar{u}_1, \bar{u}_2$. Then their difference $\tilde{u}_1 = u_1 - \bar{u}_1$, $\tilde{u}_2 = u_2 - \bar{u}_2$ satisfies the following relations for all $(z,t) \in \Omega_T^+$:

5.2 The Direct Problem: Existence and Uniqueness of a Solution

$$\tilde{u}_1(z,t) + \int_{(z+t)/2}^{t} (A(z+t-\tau)\tilde{U}(z+t-\tau,\tau))_1 \, d\tau = 0,$$
$$\tilde{u}_2(z,t) + \int_{t-z}^{t} (A(z+t-\tau)\tilde{U}(z+t-\tau,\tau))_2 \, d\tau \qquad (5.2.31)$$
$$- r_{12} \int_{(t-z)/2}^{t-z} (A(t-z-\tau)\tilde{U}(t-z-\tau,\tau))_1 \, d\tau = 0,$$

where $\tilde{U} := (\tilde{u}_1, \tilde{u}_2)^T$. Let now

$$v(t) = \begin{cases} \max_{k=1,2} \max_{0 \le z \le t} |\tilde{u}_k(z,t)|, & t \in [0, T/2], \\ \max_{k=1,2} \max_{0 \le z \le T-t} |\tilde{u}_k(z,t)|, & t \in [T/2, T]. \end{cases}$$

Then we get

$$|\tilde{u}_1(z,t)| \le 2q_0 q_{01} \int_0^t v(\tau) \, d\tau,$$
$$|\tilde{u}_2(z,t)| \le 2q_0 q_{01} \int_0^t v(\tau) \, d\tau, \quad (z,t) \in \Omega_T^+.$$

Hence,

$$v(t) \le 2q_0 q_{01} \int_0^t v(\tau) \, d\tau, \quad t \in [0, T]. \qquad (5.2.32)$$

Inequality (5.2.32) has only trivial solution $v(t) \equiv 0$. Therefore $u_1(z,t) = \bar{u}_1(z,t)$ and $u_2(z,t) = \bar{u}_2(z,t)$, i.e., the solution to (5.2.23), (5.2.24) and (5.2.25) is unique.

Now we need prove that the solution is continuously differentiable in Ω_T^+. If the coefficients satisfy the conditions $\hat{\varepsilon}(z) \in C^2[0, T/2]$, $\hat{\mu} \in C^2[0, T/2]$, $\hat{\sigma}(z) \in C^1[0, T/2]$, then one can directly prove this assertion taking derivatives of integral relations (5.2.23) and (5.2.25) with respect to z and t. In order to avoid the additional differentiation of the coefficients, we need first to rewrite these relations changing the variable τ under the integrands on ζ. Making this, we get

$$u_1(z,t) - \int_{(z+t)/2}^{z} (A(\zeta) U(\zeta, z+t-\zeta))_1 \, d\zeta = 0,$$
$$u_2(z,t) + \int_0^z (A(\zeta) U(\zeta, t-z+\zeta))_2 \, d\zeta \qquad (5.2.33)$$
$$- r_{12} \int_0^{(t-z)/2} (A(\zeta) U(\zeta, t-z-\zeta))_1 \, d\zeta$$
$$+ r_{22} \sqrt{\mu^-} \, j(t-z) = 0, \quad (z,t) \in \Omega_T^+.$$

Then we differentiate Eq. (5.2.33) to obtain integral relations for partial derivatives $\dot{u}_1, \dot{u}_2$ with respect to t of u_1 and u_2. These equations are as follows:

$$\dot{u}_1(z,t) - \int_{(z+t)/2}^{z} (A(\zeta) \dot{U}(\zeta, z+t-\zeta))_1 \, d\zeta$$
$$+ \frac{1}{2} (A((z+t)/2) U((z+t)/2, (z+t)/2))_1 = 0,$$
$$\dot{u}_2(z,t) + \int_0^z (A(\zeta) \dot{U}(\zeta, t-z+\zeta))_2 \, d\zeta \qquad (5.2.34)$$
$$- r_{12} \int_0^{(t-z)/2} (A(\zeta) \dot{U}(\zeta, t-z-\zeta))_1 \, d\zeta$$
$$- \frac{r_{12}}{2} (A((t-z)/2) U((t-z)/2, (t-z)/2))_1$$
$$+ r_{22} \sqrt{\mu^-} \, j'(t-z) = 0, \quad (z,t) \in \Omega_T^+,$$

where $\dot{U}$ is the vector-column with components $\dot{u}_1, \dot{u}_2$. Now one needs again to change variable in the integrands returning back to τ. In the result the equations take the form

$$\dot{u}_1(z,t) + \int_{(z+t)/2}^{t}(A(z+t-\tau)\dot{U}(z+t-\tau,\tau))_1 \, d\tau$$
$$+ \frac{1}{2}(A((z+t)/2)U((z+t)/2,(z+t)/2))_1 = 0,$$
$$\dot{u}_2(z,t) + \int_{t-z}^{t}(A(z-t+\tau)\dot{U}(z-t+\tau,\tau))_2 \, d\tau \qquad (5.2.35)$$
$$- r_{12} \int_{(t-z)/2}^{t-z}(A((t-z-\tau)\dot{U}((t-z-\tau,\tau))_1 \, d\tau$$
$$- \frac{r_{12}}{2}(A((t-z)/2)U((t-z)/2,(t-z)/2))_1$$
$$+ r_{22}\sqrt{\mu^-} \, j'(t-z) = 0, \quad (z,t) \in \Omega_T^+.$$

In these equations the function $U(z,z)$ is known. The Eq. (5.2.35) quite similar to the Eqs. (5.2.23) and (5.2.25). Therefore to prove the existence and uniqueness of a continuous solution in $\overline{\Omega_T^+}$ of these equations the previous methods can be used. We leave the proof of these assertins as an exercise for the reader.

Since the derivatives $\dot{u}_1(z,t), \dot{u}_2(z,t)$ exist and are continuous in $\overline{\Omega_T^+}$, then directly for the differential Eq. (5.2.14) follows an existence and continuous dependence of the derivatives of $u_1(z,t), u_2(z,t)$ with respect to z in the same domain. $\square$

The following theorem is a consequence of Theorem 5.2.1.

Theorem 5.2.2 *Let conditions of Theorem 3.2.1 hold. Assume, in addition, that additional condition $j(0) \neq 0$ holds. Then the function $f(t)$ in (5.2.19) belongs to the functional space $C^1[0,T]$ and satisfy the condition*

$$-\frac{\sqrt{\mu^-}}{\sqrt{\varepsilon^-}} < \frac{f(0^+)}{j(0)} < 0. \qquad (5.2.36)$$

Proof One needs prove only inequality (5.2.36). It follows directly from Eq. (2.2.25) that

$$f(0^+) = -\frac{r_{22}\sqrt{\mu^-}}{2\sqrt{\varepsilon^+}} j(0). \qquad (5.2.37)$$

Then one finds

$$\frac{f(0^+)}{j(0)} = -\frac{r_{22}\sqrt{\mu^-}}{2\sqrt{\varepsilon^+}} < 0.$$

Using the formula (5.2.18) for r_{22}, one obtains the estimate

$$\frac{r_{22}}{2\sqrt{\varepsilon^+}} = \frac{\sqrt{\mu^+}}{\sqrt{\varepsilon^-\mu^+} + \sqrt{\varepsilon^+\mu^-}} < \frac{1}{\sqrt{\varepsilon^-}}.$$

This implies the inequality (5.2.36).

5.3 One-Dimensional Inverse Problems

Consider the inverse problem. We assume that $h(t) \in C^1[0, T]$, $j(0) \neq 0$, and $f(t) \in C^1[0, T]$ is a given function which satisfies to the condition (5.2.36). Since only the $f(t)$ function is a given data, i.e. information, one can find at most one unknown parameters ε, μ or σ. In this context we consider two typical inverse problems.

5.3.1 Problem of Finding a Permittivity Coefficient

In this case we should accept that μ and σ are given. We assume for simplicity that $\mu(x_3) = \mu^+ > 0$ and $\sigma(x_3) = 0$ for $x_3 > 0$. In this case

$$q_1(z) = q_2(z) = \frac{\hat{\varepsilon}'(z)}{4\hat{\varepsilon}(z)} := q(z), \quad q_3(z) = 0. \tag{5.3.1}$$

Using relations (5.2.36) and (5.2.18) one can find ε^+. Elementary calculations lead to the formula

$$\varepsilon^+ = \mu^+ \left(\frac{j(0)}{f(0^+)} + \frac{\sqrt{\varepsilon^-}}{\sqrt{\mu^-}} \right)^2. \tag{5.3.2}$$

Note that the inequality (5.2.36) guaranties that ε^+ is positive. Because ε^+ has found, the value r_{12} and r_{22} become known.

In the inverse problem both functions $h(t)$ and $f(t)$ are given. Therefore one can calculate u_1^+ and u_2^+ using the conditions (5.2.17) and (5.2.19) (recall that $u_2^- = 0$ in (5.2.17)). Indeed, then the equations hold

$$u_2^+ = -r_{12} u_1^+ - r_{22} \sqrt{\mu^-} \, j(t),$$
$$u_1^+ + u_2^+ = 2\sqrt{\varepsilon^+} f(t), \quad t > 0.$$

Solving these equations we deduce:

$$u_1^+ = \frac{1}{1 - r_{12}} \left(r_{22} \sqrt{\mu^-} \, j(t) + 2\sqrt{\varepsilon^+} f(t) \right) := F_1(t),$$
$$u_2^+ = -\frac{1}{1 - r_{12}} \left(r_{22} \sqrt{\mu^-} \, j(t) + 2 r_{12} \sqrt{\varepsilon^+} f(t) \right) := F_2(t). \tag{5.3.3}$$

Note that, according (5.2.18),

$$1 - r_{12} = \frac{2\sqrt{\varepsilon^+ \mu^-}}{\sqrt{\varepsilon^- \mu^+} + \sqrt{\varepsilon^+ \mu^-}} > 0.$$

Because u_1^+ and u_2^+ are known, one can consider Eq. (5.2.14) in the domain Ω_T^+ only. These equations have the form

$$\begin{cases} \dfrac{\partial u_1}{\partial t} - \dfrac{\partial u_1}{\partial z} + q(z)(u_1 + u_2) = 0, \\ \dfrac{\partial u_2}{\partial t} + \dfrac{\partial u_2}{\partial z} - q(z)(u_1 + u_2) = 0, \quad (z,t) \in \Omega_T^+. \end{cases} \quad (5.3.4)$$

Moreover, the following conditions hold

$$u_1(z, z) = 0, \quad u_2(z, z) = -r_{22}\sqrt{\mu^-}\, j(0) \exp\left(\int_0^z q(\zeta)\, d\zeta\right) \quad (5.3.5)$$

for $z \in [0, T/2]$. The relations (5.3.3), (5.3.4) and (5.3.5) form the complete system of equations for the inverse problem. Deduce integral equations for this problem. At first, integrating Eq. (5.3.4) along the characteristic lines $\zeta + \tau = z + t$ and $\zeta - \tau = z - t$ from arbitrary point $(z, t) \in \Omega_T^+$ till to intersection with axis $\zeta = 0$ and using at the intersection points conditions (5.3.3), one gets

$$u_1(z,t) - \int_0^z q(\zeta)(u_1(\zeta, z+t-\zeta) + u_2(\zeta, z+t-\zeta))\, d\zeta = F_1(t+z),$$

$$u_2(z,t) - \int_0^z q(\zeta)(u_1(\zeta, t-z+\zeta) + u_2(\zeta, t-z+\zeta))\, d\zeta = F_2(t-z),$$

$$(z,t) \in \Omega_T^+. \quad (5.3.6)$$

At second, substituting in the first of these equations $t = z$ and using condition (5.3.5), one finds the relation

$$-\int_0^z q(\zeta)(u_1(\zeta, 2z-\zeta) + u_2(\zeta, 2z-\zeta))\, d\zeta = F_1(2z),$$

that is, an additional equation for finding $q(z)$. Differentiating it with respect to z, one gets

$$-q(z)(u_1(z,z) + u_2(z,z)) - 2\int_0^z q(\zeta)(\dot{u}_1(\zeta, 2z-\zeta) + \dot{u}_2(\zeta, 2z-\zeta))\, d\zeta = 2F_1'(2z), \quad (5.3.7)$$

Use here the relations (5.3.5). Then the first term in this equation one can rewrite as follows

$$-q(z)(u_1(z,z) + u_2(z,z)) = r_{22}\sqrt{\mu^-}\, j(0) q(z) \exp\left(\int_0^z q(\zeta)\, d\zeta\right).$$

Denote

5.3 One-Dimensional Inverse Problems

$$p(z) = q(z)\exp\left(\int_0^z q(\zeta)\,d\zeta\right) = \frac{d}{dz}\exp\left(\int_0^z q(\zeta)\,d\zeta\right).$$

Then

$$\exp\left(\int_0^z q(\zeta)\,d\zeta\right) = 1 + \int_0^z p(\zeta)\,d\zeta > 0, \qquad (5.3.8)$$

and

$$q(z) = \frac{p(z)}{1 + \int_0^z p(\zeta)\,d\zeta}. \qquad (5.3.9)$$

Dividing both sides of (5.3.7) by $r_{22}\sqrt{\mu^-}\,j(0)$, one gets

$$p(z) - \lambda_0 \int_0^z q(\zeta)(\dot{u}_1(\zeta, 2z-\zeta) + \dot{u}_2(\zeta, 2z-\zeta))\,d\zeta = p_0(z),$$

$$z \in [0, T/2], \quad (5.3.10)$$

where

$$\lambda_0 = \frac{2}{r_{22}\sqrt{\mu^-}\,j(0)}, \quad p_0(z) = \lambda_0 F_1'(2z)$$

and $q(z)$ is defined by the formula (5.3.9).

Differentiating (5.3.6) with respect to t, one obtain equations for functions $\dot{u}_1$ and $\dot{u}_2$. They have the form:

$$\dot{u}_1(z,t) - \int_0^z q(\zeta)(\dot{u}_1(\zeta, z+t-\zeta) + \dot{u}_2(\zeta, z+t-\zeta))\,d\zeta = F_1'(t+z),$$

$$\dot{u}_2(z,t) - \int_0^z q(\zeta)(\dot{u}_1(\zeta, t-z+\zeta) + \dot{u}_2(\zeta, t-z+\zeta))\,d\zeta = F_2'(t-z),$$

$$(z,t) \in \Omega_T^+ \quad (5.3.11)$$

Now we can prove the existence and uniqueness of the local solution of the inverse problem.

Theorem 5.3.1 *Let $\mu(x_3) = \mu^+ > 0$, $\sigma(x_3) = 0$, for $x_3 > 0$, and $j(t) \in C^1[0, T]$, $j(0) \neq 0$. Assume that $f(t) \in C^1[0, T]$, $T > 0$, and the condition (5.2.36) holds. Then for sufficiently small $T > 0$ there exist a unique continuously differentiable positive solution of the inverse problem.*

Proof First, we represent the Eqs. (5.3.10) and (5.3.11) in the operator form

$$\varphi = A\varphi, \qquad (5.3.12)$$

where $\varphi = (\varphi_1, \varphi_2, \varphi_3)$ is the vector-function with components defined by the formulae

$$\varphi_1 = \dot{u}_1(z, t), \quad \varphi_2 = \dot{u}_2(z, t), \quad \varphi_3 = p(z)$$

and the operator $A = (A_1, A_2, A_3)$ is given by the relations

$$\varphi_1(z, t) = \int_0^z q(\zeta)(\varphi_1(\zeta, z + t - \zeta) + \varphi_2(\zeta, z + t - \zeta)) \, d\zeta + \varphi_1^0(z, t),$$

$$\varphi_2(z, t) = \int_0^z q(\zeta)(\varphi_1(\zeta, z + t - \zeta) + \varphi_2(\zeta, z + t - \zeta)) \, d\zeta + \varphi_2^0(z, t),$$

$$\varphi_3(z) = \lambda_0 \int_0^z q(\zeta)(\varphi_1(\zeta, 2z - \zeta) + \varphi_2(\zeta, 2z - \zeta)) \, d\zeta + \varphi_3^0(z), \quad (5.3.13)$$

where $(z, t) \in \Omega_T^+$,

$$\varphi_1^0(z, t) = F_1'(t + z), \quad \varphi_2^0(z, t) = F_2'(t - z), \quad \varphi_3^0(z) = p_0(z)$$

and $q(z)$ is defined by the formula

$$q(z) = \frac{\varphi_3(z)}{1 + \int_0^z \varphi_3(\zeta) \, d\zeta}. \quad (5.3.14)$$

Then we define the ball $B(\varphi^0)$

$$\|\varphi - \varphi^0\| \leq \|\varphi^0\|,$$

centered at the element $\varphi^0 = (\varphi_1^0, \varphi_2^0, \varphi_3^0)$, in the functional space of continuous functions φ in $\overline{\Omega_T^+}$ with the norm

$$\|\varphi\| = \max_{k=1,2,3} \max_{(z,t) \in \overline{\Omega_T^+}} |\varphi_k(z, t)|.$$

Now we are going to prove that that the operator A is a contracting operator on this ball, if T is enough small.

Let φ be arbitrary element of $B(\varphi^0)$. Then $\|\varphi\| \leq 2\|\varphi^0\|$. Making estimates one finds

$$|\varphi_1(z, t) - \varphi_1^0(z, t)| \leq \int_0^z |q(\zeta)|(|\varphi_1(\zeta, z + t - \zeta)| + |\varphi_2(\zeta, z + t - \zeta)|) \, d\zeta$$

$$\leq \frac{4T \|\varphi_0\|}{1 - T\|\varphi_0\|} \|\varphi_0\|, \quad (z,t) \in \Omega_T^+.$$

We used here that

5.3 One-Dimensional Inverse Problems

$$|q(\zeta)| \leq \frac{2\|\varphi_0\|}{1 - T\|\varphi_0\|}.$$

Similarly,

$$|\varphi_2(z,t) - \varphi_2^0(z,t)| \leq \int_0^z |q(\zeta)|(|\varphi_1(\zeta, z+t-\zeta)| + |\varphi_2(\zeta, z+t-\zeta)|)\, d\zeta$$

$$\leq \frac{4T\|\varphi_0\|}{1 - T\|\varphi_0\|} \|\varphi_0\|, \quad (z,t) \in \Omega_T^+,$$

$$|\varphi_3(z) - \varphi_3^0(z)| \leq = |\lambda_0| \int_0^z |q(\zeta)|(|\varphi_1(\zeta, 2z-\zeta)| + |\varphi_2(\zeta, 2z-\zeta)|)\, d\zeta$$

$$\leq \frac{4T|\lambda_0|\|\varphi_0\|}{1 - T\|\varphi_0\|} \|\varphi_0\|, \quad z \in T/2.$$

From these estimates follows that $\|A\varphi - \varphi^0\| \leq \|\varphi^0\|$ if T satisfy the relations

$$4T \max(1, |\lambda_0|) \frac{\|\varphi^0\|}{1 - T\|\varphi^0\|} < 1, \quad T\|\varphi^0\| < 1. \tag{5.3.15}$$

This means that $A\varphi \in B(\varphi^0)$, i.e., the operator A maps the ball $B(\varphi^0)$ into itself.
Take now two arbitrary elements φ^j, $j = 1, 2$, belonging to $B(\varphi^0)$ and denote

$$\tilde{\varphi}_k = \varphi_k^1 - \varphi_k^2, \quad k = 1, 2, 3,$$

$$q^j(z) = \frac{\varphi_3^j(z)}{1 + \int_0^z \varphi_3^j(\zeta)\, d\zeta}, \quad j = 1, 2,$$

$$\tilde{q}(z) = \frac{\tilde{\varphi}_3(z)\left(1 + \int_0^z \varphi_3^2(\zeta)\, d\zeta\right) - \varphi_3^2(z) \int_0^z \tilde{\varphi}_3(\zeta)\, d\zeta}{\left(1 + \int_0^z \varphi_3^1(\zeta)\, d\zeta\right)\left(1 + \int_0^z \varphi_3^2(\zeta)\, d\zeta\right)}.$$

From (5.3.13) we find

$$\tilde{\varphi}_1(z,t) = \int_0^z [q^1(\zeta)(\tilde{\varphi}_1(\zeta, z+t-\zeta) + \tilde{\varphi}_2(\zeta, z+t-\zeta))$$
$$+ \tilde{q}(\zeta)(\varphi_1^2(\zeta, z+t-\zeta) + \varphi_2^2(\zeta, z+t-\zeta))]\, d\zeta,$$

$$\tilde{\varphi}_2(z,t) = \int_0^z [q^1(\zeta)(\tilde{\varphi}_1(\zeta, z+t-\zeta) + \tilde{\varphi}_2(\zeta, z+t-\zeta))$$
$$+ \tilde{q}(\zeta)(\varphi_1^2(\zeta, z+t-\zeta) + \varphi_2^2(\zeta, z+t-\zeta))]\, d\zeta,$$

$$\tilde{\varphi}_3(z) = \lambda_0 \int_0^z [q^1(\zeta)(\tilde{\varphi}_1(\zeta, 2z-\zeta) + \tilde{\varphi}_2(\zeta, 2z-\zeta))$$
$$+ \tilde{q}(\zeta)(\varphi_1^2(\zeta, 2z-\zeta) + \varphi_2^2(\zeta, 2z-\zeta))]\, d\zeta.$$

Then we can easily deduce the estimate:

$$|\tilde{\varphi}_1(z,t)| \leq \int_0^z \Big(|q^1(\zeta)|(|\tilde{\varphi}_1(\zeta, z+t-\zeta)| + |\tilde{\varphi}_2(\zeta, z+t-\zeta)|)$$
$$+\tilde{q}(\zeta)(|\varphi_1^2(\zeta, z+t-\zeta)| + |\varphi_2^2(\zeta, z+t-\zeta)|)\Big)d\zeta \leq a\|\tilde{\varphi}\|, \quad (z,t) \in \Omega_T^+,$$

where

$$a = 2T\|\varphi^0\|\left(\frac{1}{1-T\|\varphi^0\|} + \frac{1+2T\|\varphi^0\|}{(1-T\|\varphi^0\|)^2}\right).$$

Similarly,

$$|\tilde{\varphi}_2(z,t)| \leq a\|\tilde{\varphi}\|, \quad |\tilde{\varphi}_3(z)| \leq a|\lambda_0|\|\tilde{\varphi}\|, \quad (z,t) \in \Omega_T^+.$$

Take a positive $\rho < 1$ and choose T such small that $a \max(1, |\lambda_0|) \leq \rho$ and inequalities (5.3.15) hold. Then the operator A is contracting on the ball $B(\varphi^0)$. According the Banach's principle the Eq. (5.3.12) has a unique solution which belongs to the ball $B(\varphi^0)$. Hence, function $q(z)$ is uniquely defined by the formula (5.3.9) and it is continuous for $z \in [0, T/2]$. This means that the coefficient $\hat{\varepsilon}(z)$ is in $C^1[0, T/2]$. Moreover, this coefficient can be derived via the function $q(z)$ by the formula:

$$\hat{\varepsilon}(z) = \varepsilon^+ \exp\left(4 \int_0^z q(\zeta)\, d\zeta\right).$$

Having $\hat{\varepsilon}(z)$ we can find the correspondence between z and x_3 using formula (5.2.10), then calculate the unknown function $\varepsilon(x_3) = \hat{\varepsilon}(h^{-1}(x_3))$, where $h^{-1}(x_3)$ is the inverse of the function $h(z)$. □

Note that the problem of finding the permeability $\mu(x_3)$, when $\varepsilon(x_3) = \varepsilon^+$ is a given positive constant and $\sigma(x_3) = 0$ for $x_3 > 0$, is completely symmetric to the previous one. So, it does not require a separate investigation. Instead we consider the following problem.

5.3.2 Problem of Finding a Conductivity Coefficient

In this case we assume that ε and μ are given. For the sake of simplicity, here we assume that $\varepsilon(x_3) = \varepsilon^+ > 0$ and $\mu(x_3) = \mu^+ > 0$ for $x_3 > 0$. In this case

$$q_1(z) = q_2(z) = 0, \quad q_3(z) = \frac{\hat{\sigma}(z)}{2\varepsilon^+} := q(z), \tag{5.3.16}$$

and the Eq. (5.2.14) have the form

5.3 One-Dimensional Inverse Problems

$$\begin{cases} \dfrac{\partial u_1}{\partial t} - \dfrac{\partial u_1}{\partial z} + q(z)(u_1 + u_2) = 0, \\ \dfrac{\partial u_2}{\partial t} + \dfrac{\partial u_2}{\partial z} + q(z)(u_1 + u_2) = 0, \quad (z,t) \in \Omega_T^+, \end{cases} \quad (5.3.17)$$

which is quite similar to (5.3.4). One needs to add to these equations the boundary conditions (5.3.3) and conditions (5.3.5) along the characteristic line $t = z$. Then we obtain the integral equation for functions $u_1(z,t), u_2(z,t)$ of the form:

$$u_1(z,t) - \int_0^z q(\zeta)(u_1(\zeta, z+t-\zeta) + u_2(\zeta, z+t-\zeta))\,d\zeta = F_1(t+z),$$

$$u_2(z,t) + \int_0^z q(\zeta)(u_1(\zeta, t-z+\zeta) + u_2(\zeta, t-z+\zeta))\,d\zeta = F_2(t-z),$$

$$(z,t) \in \Omega_T^+. \quad (5.3.18)$$

Substituting in the first Eq. (5.3.18) $t = z$ and using first condition (5.3.5), we find the relation

$$\int_0^z q(\zeta)(u_1(\zeta, 2z-\zeta) + u_2(\zeta, 2z-\zeta))\,d\zeta = F_1(2z).$$

Differentiate the last equation with respect to t and denote again

$$p(z) = q(z) \exp\left(\int_0^z q(\zeta)\,d\zeta\right).$$

Then dividing both sides of the obtained relation by $r_{22}\sqrt{\mu^-}\,j(0)$, we get:

$$p(z) + \lambda_0 \int_0^z q(\zeta)(\dot{u}_1(\zeta, 2z-\zeta) + \dot{u}_2(\zeta, 2z-\zeta))\,d\zeta = p_0(z),$$

$$z \in [0, T/2], \quad (5.3.19)$$

where

$$\lambda_0 = \dfrac{2}{r_{22}\sqrt{\mu^-}\,j(0)}, \quad p_0(z) = \lambda_0 F_1'(2z) \quad (5.3.20)$$

and $q(z)$ is defined by the formula (5.3.9).

Differentiating now (5.3.18) with respect to t, we obtain the following equations for functions $\dot{u}_1$ and $\dot{u}_2$:

$$\dot{u}_1(z,t) - \int_0^z q(\zeta)(\dot{u}_1(\zeta, z+t-\zeta) + \dot{u}_2(\zeta, z+t-\zeta))\,d\zeta = F_1'(t+z),$$

$$\dot{u}_2(z,t) + \int_0^z q(\zeta)(\dot{u}_1(\zeta, t-z+\zeta) + \dot{u}_2(\zeta, t-z+\zeta))\,d\zeta = F_2'(t-z),$$

$$(z,t) \in \Omega_T^+. \quad (5.3.21)$$

The Eqs. (5.3.19) and (5.3.21) form the system of integral equations for determining the unknown functions $\dot{u}_1(z,t)$, $\dot{u}_2(z,t)$ and $p(z)$ in the domain Ω_T^+. This system differs from the system (5.3.10) and (5.3.11) by some signs under the integral terms. Therefore for this system the Contraction Mapping Principle can be applied in the space of continuous functions to conclude that this system has a unique solution, when $T > 0$ is enough small. After finding $p(z)$ we can calculate $q(z)$, using formula (5.3.9), and then find $\hat{\sigma}(z) = 2q(z)\varepsilon^+$. In this case we obtain the correspondence between z and x_3 of the form $z = x_3\sqrt{\varepsilon^+\mu^+}$. So, $\sigma(x_3) = \hat{\sigma}(x_3\sqrt{\varepsilon^+\mu^+})$. Hence, the unique solution of the inverse problem exists and is a continuous function on $[0, T/2]$, if T is small enough.

As a result the following theorem holds.

Theorem 5.3.2 *Let $\varepsilon(x_3) = \varepsilon^+ > 0$, $\mu(x_3) = \mu^+ > 0$ for $x_3 > 0$, and $j(t) \in C^1[0, T]$, $j(0) \neq 0$. Assume that $f(t) \in C^1[0, T]$, $T > 0$, and condition (5.2.37) holds. Then for sufficiently small $T > 0$ there exist a unique continuously differentiable solution to the inverse problem.*

For more general results related to inverse problems of electrodynamics we refer the book [87].

Chapter 6
Inverse Problems for Parabolic Equations

This Chapter deals with inverse coefficient problems for linear second-order $1D$ parabolic equations. We establish, first, a relationship between solutions of direct problems for parabolic and hyperbolic equations. Then using the results of Chap. 4 we derive solutions of the inverse problems for parabolic equations via the corresponding solutions of inverse problems for hyperbolic equations, using the Laplace transform. In the final part of this chapter, we study the relationship between the inverse problems for parabolic equation and inverse spectral problems.

6.1 Relationships Between Solutions of Direct Problems for Parabolic and Hyperbolic Equations

Consider the following initial-boundary value problem

$$\begin{cases} \left(\dfrac{\partial}{\partial t} - l\right) v(x,t) = 0, & (x,t) \in \mathbb{R}_+^2, \\ v|_{t=0} = 0, & v|_{x=0} = g(t), \end{cases} \qquad (6.1.1)$$

for the parabolic equation in the first quadrant $\mathbb{R}_+^2 = \{(x,t) \mid x > 0, t > 0\}$ of the plane $\mathbb{R}^2$, where

$$l := c^2(x) \frac{\partial^2}{\partial x^2} + b(x) \frac{\partial}{\partial x} + d(x).$$

Consider also the similar problem

$$\begin{cases} \left(\dfrac{\partial^2}{\partial t^2} - l\right)u(x,t) = 0, & (x,t) \in \mathbb{R}_+^2, \\ u|_{t=0} = 0, \quad u_t|_{t=0} = 0, \quad u|_{x=0} = h(t) \end{cases} \quad (6.1.2)$$

for the hyperbolic equation. It turns out that the solution of the parabolic problem (6.1.1) can be expressed via the solution of the hyperbolic problem (6.1.2), if the function $g(t)$ is related to $h(t)$ by a special way. To show this, let us define the Laplace transforms

$$\begin{aligned} \tilde{v}(x,p) &= \int_0^\infty v(x,t)\exp(-pt)\,dt, \\ \tilde{u}(x,s) &= \int_0^\infty u(x,t)\exp(-st)\,dt \end{aligned} \quad (6.1.3)$$

of the functions $v(x,t)$ and $u(x,t)$ given in (6.1.1) and (6.1.2). Here $p = p_1 + ip_2$ and $s = s_1 + is_2$ are the transform parameters and $i := \sqrt{-1}$ is the imaginary unit. These transforms exist if the functions $v(x,t)$ and $u(x,t)$ satisfy the conditions

$$|v(x,t)| \leq Ce^{\sigma t}, \quad |u(x,t)| \leq Ce^{\sigma t}, \quad x \geq 0,\ t \geq 0,$$

with the positive constant C and real σ. Then the integrals in (6.1.3) exist for $p_1 > \sigma$ and $s_1 > \sigma$. Moreover, Laplace transforms are analytic functions of p and s, respectively, in the complex half-planes $p_1 > \sigma$ and $s_1 > \sigma$.

We assume that the functions $v(x,t)$, $u(x,t)$ with derivatives with respect to x up to the second order, as well as the functions $g(t)$ and $h(t)$ admit the Laplace transform with respect to t. Denote by $\tilde{g}(p)$ and $\tilde{h}(s)$ the Laplace transforms of the input data $g(t)$ and $h(t)$ in (6.1.1) and (6.1.2), respectively. Then we can rewrite problems (6.1.1) and (6.1.2) as follows:

$$p\tilde{v}(x,p) - l\tilde{v}(x,p) = 0, \quad x > 0;\quad \tilde{v}|_{x=0} = \tilde{g}(p), \quad (6.1.4)$$
$$s^2\tilde{u}(x,s) - l\tilde{u}(x,s) = 0, \quad x > 0;\quad \tilde{u}|_{x=0} = \tilde{h}(s). \quad (6.1.5)$$

Suppose now that $s = \sqrt{p}$ and $\tilde{g}(p) = \tilde{h}(\sqrt{p})$. Then it follows from (6.1.4) and (6.1.5) that $\tilde{v}(x,p) = \tilde{u}(x,\sqrt{p})$. According the table of Laplace transforms [7] the last relations correspond to the formulae:

$$g(t) = \frac{1}{2\sqrt{\pi t^3}} \int_0^\infty h(\tau) e^{-\frac{\tau^2}{4t}} \tau\,d\tau, \quad (6.1.6)$$

$$v(x,t) = \frac{1}{2\sqrt{\pi t^3}} \int_0^\infty u(x,\tau) e^{-\frac{\tau^2}{4t}} \tau\,d\tau. \quad (6.1.7)$$

6.1 Relationships Between Solutions of Direct Problems for Parabolic ...

Note that

$$\frac{\tau}{2\sqrt{\pi t^3}}e^{-\frac{\tau^2}{4t}} = -\frac{\partial}{\partial \tau}G(\tau,t), \quad G(\tau,t) = \frac{e^{-\frac{\tau^2}{4t}}}{\sqrt{\pi t}},$$

where the function $G(\tau,t)$ solves the heat equation

$$\frac{\partial G(\tau,t)}{\partial t} = \frac{\partial^2 G(\tau,t)}{\partial \tau^2}, \quad t > 0.$$

Under the consistency condition $h(0) = 0$ and the initial condition $u(x,0) = 0$ in (6.1.5), formulae (6.1.6), (6.1.7) can be also written as follows:

$$g(t) = \int_0^\infty h'(\tau) G(\tau,t) \, d\tau, \tag{6.1.8}$$

$$v(x,t) = \int_0^\infty u_\tau(x,\tau) G(\tau,t) \, d\tau. \tag{6.1.9}$$

We use the change of variables $z = \tau^2/(4t)$ in the integral (6.1.8) to prove that

$$g(0) = h'(0). \tag{6.1.10}$$

We have:

$$g(0) := \lim_{t \to +0} \int_0^\infty h'(\tau) G(\tau,t) \, d\tau = \frac{2}{\sqrt{\pi}} \lim_{t \to +0} \int_0^\infty h'(2\sqrt{tz}) e^{-z^2} \, dz$$

$$= h'(0) \frac{2}{\sqrt{\pi}} \int_0^\infty e^{-z^2} \, dz = h'(0).$$

In the same way we can prove $v(x,0) = u(x,0) = 0$.

It can be verified directly that function $v(x,t)$ given by formula (6.1.7) satisfy the Eq. (6.1.1), if $u(x,t)$ solves the problem (6.1.2). Indeed,

$$\left(\frac{\partial}{\partial t} - l\right) v(x,t) = -\left(\frac{\partial}{\partial t} - l\right) \int_0^\infty u(x,\tau) G_\tau(\tau,t) \, d\tau$$

$$= \int_0^\infty \left(lu(x,\tau) G_\tau(\tau,t) - u(x,\tau) \frac{\partial}{\partial t} G_\tau(\tau,t)\right) d\tau$$

$$= \int_0^\infty \left(u_{\tau\tau}(x,\tau) G_\tau(\tau,t) - u(x,\tau) \frac{\partial}{\partial t} G_\tau(\tau,t)\right) d\tau$$

$$= \int_0^\infty u(x,\tau) \left(\frac{\partial^2}{\partial \tau^2} - \frac{\partial}{\partial t}\right) G_\tau(\tau,t) \, d\tau = 0.$$

Note that the integral relation (6.1.6) is invertible. Moreover, it is the Laplace transform with parameter $p = 1/(4t)$. Indeed, it can be represented in the form:

$$g(t) = \frac{1}{4\sqrt{\pi t^3}} \int_0^\infty h(\sqrt{z}) e^{-\frac{z}{4t}} \, dz. \tag{6.1.11}$$

Hence, the function $\sqrt{\pi p^{-3}} g(1/(4p))/2 := \hat{g}(p)$ is the Laplace transform of the function $h(\sqrt{t})$. Then $\hat{g}(p)$ is an analytic function for $Re(p) > 0$ and, as a result, the function $h(t)$ can be uniquely recovered from $g(t)$ by using the inverse Laplace transform

$$h(t) = \frac{1}{2\pi i} \int_{\sigma_0 - i\infty}^{\sigma_0 + i\infty} \hat{g}(p) e^{pt^2} \, dp, \quad t \geq 0, \tag{6.1.12}$$

where $\sigma_0 > 0$.

6.2 Problem of Recovering the Potential for Heat Equation

Let $v(x, t)$ be the solution of the parabolic equation

$$v_t - v_{xx} - q(x)v = 0, \quad x > 0, \, t > 0, \tag{6.2.1}$$

with the following boundary and initial conditions:

$$v|_{x=0} = g(t), \quad t \geq 0; \quad v|_{t=0} = 0, \quad x \geq 0. \tag{6.2.2}$$

For a given function $g(t)$ from some class of admissible coefficients, we will define problem (6.2.1) and (6.2.2) as the *direct problem*.

Consider the inverse *problem of recovering the potential $q(x)$ from the given trace*

$$r(t) := v_x|_{x=0}, \quad t \geq 0 \tag{6.2.3}$$

of the derivative v_x on the semi-axis $x = 0$, $t \geq 0$ of the solution $v = v(x, t; q)$.

Associate to the inverse problem (6.2.1), (6.2.2) and (6.2.3) the similar inverse problem for the hyperbolic equation, i.e. let the direct problem is given by

$$u_{tt} - u_{xx} - q(x)u = 0, \quad x > 0, \, t > 0, \tag{6.2.4}$$

$$u|_{x=0} = h(t), \, t \geq 0; \quad u|_{t=0} = 0, \, u_t|_{t=0} = 0, \, x \geq 0, \tag{6.2.5}$$

$$u_x|_{x=0} = f(t), \quad t \geq 0, \tag{6.2.6}$$

and the *Neumann type measured output data* is given by (6.2.3).

6.2 Problem of Recovering the Potential for Heat Equation

We assume here that $q(x) \in C[0, \infty)$ and the solutions to direct problems (6.2.1) and (6.2.2) and (6.2.4) and (6.2.5) admit the Laplace transforms with respect to t. Furthermore, suppose that the $h(0) = 0$ and the relationship (6.1.6) holds between the input $h(t)$ in the parabolic inverse problem (6.2.1) and (6.2.3) and the input $g(t)$ in the hyperbolic inverse problem (6.2.4) and (6.2.6). Then it is necessary that $f(0) = -h'(0)$ and functions $r(t)$ and $f(t)$ satisfy the relation

$$r(t) = \frac{1}{2\sqrt{\pi t^3}} \int_0^\infty f(\tau) e^{-\frac{\tau^2}{4t}} \tau d\tau, \quad t \geq 0. \tag{6.2.7}$$

The function $f(t)$ can be found through $r(t)$ by the formula

$$f(t) = \frac{1}{2\pi i} \int_{\sigma_0 - i\infty}^{\sigma_0 + i\infty} \hat{r}(p) e^{pt^2} dp, \quad t \geq 0, \tag{6.2.8}$$

similar to (6.1.12), where $\hat{r}(p) = \sqrt{\pi p^{-3}} r(1/(4p))/2$.

Thus, the functions $h(t)$ and $f(t)$ in the inverse problem (6.2.4)–(6.2.6) are uniquely defined by the functions $g(t)$ and $r(t)$. This means that instead of the inverse problem (6.2.1)–(6.2.3) for parabolic equation one can solve the inverse problem (6.2.4)–(6.2.6) for hyperbolic equation, to find unknown coefficient $q(x)$ for $x \geq 0$. The last inverse problem has been studied in Sect. 3.2 under the conditions that $h(0) \neq 0$ and $h(t) \in C^2([0, \infty))$. But in our case we have assumed that $h(0) = 0$. What should be done in this case? It turn out that we need to make a small change in order to reduce the present problem to the previous one. Indeed, let we suppose that $g(0) \neq 0$. Then, it follows from formula (6.1.10) that $h'(0) = g(0) \neq 0$. Differentiating relations (6.2.4), (6.2.5) and (6.2.6) with respect to t, one gets

$$\dot{u}_{tt} - \dot{u}_{xx} - q(x)\dot{u} = 0, \quad x > 0, \; t > 0, \tag{6.2.9}$$
$$\dot{u}|_{x=0} = h'(t), \; t \geq 0; \; \dot{u}|_{t=0} = 0, \; \dot{u}_t|_{t=0} = 0, \; x \geq 0, \tag{6.2.10}$$
$$\dot{u}_x|_{x=0} = f'(t), \; t \geq 0, \tag{6.2.11}$$

where $\dot{u} = u_t$. Note that the third relation (6.2.10) follows directly from Eq. (6.2.4) with $t = 0$. So, the functions $h'(t)$ and $f'(t)$ in the inverse problem (6.2.9), (6.2.10) and (6.2.11) play the same roles, as the functions $h(t)$ and $f(t)$, in the inverse problem studied in Sect. 4.2. Hence, we need suppose only that $h(t) \in C^3[0, \infty)$ in order to use the uniqueness Theorem 4.2.3 in Sect. 4.2 The following theorem is a direct consequence of this conclusion.

Theorem 6.2.1 *Assume that the function $h(t)$, which is associated with the function $g(t)$ by formula (6.1.6), satisfies the following conditions: $h(t) \in \mathbf{C}^3[0, \infty)$, $h(0) = 0$, $h'(0) \neq 0$. Suppose, in addition, that $q_1, q_2 \in C[0, \infty)$ are any two solutions of the inverse problem (6.2.1), (6.2.2) and (6.2.3) corresponding data $g(t)$ and $r(t)$. Then $q_1(x) = q_2(x)$ for $x \in [0, \infty)$.*

Remark 6.2.1 If $g(t)$ is defined by formula (6.1.6), then $g(t)$ is an analytic function for $t > 0$. Then $r(t)$ is also an analytic function of $t > 0$. Hence, uniqueness theorem for the inverse problem (6.2.1), (6.2.2) and (6.2.3) is still true, if the data $g(t)$ and $r(t)$ are given for $t \in [0, T]$ with arbitrary fixed $T > 0$.

6.3 Uniqueness Theorems for Inverse Problems Related to Parabolic Equations

Consider the following initial-boundary value problem

$$\begin{cases} v_t - (k(x)v_x)_x + q(x)v = 0, \ x > 0, \ t > 0; \\ v(x, 0) = 0, \ x \geq 0; \\ v(0, t) = 0, \ t \geq 0, \end{cases} \quad (6.3.1)$$

for the parabolic equation.

Let $v(x, t)$ solves the initial-boundary value problem (6.3.1) and, in addition, the trace of $k(x)v_x(x, t)$ is given at $x = 0$:

$$r(t) := k(0)v_x(x, t)|_{x=0}, \quad t \geq 0. \quad (6.3.2)$$

Remark that in thermal conduction $k(0)v_x(x, t)|_{x=0}$ is defined as the heat flux, where $k(x)$ is the thermal conductivity.

Here and below we assume that $k(x)$ and $q(x)$ satisfy the conditions:

$$0 < k_0 \leq k(x) \leq k_1 \leq \infty, \quad |q(x)| \leq q_0, \quad x \in [0, \infty); \quad (6.3.3)$$
$$k(x) \in C^2[0, \infty), \quad q(x) \in C[0, \infty), \quad (6.3.4)$$

where k_0, k_1, q_0 are some positive constants.

Consider the following two *inverse coefficient problems*.

ICP1: Find the unknown coefficient $q(x)$ in (6.3.1) from the given by (6.3.2) output data $r(t)$.

ICP2: Find the unknown coefficient $k(x)$ in (6.3.1) from the given by (6.3.2) output data $r(t)$.

In both cases the initial-boundary value problem (6.3.1) is defined as the direct problem.

Obviously the coefficient $k(x)$ in **ICP1** is assumed to be known as well as the coefficient $q(x)$ in **ICP2** assumed to be known. In both inverse problems we will assume that $k(0)$ is known.

6.3 Uniqueness Theorems for Inverse Problems Related to Parabolic Equations

Consider first **ICP1**. Let us try to reduce the direct problem (6.3.1) to the simplest form. For this goal, introduce the new variable y instead of x by the formula

$$y = \int_0^x \frac{d\xi}{\sqrt{k(\xi)}}. \tag{6.3.5}$$

Let $x = H(y)$ be the inverse function of $y = y(x)$ given by (6.3.5). Denote by

$$k(H(y)) = \hat{k}(y), \quad q(H(y)) = \hat{q}(y), \quad v(H(y), t) = \hat{v}(y, t).$$

Since

$$k(x)v_x(x, t) = \hat{v}_y(y, t)\sqrt{\hat{k}(y)}\Big|_{y=y(x)},$$

$$(k(x)v_x(x, t))_x = \left[\hat{v}_{yy}(y, t) + \hat{v}_y(y, t)\left(\sqrt{\hat{k}(y)}\right)_y\right]_{y=y(x)},$$

the function $\hat{v}(y, t)$ satisfies the relations

$$\hat{v}_t(y, t) - \hat{v}_{yy}(y, t) - \hat{v}_y(y, t)\left(\sqrt{\hat{k}(y)}\right)_y + \hat{q}(y)\hat{v} = 0, \quad y > 0, \ t > 0, \tag{6.3.6}$$

$$\hat{v}(y, 0) = 0, \ \hat{v}(0, t) = g(t), \ \sqrt{k(0)}\hat{v}_y(0, t) = r(t). \tag{6.3.7}$$

Introduce the new function $w(y, t)$ by the formula $\hat{v}(y, t) = S(y)w(y, t)$, where $S(y) = (k(0)/\hat{k}(y))^{1/4}$. Then $w(y, t)$ solves the problem:

$$w_t - w_{yy}(y, t) + Q(y)w = 0, \quad y > 0, \ t > 0, \tag{6.3.8}$$

$$w(y, 0) = 0, \ w(0, t) = g(t), \ \sqrt{k(0)}\left(w_y + S'(0)w\right)_{y=0} = r(t), \tag{6.3.9}$$

where $S'(0) = -\hat{k}'(0)/(4k(0))$ and $Q(y)$ is defined by the formula

$$Q(y) = q(y) - \frac{S''(y)}{S(y)} + 2\left(\frac{S'(y)}{S(y)}\right)^2$$

$$= q(y) - \left(\ln S(z)\right)'' + \left[\left(\ln S(z)\right)'\right]^2. \tag{6.3.10}$$

From the latter two relations in (6.3.9) we find

$$w_y|_{y=0} = \hat{r}(t), \quad \hat{r}(t) = \frac{r(t)}{\sqrt{k(0)}} - S'(0)g(t). \tag{6.3.11}$$

Since $S'(0)$ is known in this inverse problem the function $\hat{r}(t)$ is defined by the latter equality.

Consider now the boundary value problem for the hyperbolic equation

$$u_{tt} - u_{yy}(y,t) + Q(y)u = 0, \quad y > 0, \ t > 0, \qquad (6.3.12)$$
$$u(y,0) = 0, \ u_t(y,0) = 0, \ y > 0, \quad u(0,t) = h(t), \ t \geq 0, \qquad (6.3.13)$$

where the function $Q(y)$ is defined by (6.3.10), and the inverse problem of recovering $Q(y)$ from the data

$$u_y|_{y=0} = f(t), \quad t \geq 0. \qquad (6.3.14)$$

Assume that $h(t) \in \mathbf{C}^3[0, \infty)$, $h(0) = 0$, $h'(0) \neq 0$, and functions $g(t)$ and $h(t)$ are connected by formula (6.1.6). Then $f(0) = -h'(0)$ and

$$\hat{r}(t) = \int_0^\infty f'(\tau) G(\tau, t) \, d\tau. \qquad (6.3.15)$$

The function $f(t)$, as a trace of the normal derivative u_y on $y = 0$ to the problem (6.3.12) and (6.3.13) belongs to $\mathbf{C}^2[0, \infty)$. Then the function $\dot{u} = u_t$ satisfies the relations

$$\dot{u}_{tt} - \dot{u}_{yy} - Q(y)\dot{u} = 0, \quad y > 0, \ t > 0, \qquad (6.3.16)$$
$$\dot{u}|_{y=0} = h'(t), \ t \geq 0; \quad \dot{u}|_{t=0} = 0, \ \dot{u}_t|_{t=0} = 0, \ x \geq 0, \qquad (6.3.17)$$
$$\dot{u}_y|_{y=0} = f'(t), \quad t \geq 0. \qquad (6.3.18)$$

From (6.3.16) and (6.3.17) follows that $\dot{u}(0,0) = h'(0)$. Therefore, $f'(0) = -h''(0)$ and $\hat{r}(0) = f'(0) = -h''(0)$. Since

$$\hat{r}(0) = \frac{r(0)}{\sqrt{k(0)}} - S'(0)g(0), \quad g(0) = h'(0),$$

we can find $S'(0)$ by the formula

$$S'(0) = \frac{h''(0)}{h'(0)} + \frac{r(0)}{\sqrt{k(0)} h'(0)}. \qquad (6.3.19)$$

On the over hand, $S'(0) = -\hat{k}'(0)/(4k(0))$. Thus, the formula (6.3.19) gives the necessary condition of solvability to inverse problem of finding $q(x)$ from the data $g(t)$ and $r(t)$, it means, that the value $h''(0)$ must be taken from the relation (6.3.19).

For the inverse problem (6.3.16), (6.3.17) and (6.3.18) the uniqueness theorem holds. Hence, the inverse problem of finding $Q(y)$ for Eqs.(6.3.8), (6.3.9) and (6.3.11) has at most one solution. Since $k(x)$ is known, then $y = y(x)$ given by (6.3.5) as well as $S(y)$ are also known. So, we can find $q(y)$ from given $Q(y)$ and $S(y)$. Therefore we can determine and $q(x) = q(y(x))$. Taking into account the Remark 4.3.1 we come to the following uniqueness theorem.

6.3 Uniqueness Theorems for Inverse Problems Related to Parabolic Equations

Theorem 6.3.1 *Let conditions 6.3.3 hold. Assume that the function $h(t)$, which is associated with the function $g(t)$ by formula (6.1.6), satisfies the following conditions: $h(t) \in \mathbf{C}^3[0, \infty)$, $h(0) = 0$, $h'(0) \neq 0$ and $h''(0)$ satisfies the relation (6.3.19). Denote by $q, q_2 \in C[0, \infty)$ two solutions of **ICP1**, defined by (6.3.1) and (6.3.2), with data $g(t)$ and $r(t)$, given for $t \in [0, T]$, $T > 0$. Then $q_1(x) = q_2(x)$ for $x \in [0, \infty)$.*

Consider now **ICP2**. Assume now that $q(x) = q_0$ is a given constant and $k(0) > 0$ is known. We reduce the **ICP2** to Eqs. (6.3.8) and (6.3.11). Since in this case $\hat{k}(y)$ is unknown, $S'(0)$ in (6.3.11) is also unknown. Let us explain how to find $S'(0)$. Consider again the problem (6.3.12) and (6.3.13). Then the relation (6.3.19) holds and defines $S'(0)$. So the function $\hat{r}(t)$ is become known and Eq. (6.3.15) uniquely determines $f'(t)$. As in the previous case, solving the inverse problem for the hyperbolic equation we can find the function $Q(y)$. Then using (6.3.10) we obtain the second order differential equation for function $\ln S(y)$:

$$Q(y) - q_0 = -\big(\ln S(z)\big)'' + \big[\big(\ln S(z)\big)'\big]^2 \qquad (6.3.20)$$

with given Cauchy data $S(0)$ and $S'(0)$. From this equation and initial data the function $\ln S(y)$ is uniquely determined for all $x \in [0, \infty)$. After this we can find $\hat{k}(y)$, $y = y(x)$ and $k(x) = \hat{k}(y(x))$. Hence the following theorem holds.

Theorem 6.3.2 *Let $h(t) \in \mathbf{C}^3[0, \infty)$, $h(0) = 0$, $h'(0) \neq 0$, and functions $g(t)$ and $h(t)$ are connected by formula (6.1.6). Let, moreover, $q(x) = q_0$ be a given constant and $k(0) > 0$ be known and $k_1(x)$ and $k_2(x)$ be two solutions to the problem (6.3.1) and (6.3.2) with data $g(t)$ and $r(t)$ given for $t \in [0, T]$, $T > 0$. Then $k_1(x) = k_2(x)$ for $x \in [0, \infty)$.*

6.4 Relationship Between the Inverse Problem and Inverse Spectral Problems for Sturm-Liouville Operator

In this section we will analyze the relations between the inverse problems for parabolic equation and inverse spectral problems.

Let $v(x, t)$ be the solution of the parabolic equation

$$v_t - v_{xx} + q(x)v = g(x, t), \quad 0 < x < 1, \ t > 0, \qquad (6.4.1)$$

with given function $g(x, t)$ and given the initial and boundary data

$$v_x|_{x=0} = 0, \ v|_{x=1} = 0, \quad t \geq 0; \quad v|_{t=0} = 0, \quad x \geq 0. \qquad (6.4.2)$$

Our aim is to show that problem (6.4.1) and (6.4.2) is related to the problem of finding eigenvalues for the ordinary differential equation

$$-y''(x) + q(x)y(x) = \lambda y(x), \quad 0 < x < 1, \tag{6.4.3}$$

with boundary data

$$y'(0) = 0, \quad y(1) = 0. \tag{6.4.4}$$

Remember that λ is called an *eigenvalue* of the differential operator $-y''(x) + q(x)y(x)$ subject to the boundary conditions (6.4.4) provided there exists a function $u(x)$, not identically zero, solving problem (6.4.3) and (6.4.4). The solution $u(x)$ is called the corresponding *eigenfunctions*. It is well known that the eigenfunctions of the differential operator form countable set λ_n, $n = 1, 2, \ldots$, with the unique concentration point at infinity and the eigenfunctions $y_n(x)$, $n = 1, 2, \ldots$, are dense in the space $L_2(0, 1)$. Note that all $\lambda_n > 0$, if $q(x) \geq 0$.

Applying the Fourier method to problem (6.4.1) and (6.4.2) we can construct the solution in the form

$$v(x,t) = \sum_{n=1}^{\infty} v_n(t) y_n(x), \tag{6.4.5}$$

where $v_n(t)$ are solutions to the Cauchy problem

$$v'_n + \lambda_n v_n = g_n(t), \quad v_n(0) = 0, \quad n = 1, 2, \ldots, \tag{6.4.6}$$

and $g_n(t)$ are the Fourier coefficients of $g(x,t)$ with respect to the eigenfunctions $y_n(x)$, $n = 1, 2, \ldots$, i.e.,

$$g_n(t) = \frac{1}{\|y_n\|^2} \int_0^1 g(x,t) y_n(x) dx, \quad n = 1, 2, \ldots. \tag{6.4.7}$$

The solutions to the problem (6.4.6) have the form

$$v_n(t) = \int_0^t g_n(\tau) e^{-\lambda_n(t-\tau)} d\tau, \quad n = 1, 2, \ldots. \tag{6.4.8}$$

Thus, the solution to the problem (6.4.1) and (6.4.2) is given by the formulae (6.4.5), (6.4.7), (6.4.8). From them follows that

$$v(x,t) = \sum_{n=1}^{\infty} \frac{y_n(x)}{\|y_n\|^2} \int_0^t e^{-\lambda_n(t-\tau)} \int_0^1 g(x,\tau) y_n(x) dx d\tau. \tag{6.4.9}$$

6.4 Relationship Between the Inverse Problem and Inverse Spectral Problems ...

We can represent formula (6.4.8) in an other form. To this end, denote the solution the equation (6.4.3) with the Cauchy data

$$y(0) = 1, \quad y'(0) = 0 \qquad (6.4.10)$$

by $y(x, \lambda)$. Then λ_n satisfy the conditions

$$y(1, \lambda_n) = 0, \quad n = 1, 2, \ldots. \qquad (6.4.11)$$

Denote by $y_n(x) = y(x, \lambda_n)$.

Introduce the *spectral function* $\rho(\lambda)$, $\lambda \in (-\infty, \infty)$, by the formula

$$\rho(\lambda) = \begin{cases} 0, & \text{if } \lambda < \lambda_1, \\ \sum_{n=1}^{k} \dfrac{1}{\|y_n\|^2}, & \text{if } \lambda_k < \lambda < \lambda_{k+1}. \end{cases} \qquad (6.4.12)$$

Then formula (6.4.9) can be written as follows

$$v(x, t) = \int_{-\infty}^{\infty} y(x, \lambda) \int_{0}^{t} e^{-\lambda(t-\tau)} \int_{0}^{1} g(\xi, \tau) y(\xi, \lambda) d\xi d\tau d\rho(\lambda), \qquad (6.4.13)$$

where the integral with respect to λ should be understood as a Stieltjes integral or in the sense of distributions.

For the direct problem (6.4.1) and (6.4.2) with the given function $g(x, t)$, we consider the inverse problem of recovering the potential $q(x)$ from a given trace of the solution $v(x, t)$ on semi-axis $x = 0, t \geq 0$, i.e.,

$$r(t) := v|_{x=0}, \quad t \geq 0. \qquad (6.4.14)$$

It is easy to derive the function $r(t)$ via the spectral function $\rho(\lambda)$. Indeed, by $y(0, \lambda) = 1$, we conclude:

$$r(t) = \int_{-\infty}^{\infty} \int_{0}^{t} e^{-\lambda(t-\tau)} \int_{0}^{1} g(\xi, \tau) y(\xi, \lambda) d\xi d\tau d\rho(\lambda), \quad t \geq 0. \qquad (6.4.15)$$

Consider now case when $g(x, t) = \delta(x + 0)\delta(t + 0)$. Here $\delta(x + 0)$ is the Dirac-delta function located at the point $x = +0$. Then two inner integrals in formula (6.4.15) are calculated explicitly and we obtain more simple formula:

$$r(t) = \int_{-\infty}^{\infty} e^{-\lambda t} d\rho(\lambda), \quad t \geq 0. \qquad (6.4.16)$$

This equality uniquely defines the spectral function $\rho(\lambda)$. Then for the direct problem of finding eigenvalues and eigenfunctions related to Eqs. (6.4.3) and (6.4.4) we can consider the inverse spectral problem: given $\rho(\lambda)$ find $q(x)$. This problem is equivalent to the inverse problem for the parabolic equation. Indeed, given $r(t)$ we find $\rho(\lambda)$ and vice versa given $\rho(\lambda)$ we find $r(t)$. For the inverse spectral problem is well known uniqueness theorem stated by V.A. Marchenko first in [65] (see also, [66]).

Theorem 6.4.1 *The potential $q(x) \in C[0, 1]$ is uniquely recovered by the spectral function $\rho(\lambda)$.*

As a corollary, we obtain the uniqueness theorem for the inverse problem (6.4.1), (6.4.2), (6.4.14).

Theorem 6.4.2 *Let $g(x, t) = \delta(x + 0)\delta(t + 0)$ and $q(x) \in C[0, 1]$. Then $q(x)$ is uniquely determined by the given function $r(t)$.*

Note that the assertions of both theorems remains true if the finite interval $[0, 1]$ is replaced on semi-infinite interval $[0, \infty)$.

6.5 Identification of a Leading Coefficient in Heat Equation: Dirichlet Type Measured Output

In this section we consider the problem of determining the space-dependent thermal conductivity $k(x)$ of the one-dimensional heat equation $u_t(x, t) = (k(x)u_x(x, t))_x$ from boundary measurements in the finite domain $\Omega_T = \{(x, t) \in \mathbb{R}^2 : 0 < x < l, \ 0 < t \leq T\}$.

Consider the inverse problem of identifying the leading coefficient $k(x)$ in

$$\begin{cases} u_t(x, t) = (k(x)u_x(x, t))_x, & (x, t) \in \Omega_T, \\ u(x, 0) = 0, & 0 < x < l, \\ -k(0)u_x(0, t) = g(t), & u_x(l, t) = 0, \ 0 < t < T, \end{cases} \quad (6.5.1)$$

from the measured temperature $f(t)$ at the left boundary $x = 0$ of a nonhomogeneous rod:

$$f(t) := u(0, t), \quad t \in [0, T]. \quad (6.5.2)$$

Note that problem (6.5.1) is a simplest $1D$ model of heat conduction in a rod occupying the interval $(0, l)$. The Neumann condition $-k(0)u_x(0, t) = g(t)$ in the direct problem (6.5.1) means that the heat flow $g(t)$ is prescribed at the left boundary of the rod as an input datum for all $t \in [0, T]$.

We assume that the functions $k(x)$ and $g(t)$ satisfy the following conditions:

$$\begin{cases} k \in H^1(0, l), \ 0 < c_0 \leq k(x) \leq c_1 < \infty; \\ g \in H^1(0, T), \ g(t) > 0, \ \text{for all } t \in (0, T) \text{ and } g(0) = 0. \end{cases} \quad (6.5.3)$$

The second condition in (6.5.3) means heating of a rod at the left end, that is heat flux at $x = 0$ is positive: $g(t) > 0$ for all $t \in (0, T)$ and is zero at an initial time $t = 0$, i.e. $g(0) = 0$.

Under conditions (6.5.3) the initial boundary value problem (6.5.1) has the unique regular weak solution

$$\begin{cases} u \in L^\infty(0, T; H^2(0, l)), \\ u_t \in L^\infty(0, T; L^2(0, l)) \cap L^2(0, T; H^1(0, l)), \\ u_{tt} \in L^2(0, T; H^{-1}(0, l)), \end{cases} \qquad (6.5.4)$$

with improved regularity ([24], Sect. 7.1, Theorem 5).

In next subsection we derive some important properties of the solution $u(x, t)$ of the direct problem (6.5.1), in particular the output $u(0, t; k)$. Then we introduce an input-output operator and reformulate the *inverse coefficient problem* (6.5.1) and (6.5.2) as a nonlinear operator equation. We prove in Sect. 6.5.2 that the input-output operator is compact and Lipschitz continuous which allows to prove an existence of a minimizer of the corresponding regularized functional. In Sect. 6.5.3 we introduce an adjoint problem and derive an integral relationship relating the change $\delta k(x) := k_1(x) - k_2(x)$ in the coefficient $k(x)$ to the change of the output $\delta u(x, t) = u(0, t; k_1) - u(0, t; k_2)$. This permits to obtain explicit formula for the Fréchet gradient $J'(k)$ of the Tikhonov functional via the solutions of the direct and adjoint problems. This approach, defined as an adjoint problem approach, has been proposed by Cannon and DuChateau [16, 22] and then developed in [37]. Furthermore, these results also constitute a theoretical framework of the numerical algorithm to recover the unknown coefficient. Some numerical examples applied to severely ill-posed benchmark problems, presented in the final subsection Sect. 6.5.4 demonstrate accurate reconstructions by the CG-algorithm.

6.5.1 Some Properties of the Direct Problem Solution

The theorem below shows that sign of the input datum $g(t)$ has a significant impact on the solution of the direct problem (6.5.1).

Theorem 6.5.1 *Let conditions* (6.5.3) *hold. Then* $g(t) > 0$ *for all* $t \in (0, T)$, *implies* $u_x(x, t) \leq 0$, *for all* $(x, t) \in \Omega_T$.

Proof Let $\varphi \in C_0^{2,1}(\Omega_T) \cup C_0(\overline{\Omega_T})$ be an arbitrary smooth function. Multiply both sides of the parabolic equation (6.5.1) by $\varphi_x(x, t)$, integrate on Ω_T and then perform integration by parts multiple times. After elementary transformations we get:

$$\iint_{\Omega_T} u_x(\varphi_t + k(x)\varphi_{xx}) \, dxdt$$
$$= \int_0^T (u\varphi_t)_{x=0}^{x=l} \, dt + \int_0^T (k(x)u_x\varphi_x)_{x=0}^{x=l} \, dt - \int_0^l (u\varphi_x)_{t=0}^{t=T} \, dx. \qquad (6.5.5)$$

Now we require that the function $\varphi(x, t)$ is chosen to be the solution of the following backward parabolic problem:

$$\begin{cases} \varphi_t + k(x)\varphi_{xx} = F(x,t), & (x,t) \in \Omega_T, \\ \varphi(x, T) = 0, & x \in (0, l), \\ \varphi(0, t) = 0, \ \varphi(l, t) = 0, & t \in (0, T), \end{cases} \tag{6.5.6}$$

where an arbitrary continuous function $F(x, t)$ which will be defined below. Then taking into account the homogeneous initial and boundary conditions given in (6.5.1) and (6.5.6) with the flux condition $-k(0)u_x(0, t) = g(t)$, in the integral identity (6.5.5) we get:

$$\iint_{\Omega_T} u_x(x, t) F(x, t) dx dt = \int_0^T g(t) \varphi_x(0, t) dt. \tag{6.5.7}$$

We apply the maximum principle to the backward in time parabolic problem (6.5.6). For this aim we require that the arbitrary function $F(x, t)$ satisfies the condition $F(x, t) > 0$ for all $(x, t) \in \Omega_T$. Then $\varphi(x, t) < 0$ on Ω_T. This, with the boundary condition $\varphi(0, t) = 0$, implies

$$\varphi_x(0, t) := \lim_{h \to 0^+} \frac{\varphi(h, t) - \varphi(0, t)}{h} \leq 0.$$

On the other hand, by the condition $g(t) > 0$ we conclude that the right hand side of (6.5.7) is non-positive, so

$$\iint_{\Omega_T} u_x(x, t) F(x, t) dx dt \leq 0, \quad \text{for all } F(x, t) > 0.$$

This implies that $u_x(x, t) \leq 0$ for all (x, t) in Ω_T. □

Remark 6.5.1 The result of the above theorem has a precise physical meaning. By definition, $g(t) := -k(0)u_x(0, t)$ is the heat flux at $x = 0$. The sign minus here means that, by convention, the heat flows in the positive x-direction, i.e. from regions of higher temperature to regions of lower temperature. The condition $g(t) > 0$, $t \in (0, T)$ implies that the heat flux at $x = 0$ is positive. Theorem 6.5.1 states that in the absence of other heat sources, the positive heat flux $g(t) > 0$ at the left end $x = 0$ of a rod results the nonnegative flux at all points $x \in (0, l)$ of a rod, that is $-k(x)u_x(x, t) \geq 0$.

Corollary 6.5.1 *Assume that, in addition to conditions of Theorem 6.5.1, the solution of the direct problem belongs to $C^{1,0}(\overline{\Omega}_T)$. Then there exists such $\epsilon = \epsilon(t) > 0$ that $u_x(x, t) < 0$, for all $(x, t) \in \Omega_T^\epsilon$, where $\Omega_T^\epsilon := \{(x, t) \in \Omega_T : 0 < x < \epsilon(t), \ 0 < t \leq T\}$ is the triangle with two rectilinear sides $\ell_1 := \{(x, t) \in \overline{\Omega_T^\epsilon} : x = 0, \ t \in [0, T]\}$, $\ell_2 := \{(x, t) \in \overline{\Omega_T^\epsilon} : x \in (0, \epsilon(T)], \ t = T\}$ and the curvilinear side*

$\ell_3 := \{(x, t) \in \overline{\Omega_T^\epsilon} : x = \epsilon(t), \ t \in (0, T)\}$. *The parameter $\epsilon(t) > 0$ depending on $t \in (0, T]$ satisfies the condition $\epsilon(0) = 0$.*

Proof Indeed, since $u_x(x, t)$ is a continuous function of $x \in (0, l)$ and $u_x(0, t) < 0$ for all $t \in (0, T)$, it remains negative throughout the right neighborhood Ω_T^ϵ of $x = 0$, that is $u_x(x, t) < 0$ for all $(x, t) \in \Omega_T^\epsilon$. Since $u_x(0, 0) = 0$, this neighborhood is the above defined triangle Ω_T^ϵ with the bottom vertex at the origin $(x, t) = (0, 0)$. □

6.5.2 Compactness and Lipschitz Continuity of the Input-Output Operator. Regularization

Let us define the *set of admissible coefficients*

$$\mathcal{K} := \{k \in H^1(0, l) : 0 < c_0 \leq k(x) \leq c_1 < \infty\}. \tag{6.5.8}$$

For a given coefficient $k \in \mathcal{K}$ we denote by $u = u(x, t; k)$ the solution of the direct problem (6.5.1). Then introducing the input-output operator or *coefficient-to-data mapping*

$$\begin{cases} \Phi[k] := u(x, t; k)|_{x=0^+}, \\ \Phi[\cdot] : \mathcal{K} \subset H^1(0, l) \mapsto L^2(0, T), \end{cases} \tag{6.5.9}$$

we can reformulate the *inverse coefficient problem* (6.5.1) and (6.5.2) in the following nonlinear operator equation form:

$$\Phi[k](t) = f(t), \quad t \in (0, T], \tag{6.5.10}$$

where $f(t)$ is the noise free measured output and $u(0, t; k)$ is the output, corresponding to the coefficient $k \in \mathcal{K}$. Therefore the inverse coefficient problem with the given measured output datum can be reduced to the solution of the nonlinear equation (6.5.10) or to inverting the nonlinear input-output operator defined by (6.5.9).

Lemma 6.5.1 *Let conditions* (6.5.3) *hold. Then the input-output operator* $\Phi[\cdot] : \mathcal{K} \subset H^1(0, l) \mapsto L^2(0, T)$ *defined by* (6.5.9) *is a compact operator.*

Proof Let $\{k_m\} \subset \mathcal{K}$, $m = \overline{1, \infty}$, be a bounded in $H^1(0, l)$-norm sequence of coefficients. Denote by $\{u_m(x, t)\}$, $u_m := u(x, t; k_m)$, the sequence of corresponding regular weak solutions of the direct problem (6.5.1). Then $\{u(0, t; k_m\}$ is the sequence of outputs. We need to prove that this sequence is a relatively compact subset of $L^2(0, T)$ or, equivalently, the sequence $\{u_m(0, t)\}$ is bounded in the norm of the Sobolev space $H^1(0, T)$.

To estimate the norm $\|u_{m,t}(0,\cdot)\|_{L^2(0,T)}$ we use inequality (B.2.2) in Appendix B.2:

$$\int_0^T u_{m,t}^2(0,t)dt$$
$$\leq \frac{2}{l}\int_0^T \left[\operatorname*{ess\,sup}_{[0,T]} \int_0^l u_{m,t}^2(x,t)dx\right]dt + 2l\int_0^T\int_0^l u_{m,xt}^2(x,t)dxdt. \quad (6.5.11)$$

Now we use a priori estimates (B.2.4) and (B.2.5) for the regular weak solution, given in Appendix B.2 to estimate the right hand side integrals. We have:

$$\operatorname*{ess\,sup}_{[0,T]} \int_0^l u_{m,t}^2(x,t)dx \leq C_1^2 \|g'\|_{L^2(0,T)}^2,$$
$$\iint_{\Omega_T} u_{m,xt}^2 dx dt \leq C_2^2 \|g'\|_{L^2(0,T)}^2$$

where the constant $c_0 > 0$ is the lower bound on the coefficient $k(x)$ and

$$\begin{cases} C_1 = (l/c_0)^{1/2}\exp(2Tc_0/l^2), \\ C_2 = \left(2TC_1^2/l^2 + 1/c_0^2\right)^{1/2} \end{cases} \quad (6.5.12)$$

are the same constants defined by (B.1.4) in Appendix B.2. Substituting these estimates into (6.5.11) we conclude that

$$\int_0^T u_{m,t}^2(0,t)dt \leq C_3^2 \|g'\|_{L^2(0,T)}^2, \quad (6.5.13)$$

where the constant $C_3 > 0$ is defined as follows:

$$C_3^2 = (2T/l)C_1^2 + 2lC_2^2. \quad (6.5.14)$$

Using a priori estimate (B.1.6) given in Appendix B.1, in the same way we can prove that

$$\int_0^T u_m^2(0,t)dt \leq C_3^2 \|g\|_{L^2(0,T)}^2, \quad (6.5.15)$$

where $C_3 > 0$ is the constant defined by (6.5.14).

It follows from estimates (6.5.13) and (6.5.15) that the sequence of outputs $\{u_m(0,t)\}$ is uniformly bounded in the norm of $H^1(0,T)$. By the compact imbedding $H^1(0,T) \hookrightarrow L^2(0,T)$. This implies that this sequence is a precompact in $L^2(0,T)$. Therefore, the input-output operator Φ transforms each bounded in $H^1(0,l)$ sequence of coefficients $\{k_m(x)\}$ to the precompact in $L^2(0,l)$ sequence of outputs $\{u_m(0,t)\}$. This means that Φ is a compact operator. This completes the proof of the lemma. □

6.5 Identification of a Leading Coefficient in Heat Equation: Dirichlet ...

As a consequence of Lemma 6.5.1 we conclude that *the inverse coefficient problem* (6.5.1) *and* (6.5.2) *is ill-posed*.

Lemma 6.5.2 *Let conditions* (6.5.3) *hold. Denote by* $u(x, t; k_1)$ *and* $u(x, t; k_2)$ *the regular weak solutions of the direct problem* (6.5.1) *corresponding to the admissible coefficients* $k_1, k_2 \in \mathcal{K}$, *respectively. Then the following estimate holds:*

$$\|u(0,t;k_1) - u(0,t;k_2)\|_{L^2(0,T)}^2$$
$$\leq M_c C_2^2 \|g\|_{L^2(0,l)}^2 \|k_1 - k_2\|_{C[0,l]}^2. \quad (6.5.16)$$

where $M_c = (2T/l + l/2)/c_0 > 0$ *and* $C_2 > 0$ *is the constant defined in* (6.5.12).

Proof For simplicity, we denote by $v(x, t) := u(x, t; k_1) - u(x, t; k_2)$. Then the function $v(x, t)$ solves the following initial boundary value problem:

$$\begin{cases} v_t = (k_1(x)v_x)_x + (\delta k(x)u_{2x})_x, & (x, t) \in \Omega_T, \\ v(x, 0) = 0, & 0 < x < l, \\ -k_1(0)v_x(0, t) = \delta k(0)u_{2x}(0, t), & v_x(l, t) = 0, \quad 0 < t < T, \end{cases} \quad (6.5.17)$$

where $u_i(x, t) := u(x, t; k_i)$, $i = 1, 2$ and $\delta k(x) = k_1(x) - k_2(x)$.

To estimate the norm $\|v(0, \cdot)\|_{L^2(0,T)} := \|\delta u(0, \cdot)\|_{L^2(0,T)}$ we use first the identity

$$v^2(0,t) = \left(v(x,t) - \int_0^x v_\xi(\xi,t)d\xi\right)^2, \quad v \in L^2(0,T; H^1(0,l)).$$

We apply to the right hand side the inequality $(a - b)^2 \leq 2(a^2 + b^2)$ and then the Hölder inequality. Integrating then on $[0, T]$ we obtain the following inequality:

$$\int_0^T v^2(0,t)dt \leq 2\int_0^T v^2(x,t)dt + 2x \int_0^T \int_0^l v_x^2(x,t)dxdt.$$

Integrating again on $[0, l]$ and then dividing by $l > 0$ both sides we arrive at the inequality:

$$\int_0^T v^2(0,t)dt \leq \frac{2}{l} \iint_{\Omega_T} v^2(x,t)dxdt + l \iint_{\Omega_T} v_x^2(x,t)dxdt. \quad (6.5.18)$$

This inequality shows that to prove the lemma we need to estimate the right hand side norms $\|v\|_{L^2(\Omega_T)}$ and $\|v_x\|_{L^2(\Omega_T)}$ via the norm $\|\delta k\|_{H^1(0,l)}$. To do this, we use now the standard L^2-energy estimates for the weak solution of the initial boundary value problem (6.5.17).

Multiplying both sides of the parabolic equation (6.5.17) by $v(x, t)$, integrating on $\Omega_t = \{(x, \tau) \in \mathbb{R}^2 : 0 < x < l, \ 0 < \tau \leq t, \ t \in (0, T]\}$ and then using the formulas for integration by parts we obtain the following identity:

$$\frac{1}{2}\int_0^l v^2(x,t)dt + \iint_{\Omega_t} k_1(x)v_x^2(x,\tau)d\tau dx - \int_0^t (k_1(x)v_x(x,\tau)v(x,\tau))_{x=0}^{x=l} d\tau$$
$$= -\iint_{\Omega_t} \delta k(x)u_{2x}(x,\tau)v_x(x,\tau)d\tau dx + \int_0^t (\delta k(x)u_{2x}(x,\tau)v(x,\tau))_{x=0}^{x=l} d\tau,$$

for all $t \in [0, T]$. The terms under the last left and right hand side integrals are zero at $x = l$ due to the homogeneous boundary conditions in (6.5.17). Taking into account the Neumann boundary condition $-k_1(0)\delta u_x(0, t) = \delta k(0)u_x(0, t; k_2)$ in the last left hand side integral we deduce that this term and the term at $x = 0$ under the last right hand side integral are mutually exclusive. Then the above integral identity becomes

$$\int_0^l v^2(x,t)dt + 2\iint_{\Omega_t} k_1(x)v_x^2(x,\tau)d\tau dx$$
$$= -2\iint_{\Omega_t} \delta k(x)u_{2x}(x,\tau)v_x(x,\tau)d\tau dx, \quad (6.5.19)$$

for a.e. $t \in [0, T]$.

The first consequence of the energy identity (6.5.19) is the inequality

$$c_0 \iint_{\Omega_T} v_x^2(x,t)dtdx \leq \|\delta k\|_{C[0,l]} \left|\iint_{\Omega_T} u_{2x}(x,\tau)v_x(x,t)dtdx\right|,$$

since $k_1(x) \geq 2c_0$. Applying the Hölder inequality to the right hand side integral and then dividing both sides by the norm $\|v_x\|_{L^2(0,T;L^2(0,l))}$, we deduce from this identity the estimate for the L^2-norm of the gradient of the weak solution of problem (6.5.17):

$$\|v_x\|_{L^2(0,T;L^2(0,l))} \leq \frac{1}{2c_0} \|\delta k\|_{C[0,l]} \|u_{2x}\|_{L^2(0,T;L^2(0,l))}, \ c_0 > 0. \quad (6.5.20)$$

Now we use again the energy identity (6.5.19) to estimate the first right hand side integral in (6.5.18). We have:

$$\int_0^l v^2(x,t)dx \leq 2\|\delta k\|_{C[0,l]} \|u_{2x}\|_{L^2(0,T;L^2(0,l))} \|v_x\|_{L^2(0,T;L^2(0,l))}.$$

6.5 Identification of a Leading Coefficient in Heat Equation: Dirichlet ...

Integrating both sides on $[0, T]$ and using estimate (6.5.20) for the right hand side norm $\|v_x\|_{L^2(0,T;L^2(0,l))}$ we conclude:

$$\iint_{\Omega_T} v^2(x,t)dxdt \leq \frac{T}{c_0} \|\delta k\|^2_{C[0,l]} \|u_{2x}\|^2_{L^2(0,T;L^2(0,l))}.$$

Substituting this with estimate (6.5.20) in (6.5.18) we obtain:

$$\int_0^T v^2(0,t)dx \leq \frac{1}{c_0}\left(\frac{2T}{l} + \frac{l}{2}\right) \|\delta k\|^2_{C[0,l]} \|u_{2x}\|^2_{L^2(0,T;L^2(0,l))}. \quad (6.5.21)$$

For the gradient norm $\|u_{2x}\|_{L^2(0,T;L^2(0,l))}$ of the solution $u(x,t;k_2)$ of the direct problem (6.5.1) with $k(x) = k_2(x)$ we use the estimate

$$\iint_{\Omega_T} u_{2x}^2(x,\tau)dxdt \leq C_2^2 \int_0^T g^2(t)dt, \; C_2 > 0$$

given by (B.1.12) in Appendix B.2. Taking into account this in (6.5.21) we finally arrive at the required estimate (6.5.16). This completes the proof of the lemma. $\square$

The main consequence of this lemma is the Lipschitz continuity of the nonlinear input-output operator $\Phi[\cdot] : \mathcal{K} \subset H^1(0,l) \mapsto L^2(0,T)$.

Corollary 6.5.2 *Let conditions of Lemma 6.5.2 hold. Then the input-output operator defined by (6.5.9) is Lipschitz continuous,*

$$\|\Phi[k_1] - \Phi[k_2]\|_{L^2(0,T)} \leq L_0 \|k_1 - k_2\|_{C[0,l]}, \; k_1, k_2 \in \mathcal{K}, \quad (6.5.22)$$

with the Lipschitz constant

$$L_0 = ((2T/l + l)/c_0)^{1/2} C_2 \|g\|_{L^2(0,T)} > 0. \quad (6.5.23)$$

Remark 6.5.2 By the condition $k \in H^1(0,l)$ the function $k(x)$ can be identified with a continuous function and the estimate $\|k\|_{C[0,l]} \leq C_l \|k\|_{H^1(0,l)}$ holds. This implies that the Lipschitz continuity of the input-output mapping remains valid also in the natural norm of $k(x)$:

$$\|\Phi[k_1] - \Phi[k_2]\|_{L^2(0,T)} \leq L_0 C_l \|k_1 - k_2\|_{H^1(0,l)}, \; C_l > 0. \quad (6.5.24)$$

Due to measurement error in the output data the exact equality in the operator equation

$$\Phi[k](t) = f^\delta(t), \; t \in (0, T], \quad (6.5.25)$$

is not possible in practice, where $f^\delta \in L^2(0, T)$ is the noisy data: $\|f - f^\delta\|_{L^2(0,T)} \leq \delta$, $\delta > 0$. Hence, one needs to introduce the regularized form

$$J_\alpha(k) = \frac{1}{2}\left\|\Phi[k] - f^\delta\right\|^2_{L^2(0,T)} + \frac{\alpha}{2}\|k'\|^2_{L^2(0,T)}, \quad k \in \mathcal{K} \quad (6.5.26)$$

of the Tikhonov functional

$$J(k) = \frac{1}{2}\left\|\Phi[k] - f^\delta\right\|^2_{L^2(0,T)}, \quad k \in \mathcal{K}$$

and consider inverse coefficient problem (6.5.1) and (6.5.2) as a minimum problem for the functional (6.5.26) on the set of admissible coefficients $\mathcal{K}$. Note that the regularization (6.5.26) including the term $\|k'\|_{L^2(0,T)}$ is referred as the *Tikhonov regularization with a Sobolev norm* or as higher-order Tikhonov regularization (see [2]).

Having compactness and continuity of the input-output operator, we can prove an existence of a minimizer $k^\delta_\alpha \in \mathcal{K}$.

Theorem 6.5.2 *Let conditions (6.5.3) hold. Then for any $\alpha > 0$, the functional (6.5.26) attains a minimizer $k^\delta_\alpha \in \mathcal{K}$.*

Proof Since the input-output operator $\Phi[\cdot] : \mathcal{K} \subset H^1(0,l) \mapsto L^2(0,T)$ is compact and continuous, it is weakly continuous. This implies that the functional $J_\alpha(k)$ is weakly lower semi-continuous, also coercive and bounded from below. This guarantees the existence of a minimizer. □

Let us assume now that the input-output operator is injective. In this case we may apply the regularization theory for nonlinear inverse problems given in ([23], Chap. 11) to guarantee the convergence of subsequences to a minimum norm solution. Remark that assuming the injectivity of the input-output operator we assume that solution of the inverse problem (6.5.1) and (6.5.2) is unique.

Theorem 6.5.3 *Let conditions (6.5.3) hold. Assume that the input-output operator $\Phi[\cdot] : \mathcal{K} \subset H^1(0,l) \mapsto L^2(0,T)$ is injective. Denote by $\{f^{\delta_m}\} \subset L^2(0,T)$ a sequence of noisy data satisfying the conditions $\|f - f^{\delta_m}\|_{L^2(0,T)} \leq \delta_m$ and $\delta_m \to 0$, as $m \to \infty$. If*

$$\alpha_m := \alpha(\delta_m) \to 0 \quad \text{and} \quad \frac{\delta_m^2}{\alpha(\delta_m)} \to 0, \quad \text{as } \delta_m \to 0, \quad (6.5.27)$$

then the regularized solutions $k^{\delta_m}_{\alpha_m} \in \mathcal{K}$ converge to the best approximate solution $u^\dagger := A^\dagger f$ of equation (6.5.10), as $m \to \infty$.

For the general case this theorem with proof is given in ([23] Chap. 11, Theorem 11.3).

Remark that this theorem is an analogue of Theorem 2.5.2, Chap. 2 for the case of the nonlinear input-output operator.

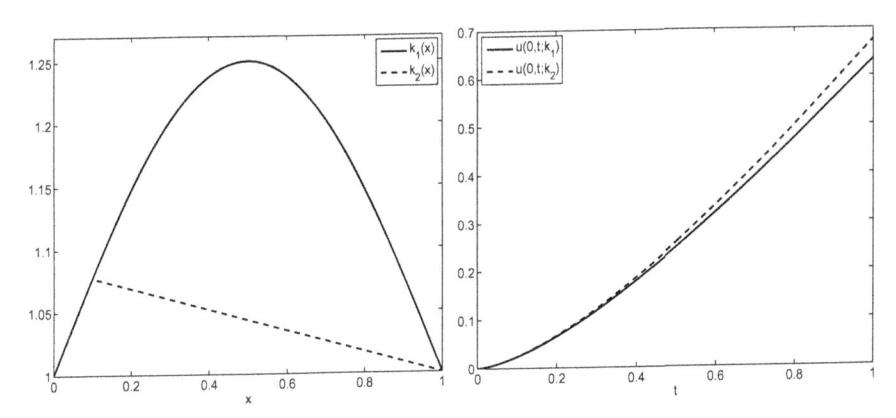

6.5 Identification of a Leading Coefficient in Heat Equation: Dirichlet ...

Fig. 6.1 Coefficients $k_i(x)$ (*left figure*) and the corresponding outputs $u(0, t; k_i)$, $i = 1, 2$ (*right figure*)

As mentioned, Lemma 6.5.1 implies that the inverse coefficient problem (6.5.1) and (6.5.2) is ill-posed. Furthermore, the following example providing further insights into the severely ill-posedness of this inverse problem shows that very small changes in the measured output $u(0, t; k)$ can lead to unacceptably large perturbations in the coefficient $k(x)$.

The outputs $u(0, t; k_i)$, $i = 1, 2$, plotted in the right Fig. 6.1, are obtained from the finite-element solution of the direct problem (6.5.1) for the following coefficients (Fig. 6.1):

$$k_1(x) = 1 + 0.25 \sin(\pi x),$$

$$k_2(x) = \begin{cases} 1 + (k_1(\xi) - 1) \frac{x}{\xi}, & x \in [0, \xi], \\ 1 + (k_1(\xi) - 1) \frac{1-x}{1-\xi}, & x \in (\xi, 1], \ \xi = 0.1, \end{cases}$$

with $g(t) = t, t \in [0, T]$. The figures illustrate high sensitivity of the inverse problem to changes in output datum. That is, almost indistinguishable outputs may correspond to quite different coefficients. Thus solution of the inverse problem is highly sensitive to noise and this is a reason that inverse coefficient problems are extremely difficult to solve numerically.

6.5.3 Integral Relationship and Gradient Formula

We derive now an *integral relationship* relating the change $\delta k(x) := k_1(x) - k_2(x)$ in the coefficient to the change of the output $\delta u(x, t) = u(0, t; k_1) - u(0, t; k_2)$ corresponding to the coefficients $k_1, k_2 \in \mathcal{K}$.

Lemma 6.5.3 *Let conditions* (6.5.3) *hold. Denote by* $u_m(x, t) := u(x, t; k_m)$ *and* $u_m(0, t) = u(0, t; k_m)$ *the solutions of the direct problem* (6.5.1) *and outputs, corresponding to the given admissible coefficients* $k_m \in \mathcal{K}$, $m = 1, 2$. *Then the following integral relationship holds:*

$$\int_0^T [u_1(0, t) - u_2(0, t)] q(t) dt = -\iint_{\Omega_T} \delta k(x) u_{2x}(x, t) \varphi_x(x, t) dx dt, \quad (6.5.28)$$

where $\delta k(x) = k_1(x) - k_2(x)$ *and the function* $\varphi(x, t) = \varphi(x, t; q)$ *solves the adjoint problem*

$$\begin{cases} \varphi_t + (k_1(x)\varphi_x)_x = 0, & (x, t) \in (0, l) \times [0, T), \\ \varphi(x, T) = 0, & x \in (0, l), \\ -k_1(0)\varphi_x(0, t) = q(t), & \varphi_x(l, t) = 0, \quad t \in (0, T), \end{cases} \quad (6.5.29)$$

with an arbitrary datum $q \in H^1(0, T)$ *satisfying the consistency condition* $q(T) = 0$.

Proof The function $\delta u(x, t) := u_1(x, t) - u_2(x, t)$ solves the following initial boundary value problem

$$\begin{cases} \delta u_t - (k_1(x)\delta u_x)_x = (\delta k(x) u_{2x}(x, t))_x, & (x, t) \in \Omega_T, \\ \delta u(x, 0) = 0; & x \in (0, l), \\ -k_1(0)\delta u_x(0, t) = \delta k(0) u_{2x}(0, t), & \delta u_x(l, t) = 0, \quad t \in (0, T), \end{cases} \quad (6.5.30)$$

where $u_{2x}(x, t) := u_x(x, t; k_2)$. Multiply both sides of Eq. (6.5.30) by an arbitrary function $\varphi \in L^\infty(0, T; H^2(0, l))$, with $\varphi_t \in L^\infty(0, T; L^2(0, l)) \cap L^2(0, T; H^1(0, l))$, integrate on Ω_T and use integration by parts formula multiple times. Then we get:

$$\int_0^l (\delta u(x, t) \varphi(x, t))\big|_{t=0}^{t=T} dx$$

$$- \int_0^T (k_1(x)\delta u_x(x, t)\varphi(x, t) - k_1(x)\delta u(x, t)\varphi_x(x, t))\big|_{x=0}^{x=l} dt$$

$$- \iint_{\Omega_T} \delta u(x, t) [\varphi_t(x, t) + (k_1(x)\varphi_x(x, t))_x] dx dt$$

$$= -\iint_{\Omega_T} \delta k(x) u_{2x}(x, t) \varphi_x(x, t) dx dt$$

$$+ \left(\delta k(x) \int_0^T u_{2x}(x, t)\varphi(x, t) dt\right)\bigg|_{x=0}^{x=l}.$$

The first left hand side integral is zero due to the initial and final conditions in (6.5.30) and (6.5.29). At $x = l$, the terms under the second left hand side and

6.5 Identification of a Leading Coefficient in Heat Equation: Dirichlet ...

the last right hand side integrals drop out due to the homogeneous Neumann conditions. The third left hand side integral is also zero due to the adjoint equation $\varphi_t(x,t) + (k_1(x)\varphi_x(x,t))_x = 0$. Taking into account the nonhomogeneous Neumann conditions in (6.5.29) and (6.5.30) we conclude that

$$-\int_0^T \delta k(0) u_{2x}(0,t)\varphi(0,t)dt + \int_0^T \delta u(0,t) q(t) dt$$
$$= -\iint_{\Omega_T} \delta k(x) u_{2x}(x,t)\varphi_x(x,t) dx dt - \int_0^T \delta k(0) u_{2x}(0,t)\varphi(0,t) dt.$$

Since the first left hand side and the last right side integrals are equal, we arrive at the required integral relationship. □

Now we use the integral relationship (6.5.28) to derive the Fréchet gradient $J'(k)$ of the Tikhonov functional

$$J(k) = \frac{1}{2} \|\Phi[k] - f\|_{L^2(0,l)}^2, \quad k \in \mathcal{K}, \tag{6.5.31}$$

where $f \in L^2(0,l)$ is the noise free measured output. Let $u(x,t;k)$ and $u(x,t;k+\delta k)$ be the solutions of the direct problem (6.5.1) corresponding to the coefficients $k, k + \delta k \in \mathcal{K}$. Calculating the increment $\delta J(k) := J(k+\delta k) - J(k)$ of functional (6.5.31) we get:

$$\delta J(k) = \int_0^T [u(0,t;k) - f(t)]\delta u(0,t;k) dt + \frac{1}{2}\|\delta u(0,\cdot;k)\|_{L^2(0,T)}^2, \tag{6.5.32}$$

for all $k, k + \delta k \in \mathcal{K}$. Now we use Lemma 6.5.3, taking $k_1(x) = k(x) + \delta k(x)$ and $k_2(x) = k(x)$. Then $\delta u(0,t;k) := u(0,t;k+\delta k) - u(0,t;k)$. Choosing the arbitrary input $q \in H^1(0,T)$ in the adjoint problem (6.5.28) as $q(t) := -[u(0,t;k) - f(t)]$ we deduce from the integral relationship (6.5.28) that

$$\int_0^T [u(0,t;k) - f(t)]\delta u(0,t;k) dt$$
$$= \iint_{\Omega_T} \delta k(x) u_x(x,t;k) \varphi_x(x,t;k+\delta k) dx dt, \tag{6.5.33}$$

where the function $\varphi(x,t;k_1)$, $k_1(x) := k(x) + \delta k(x)$ is the solution of the adjoint problem

$$\begin{cases} \varphi_t + (k_1(x)\varphi_x)_x = 0, \ (x,t) \in (0,l) \times [0,T), \\ \varphi(x,T) = 0, \ x \in (0,l), \\ -k_1(0)\varphi_x(0,t) = -[u(0,t;k) - f(t)], \ \varphi_x(l,t) = 0, \ t \in (0,T). \end{cases} \tag{6.5.34}$$

Using the integral relationship (6.5.33) in the right hand side of the increment formula (6.5.32) we deduce that

$$\delta J(k) = \int_0^l \left(\int_0^T u_x(x,t;k) \varphi_x(x,t;k+\delta k) dt \right) \delta k(x) dx$$

$$+ \frac{1}{2} \int_0^T [\delta u(0,t;k)]^2 dt, \qquad (6.5.35)$$

Estimate (6.5.16) implies that

$$\|\delta u(0,t;k)\|_{L^2(0,T)}^2 \leq M_c C_2^2 \|g\|_{L^2(0,l)}^2 \|\delta k\|_{C[0,l]}^2, \quad M_c, C_2 > 0, \qquad (6.5.36)$$

i.e. the second right hand side integral in (6.5.35) is of the order $\mathcal{O}\left(\|\delta k\|_{C[0,l]}^2\right)$.

Corollary 6.5.3 *For the Fréchet gradient $J'(k)$ of the Tikhonov functional (6.5.31) corresponding to the inverse coefficient problem (6.5.1) and (6.5.2) the following gradient formula holds:*

$$J'(k)(x) = \int_0^T u_x(x,t;k) \varphi_x(x,t;k) dt, \quad k \in \mathcal{K}, \qquad (6.5.37)$$

where $u(x,t;k)$ and $\varphi(x,t;k)$ are the solutions of the direct problem (6.5.1) and the adjoint problem (6.5.34) corresponding to the coefficient $k \in \mathcal{K}$.

By definition (6.5.26), the gradient formula for the regularized form of the Tikhonov functional is as follows:

$$J'_\alpha(k)(x) = (u_x(x,\cdot;k), \varphi_x(x,\cdot;k))_{L^2(0,T)} + \alpha k'(x), \quad k \in \mathcal{K}, \qquad (6.5.38)$$

for a.e. $x \in [0,l]$.

6.5.4 Reconstruction of an Unknown Coefficient

Having now the explicit gradient formulae (6.5.37) and (6.5.38) we may use the Conjugate Gradient Algorithm (CG-algorithm) to the inverse coefficient problem (6.5.1) and (6.5.2). However, the version of this algorithm described in Sect. 2.4 cannot be used in this case, since the formula

$$\beta_n := \frac{\left(J'(p^{(n)}), p^{(n)}\right)}{\|\Phi p^{(n)}\|^2}$$

6.5 Identification of a Leading Coefficient in Heat Equation: Dirichlet ...

for the descent direction parameter β_n will not work. The reason is that the input-output mapping corresponding to the inverse problem (6.5.1) and (6.5.2) in defined only for positive functions $0 < c_0 \leq k(x) \leq c_1$ and the function $p^{(n)}(x)$ in $\|\Phi p^{(n)}\|^2$ may not be positive. As an equivalent alternative, the descent direction parameter $\beta_n > 0$ will be defined from the minimum problem

$$F_n(\beta_n) := \inf_{\beta > 0} F_n(\beta), \quad F_n(\beta) := J(k^{(n)} - \beta J'(k^{(n)})), \tag{6.5.39}$$

for each $n = 0, 1, 2, \ldots$, as in Lemma 3.4.4 of Sect. 3.4.

Thus, the following version of the CG-algorithm is used below in numerical solving of the inverse coefficient problem (6.5.1) and (6.5.2).

Step 1. For $n = 0$ choose the initial iteration $k^{(0)}(x)$.
Step 2. Compute the initial descent direction $p^{(0)}(x) := J'(k^{(0)})(x)$.
Step 3. Find the *descent direction parameter* from (6.5.39).
Step 4. Find next iteration $k^{(n+1)}(x) = k^{(n)}(x) - \beta_n p^{(n)}(x)$ and compute the convergence error

$$e(n; k^{(n)}; \delta) := \|f^\delta - u(0, t; k^{(n)})\|_{L^2(0,T)}$$

Step 5. If the *stopping condition*

$$e(n; k^{(n)}; \delta) \leq \tau_M \delta < e(n; k^{(n-1)}; \delta), \quad \tau_M > 1, \; \delta > 0 \tag{6.5.40}$$

holds, then go to Step 7.
Step 6. Set $n := n + 1$ and compute

$$\begin{cases} p^{(n)}(x) := J'(k^{(n)})(x) + \gamma_n p^{(n-1)}(x), \\ \gamma_n = \frac{\|J'(k^{(n)})\|^2}{\|J'(k^{(n-1)})\|^2} \end{cases}$$

and go to Step 3.
Step 7. Stop the iteration process.

In the case when the algorithm is applied to the regularized form of the Tikhonov functional (6.5.26), which gradient is defined by formula (6.5.38), one needs to replace in the above algorithm $J(k^{(n)})$ and $J'(k^{(n)})$ with $J_\alpha(k^{(n)})$ and $J'_\alpha(k^{(n)})$, respectively.

For discretization and numerical solution of the direct problem (6.5.1) as well as the adjoint problem (6.5.34) the finite element algorithm with piecewise quadratic Lagrange basis functions, introduced in Sect. 3.4.1, is used. These schemes with composite numerical integration formula are also used for approximating the spatial derivatives $u_x(x; t; k)$ and $\varphi_x(x; t; k)$ in the gradient formulas (6.5.37) and (6.5.38).

In the examples below the finer mesh with the mesh parameters $N_x = 201$ and $N_t = 801$ are used to generate the noise free synthetic output data. For this mesh the computational noise level is estimated as 10^{-4}. With this accuracy the synthetic output data f is assumed noise-free. Coarser mesh with the parameters $N_x = 51$ and $N_t = 101$ is used in the numerical solution of the inverse problem, to avoid of inverse crime.

The noisy output data f^δ, with $\|f - f^\delta\|_{L^2_h(0,T)} = \delta$, is generated by employing the "randn" function in MATLAB, that is,

$$u^\delta_{T,h}(x) = u_{T,h}(x) + \gamma \|u_{T,h}\|_{L^2_h(0,l)} \text{randn}(N),$$

where $\gamma > 0$ is the MATLAB noise level. Remark that $\gamma > 0$ and $\delta > 0$ are of the same order. The parameter $\tau_M > 1$ in the stopping condition (6.5.40) is taken below as $\tau_M = 1.05 \div 1.07$, where n is the iteration number. We will employ also the accuracy error defined as

$$E(n; k^{(n)}; \delta) := \|k - k^{(n)}\|_{L^2(0,T)},$$

for performance analysis of the CG-algorithm.

In the examples below, we represent attempts to capture performance characteristics of the CG-algorithm not only in the case when the initial datum and the Neumann boundary datum at $x = l$ are homogeneous, but also in the case when the direct problem is given in the general form:

$$\begin{cases} u_t(x,t) = (k(x)u_x(x,t))_x, & (x,t) \in \Omega_T, \\ u(x,0) = h(x), & 0 < x < l, \\ -k(0)u_x(0,t) = g_0(t), & -k(0)u_x(l,t) = g_1(t), \quad 0 < t < T. \end{cases} \quad (6.5.41)$$

In this case the input data must satisfy the following consistency conditions hold:

$$\begin{cases} g_0(0) = -k(0)h'(0), \\ g_1(0) = -k(l)h'(l), \end{cases}$$

due to the regularity of the solution. Since the fluxes $g_0(t)$ and $g_1(t)$ at the endpoints initially are assumed to be positive and since $k(x) \geq c_1 > 0$, the above consistency conditions imply that

$$\begin{cases} g_0(t) > 0, \ g_1(t) > 0, & \text{for all } t \in [0, T], \\ h'(0) < 0, \ h'(l) < 0. \end{cases} \quad (6.5.42)$$

On the other hand, these conditions allow to find the values $k(0)$ and $k(1)$ of the unknown coefficients at the endpoints:

$$\begin{cases} k(0) = -g_0(0)/h'(0), & h'(0) < 0 \\ k(l) = -g_1(0)/h'(l), & h'(l) < 0. \end{cases} \quad (6.5.43)$$

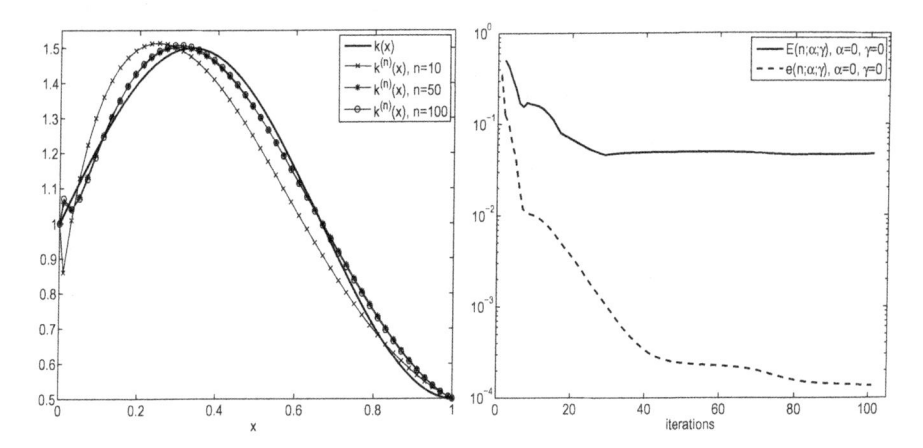

6.5 Identification of a Leading Coefficient in Heat Equation: Dirichlet ...

Fig. 6.2 The reconstructed coefficients for different values of the iteration number n (*left figure*) and behaviour of the convergence and accuracy errors depending on n (*right figure*): noise free data, for Example 6.5.1

This, in turn, permits to use the function $k^{(0)}(x) = (k(0)(l-x) + k(l)x)/l$, i.e. the linear approximation of $k(x)$, as an initial iteration in the CG-algorithm, also. Otherwise, i.e. in the case when $g_1(t) \equiv 0$, $h(x) \equiv 0$ and $g_0(t)$ satisfies conditions (6.5.3), i.e. $g_0(t) > 0$, for all $t \in (0, T)$ and $g_0(0) = 0$, any constant function $k^{(0)}(x) = k_0$, $c_0 \le k_0 \le c_1$ can be taken as an initial iteration, where $c_0 > 0$ and $c_1 > 0$ are lower and upper bounds for the coefficient $k(x)$. Note that an initial iteration has indistinguishable effect on the reconstruction quality, as computational experiments show. Only the attainability of the stopping criterion (6.5.40) becomes faster. Finally, remark that reasonable values for the parameter of regularization $\alpha > 0$ in the examples below were determined by numerical experiments on carefully chosen examples, of course, taking into account the convergence conditions (6.5.27).

Example 6.5.1 Performance characteristics of the CG-algorithm: noise free output data

The synthetic output datum $f(t)$ in this example is generated from the numerical solution of the direct problem (6.5.1) for the given $k(x) = 1 + 0.5\sin(3\pi x/2)$, $x \in [0, 1]$, with the source $g(t) = 20\exp(-300(t - 0.25)^2)$, $t \in [0, T]$, $T = 0.5$, where $g(0) \approx 1.44 \times 10^{-7}$. The initial iteration here is taken as $k^{(0)}(x) = 1, x \in [0, l]$.

The right Fig. 6.2 shows the behaviour of the convergence error (bottom curve) and the accuracy error (upper curve) depending on the iteration number n. It is clearly seen that the accuracy error $E(n; k^{(n)}; \delta)$ remains almost the same after $30 \div 35$th iterations, although the convergence error $e(n; k^{(n)}; \delta)$ still decreased after 80 iterations. For some values of the iteration number n these errors are reported in Table 6.1. The reconstructed coefficients $k^{(n)}(x)$ from noise free data are plotted in the left Fig. 6.2, for the iteration numbers $n = 10; 50; 100$. It is seen that after 50 iterations the

Table 6.1 Errors depending on the number of iterations n: noise free output data ($\delta = 0$), for Example 6.5.1

n	$e(n; \alpha; \gamma)$	$E(n; \alpha; \gamma)$
25	1.90×10^{-3}	5.26×10^{-2}
50	2.39×10^{-4}	4.94×10^{-2}
75	1.75×10^{-4}	4.70×10^{-2}
100	1.36×10^{-4}	4.71×10^{-2}

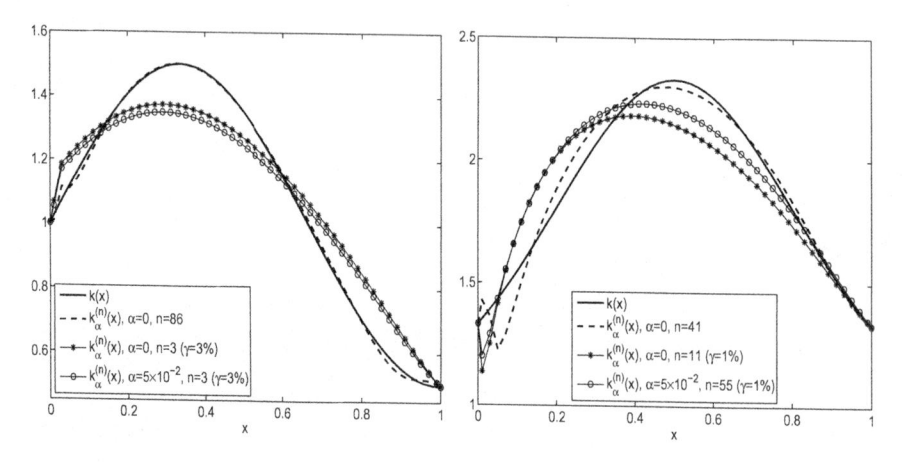

Fig. 6.3 The reconstructed coefficient from noisy data with and without regularization: *left figure*, for Example 6.5.2 and the *right figure*, for Example 6.5.3

reconstructions are almost the same. In all these reconstructions the CG-algorithm is applied without regularization ($\alpha = 0$). □

Example 6.5.2 Reconstruction of an unknown coefficient from noisy data with and without regularization

In this example the reconstruction of the function $k(x) = 1 + 0.5\sin(3\pi x/2)$, $x \in [0, 1]$, is considered in the case when the inputs in the direct problem (6.5.41) are $g_0(t) = 10\exp(-300(t - 0.04)^2)$, $g_0(t) = 20\exp(-300(t - 0.02)^2)$, $t \in [0, T]$, $T = 0.5$, and $h(x) = Ax^2 + Bx$, $x \in [0, 1]$. For the Neumann data $g_0(0) \approx 6.1878$ and $g_1(0) \approx 17.7384$, and the parameters A and B in initial datum $h(x)$ is chosen from the consistency conditions: $h'(0) = -g_0(0)/k(0)$, $h'(1) = -g_1(0)/k(1)$. We have: $h(x) = (1/2)[h'(1) - h'(0)]x^2 + h'(0)x$. The CG-algorithm is applied to the inverse problem defined by (6.5.41) and (6.5.2) with the initial iteration $k^{(0)}(x) = (k(0)(l - x) + k(l)x)/l$. The reconstructed coefficients from noise free and noisy data are plotted in the left Fig. 6.3. In the noise-free case the reconstructed coefficient almost coincides with the exact one. Influence of the parameter of regularization $\alpha > 0$ on accuracy of reconstruction is seen from the plots in the left Fig. 6.3.

6.5 Identification of a Leading Coefficient in Heat Equation: Dirichlet ...   191

Table 6.2 Errors obtained for noise free and noisy output data, for Example 6.5.2

δ, α	iteration	$e(n; \alpha; \gamma)$	$E(n; \alpha; \gamma)$
$\delta = 0$	86	1.40×10^{-3}	4.28×10^{-2}
$\delta = 3\%, \alpha = 0$	3	9.75×10^{-2}	1.46×10^{-1}
$\delta = 3\%,$ $\alpha = 5.0 \times 10^{-2}$	3	1.03×10^{-1}	1.59×10^{-1}

Table 6.3 Errors in the reconstruction of Gaussian, for Example 6.5.3

δ, α	n	$e(n; \alpha; \gamma)$	$E(n; \alpha; \gamma)$
$\delta = 0$	41	6.70×10^{-3}	2.05×10^{-1}
$\delta = 1\%, \alpha = 0$	9	1.89×10^{-2}	1.56×10^{-1}
$\delta = 1\%,$ $\alpha = 5.0 \times 10^{-2}$	55	1.95×10^{-2}	2.98×10^{-1}

The values of the convergence and accuracy errors are given in Table 6.2. In the case of noisy data the iteration numbers n in this table is determined by the stopping condition (6.5.40) with $\tau_M = 1.05$. □

Example 6.5.3 Reconstruction of Gaussian from noisy data with and without regularization

In this final example, we use input data from the previous example for the reconstruction of the $k(x) = 1 + exp(-(x-0.5)^2/(2\sigma^2))/(\sigma\sqrt{2\pi})$, with mean $\sigma = 0.3$, in the inverse coefficient problem defined by (6.5.41) and (6.5.2). The constant $k^{(0)}(x) = 2$ is used as an initial iteration. The reconstructed coefficients from noise free and noisy data are plotted in the right Fig. 6.3. Table 6.3 reports the values of the convergence and accuracy errors with the iteration numbers n. □

We remark finally that, in all examples the regularization effect on the accuracy of the obtained reconstructions was negligible. That is, an accuracy of the numerical solution obtained by CG-algorithm has slightly improved due to the regularization, but at the expense of the number of iterations, as Table 6.3 shows.

6.6 Identification of a Leading Coefficient in Heat Equation: Neumann Type Measured Output

In the preceding section we studied the inverse coefficient problem when Dirichlet data is given as a measured output. However, the most significant case in applications arises when the heat flux $f(t) := k(0)u_x(0, t; k)$, i.e. the *Neumann type datum* is given an available *measured output*. This is an important, but more complicated case since the *output* $k(0)u_x(0, t; k)$ contains the derivative which is a source of an additional ill-posedness. While the study of coefficient identification problems for evolution equations with Dirichlet type measured output is comprehensive enough, to the best of our knowledge, only a few results are known for these problems with Neumann type measured output. Applying the methodology given in §5.5 to the inverse coefficient problem for heat equation with Neumann type measured output, we will also emphasize some distinctive features of this problem.

Consider the inverse problem of identifying the leading coefficient $k(x)$ in

$$\begin{cases} u_t(x,t) = (k(x)u_x(x,t))_x + F(x,t), & (x,t) \in \Omega_T, \\ u(x,0) = u_0(x), & 0 < x < l, \\ u(0,t) = 0, \ u_x(l,t) = 0, & 0 < t < T, \end{cases} \quad (6.6.1)$$

from the *Neumann type measured output* $f(t)$ at the left boundary $x = 0$ of a nonhomogeneous rod:

$$f(t) := k(0)u_x(0,t), \quad t \in [0,T]. \quad (6.6.2)$$

At the first stage, to guarantee, for instance, the existence of the regular weak solution of the direct problem (6.6.1) assume that the functions $k(x)$, $F(x,t)$ and $u_0(x)$ satisfy the following conditions:

$$\begin{cases} k \in H^1(0,l), \ 0 < c_0 \le k(x) \le c_1 < \infty; \\ F_t \in L^2(0,T; L^2(0,l)), \ \exists F(\cdot, 0^+) \in L^2(0,l); \\ F(x,t) > 0, \ (x,t) \in \Omega_T; \\ u_0 \in H^2(0,l), \ u_0(0) = 0, \ u_0'(l) = 0. \end{cases} \quad (6.6.3)$$

The condition $F(x,t) > 0$ here means that heating within the rod is considered. $u_0(0) = 0$ and $u_0'(l) = 0$ are the consistency conditions.

Under conditions (6.6.3) the initial boundary value problem (6.6.1) has the unique regular weak solution

$$\begin{cases} u \in L^\infty(0,T; H^2(0,l)), \\ u_t \in L^\infty(0,T; L^2(0,l)) \cap L^2(0,T; H^1(0,l)), \\ u_{tt} \in L^2(0,T; H^{-1}(0,l)), \end{cases} \quad (6.6.4)$$

with improved regularity ([24], Sect. 7.1, Theorem 5).

Initially, we define the *set of admissible coefficients* as in Sect. 6.5.2:

$$\mathcal{K} := \{k \in H^1(0,l) : 0 < c_0 \le k(x) \le c_1 < \infty\}. \quad (6.6.5)$$

We will show below that for the considered inverse problem (6.6.1) and (6.6.2) *this set of admissible coefficients is not enough due to the Neumann type measured output* (6.6.2). Specifically, we will prove that, different from the inverse coefficient problem considered in the preceding section, even regular weak solution is not enough for solvability of inverse coefficient problem with Neumann type measured output.

6.6 Identification of a Leading Coefficient in Heat Equation: Neumann ...

For a given coefficient $k \in \mathcal{K}$ we denote by $u = u(x, t; k)$ the solution of the direct problem (6.6.1) and introduce the *input-output operator*

$$\begin{cases} \Psi[k](t) := (k(x)u_x(x, t; k))_{x=0^+}, \\ \Psi[\cdot] : \mathcal{K} \subset H^1(0, l) \mapsto L^2(0, T). \end{cases} \quad (6.6.6)$$

Then we can reformulate the *inverse coefficient problem* (6.6.1) and (6.6.2) in the following nonlinear operator equation form:

$$\Psi[k](t) = f(t), \quad t \in (0, T]. \quad (6.6.7)$$

Taking into account that the output $f(t)$ contains measurement error, we introduce the regularized form

$$J_\alpha(k) = \frac{1}{2} \|\Psi[k] - f\|^2_{L^2(0,T)} + \frac{\alpha}{2} \|k'\|^2_{L^2(0,T)}, \quad k \in \mathcal{K} \quad (6.6.8)$$

of the Tikhonov functional

$$J(k) = \frac{1}{2} \|\Psi[k] - f\|^2_{L^2(0,T)}, \quad k \in \mathcal{K}, \quad (6.6.9)$$

where $\alpha > 0$ is the parameter of regularization.

Therefore, for a noisy measured output the inverse coefficient problem (6.6.1) and (6.6.2) can only be considered as a minimum problem for the regularized functional (6.6.8) on the set of admissible coefficients $\mathcal{K}$.

6.6.1 Compactness of the Input-Output Operator

We need some auxiliary results related to the regular weak solution of the direct problem. Although some of the estimates below can be derived from those given in Sect. 5.2 and in Appendix B.2, we derive them here for completeness.

Lemma 6.6.1 *Let conditions (6.5.3) hold. Then for the regular weak solution of the parabolic problem (6.6.1) the following estimates hold:*

$$\|u_t\|^2_{L^2(0,l)} \leq \left(\|F_t\|^2_{L^2(0,T;L^2(0,l))} + 2C_1^2 \right) e^t, \quad t \in [0, T], \quad (6.6.10)$$

$$\|u_{xt}\|^2_{L^2(0,T;L^2(0,l))} \leq \frac{1}{2c_0} \left(\|F_t\|^2_{L^2(0,T;L^2(0,l))} + 2C_1^2 \right) e^T, \quad (6.6.11)$$

where

$$C_1^2 = \|F(\cdot, 0^+)\|^2_{L^2(0,l)} + \|(k\, u_0')'\|^2_{L^2(0,l)}. \quad (6.6.12)$$

Proof Differentiate (formally) Eq. (6.6.1) with respect to $t \in (0, T)$, multiply both sides by $u_t(x, t)$, integrate on $\Omega_t := (0, l) \times (0, t)$ and use the integration by parts formula. Taking then into account the initial and boundary conditions we obtain the following integral identity:

$$\int_0^l |u_t|^2 dx + 2 \int_0^t \int_0^l k(x)|u_{x\tau}|^2 dx d\tau$$
$$= 2 \int_0^t \int_0^l F_\tau u_\tau dx d\tau + \int_0^l |u_t(x, 0^+)|^2 dx, \ t \in [0, T]$$

We use the limit equation $u_t(x, 0^+) = (k(x)u_x(x, 0^+))_x + F(x, 0^+)$ to estimate the last right hand side integral. Squaring both sides of this equation, using the identity $(a+b)^2 \leq 2a^2 + 2b^2$ and then integrating over $[0, l]$ we deduce:

$$\int_0^l |u_t(x, 0^+)|^2 dx \leq 2 \int_0^l \left((k(x)u_0')'\right)^2 dx + 2 \int_0^l |F(x, 0^+)|^2 dx.$$

Substituting this in above integral identity we conclude that

$$\int_0^l |u_t|^2 dx + 2 \int_0^t \int_0^l k(x)|u_{x\tau}|^2 dx d\tau \leq 2 \int_0^t \int_0^l F_\tau u_\tau dx d\tau + 2C_1^2, \quad (6.6.13)$$

for all $t \in [0, T]$, where the constant $C_1 > 0$ is defined by (6.6.12).

The first consequence of (6.6.13) is the inequality

$$\int_0^l |u_t|^2 dx \leq 2 \int_0^t \int_0^l F_\tau u_\tau dx d\tau + 2C_1^2,$$

which implies:

$$\int_0^l |u_t|^2 dx \leq \int_0^t \int_0^l |u_t|^2 dx d\tau + \int_0^t \int_0^l |F_\tau|^2 dx d\tau + 2C_1^2, \ t \in [0, T].$$

Applying the Gronwall-Bellman inequality after elementary transformations we arrive at the required first estimate (6.6.10).

As a second consequence of (6.6.13) we get the inequality:

$$c_0 \int_0^T \int_0^l |u_{xt}|^2 dx dt \leq \int_0^T \int_0^l F_t u_t dx dt + C_1^2,$$

which by the inequality $ab \leq (a^2 + b^2)/2$ yields:

$$c_0 \int_0^T \int_0^l |u_{xt}|^2 dx dt \leq \frac{1}{2} \int_0^T \int_0^l |u_t|^2 dx dt + \frac{1}{2} \int_0^T \int_0^l |F_t|^2 dx dt + C_1^2.$$

6.6 Identification of a Leading Coefficient in Heat Equation: Neumann ...

The required second estimate (6.6.11) easily is obtained from this inequality after elementary transformations, taking into account estimate (6.6.10). □

As will be seen below, we need to estimate the norm $\|u_{tt}\|^2_{L^2(0,T;L^2(0,l))}$. This requires higher regularity of the regular weak solution, in particular, $u_{tt} \in L^2(0, T; L^2(0, l))$. If, in addition to conditions (6.6.3), the condition $u_0 \in H^3(0, l) \cap \mathcal{V}(0, l)$ holds, then by Theorem 6 in ([24], Sect. 7.1) the weak solution $u \in L^2(0, T; H^4(0, l)) \cap L^2(0, T; \mathcal{V}(0, l))$, $u_t \in L^\infty(0, T; H^2(0, l))$, $u_{tt} \in L^2(0, T; L^2(0, l))$ of the parabolic problem (6.6.1) with higher regularity exists and unique.

Lemma 6.6.2 *Let conditions (6.6.3) hold. Assume, in addition, that the inputs $k(x)$, $F(x, t)$ and $u_0(x)$ satisfy the following conditions:*

$$\begin{cases} k \in H^2(0, l); \\ \exists\, F_x(\cdot, 0^+) \in L^2(0, l); \\ u_0 \in H^3(0, l) \cap \mathcal{V}(0, l). \end{cases} \quad (6.6.14)$$

Then for the regular weak solution of the parabolic problem (6.6.1) the following estimate holds:

$$\|u_{tt}\|^2_{L^2(0,T;L^2(0,l))} \leq \|F_t\|^2_{L^2(0,T;L^2(0,l))} + 4c_1 C_2^2 \quad (6.6.15)$$

$$\|u_{xt}\|^2_{L^2(0,T;L^2(0,l))} \leq \frac{1}{c_0}\|F_t\|^2_{L^2(0,T;L^2(0,l))} + \frac{4c_1}{c_0} C_2^2. \quad (6.6.16)$$

where $c_0, c_1 > 0$ are the constants defined in (6.6.3) and

$$C_2^2 = \|F_x(\cdot, 0^+)\|^2_{L^2(0,l)} + \|(k\, u_0')''\|^2_{L^2(0,l)}. \quad (6.6.17)$$

Proof Differentiating equation (6.6.1) with respect to $t \in (0, T)$, multiplying both sides by $u_{tt}(x, t)$, integrating over $\Omega_t := (0, l) \times (0, t]$ and using then the integration by parts formula we obtain:

$$2\iint_{\Omega_t} |u_{\tau\tau}|^2 dx dt + \int_0^l k(x)|u_{xt}|^2 dx$$
$$= 2\iint_{\Omega_t} F_\tau u_{\tau\tau} dx + 2\int_0^t (k(x) u_{x\tau} u_{\tau\tau})\big|_{x=0}^{x=l}\, d\tau + \int_0^l k(x)|u_{xt}(x, 0^+)|^2 dx.$$

By the homogeneous boundary conditions (6.6.1) the second right hand side integral is zero. Using the inequality $2ab \leq a^2 + b^2$ in the first right hand side integral after transformation we deduce:

$$\int_0^t \int_0^l |u_{\tau\tau}|^2 dx d\tau + c_0 \int_0^l |u_{xt}|^2 dx$$
$$\leq \int_0^T \int_0^l |F_t|^2 dx d\tau + 2\int_0^l k(x)|u_{xt}(x, 0^+)|^2 dx. \quad (6.6.18)$$

To estimate the second right hand side integral we use conditions (6.6.14). We have:

$$\int_0^l k(x)|u_{xt}(x,0^+)|^2 dx \le c_1 \int_0^l \left[(k(x)u_x(x,0^+))_{xx} + F_x(x,0^+)\right]^2$$
$$\le 2c_1 \int_0^l \left(k(x)u_0'(x)''\right)^2 dx + 2c_1 \int_0^l |F_x(x,0^+)|^2 dx.$$

With (6.6.18), this implies the required first estimate (6.6.15) with the constant $C_2 > 0$ defined by (6.6.17).

The second estimate (6.6.16) can be obtained in the same way. □

Now we are able to prove the compactness lemma.

Lemma 6.6.3 *Let conditions (6.6.3) and (6.6.14) hold. Then the input-output operator* $\Phi[\cdot] : \mathcal{K}_c \subset H^2(0,l) \mapsto L^2(0,T)$ *defined by (6.6.6) is a compact operator on the set of admissible coefficients*

$$\mathcal{K}_c := \{k \in H^2(0,l) : 0 < c_0 \le k(x) \le c_1 < \infty\}. \quad (6.6.19)$$

Proof Let $\{k_m\} \subset \mathcal{K}_c$, $m = \overline{1,\infty}$, be a bounded sequence of coefficients in $\mathcal{K}_c$. Denote by $\{u_m(x,t)\}$, $u_m(x,t) := u(x,t;k_m)$, the sequence of corresponding regular weak solutions of the direct problem (6.6.1). Then $\{k_m(0)u_x(0,t;k_m)\}$ is the sequence of outputs. We need to prove that this sequence is a relatively compact subset of $L^2(0,T)$ or, equivalently, the sequence $\{k_m(0)u_m(0,t)\}$ is bounded in the norm of the Sobolev space $H^1(0,T)$.

By the homogeneous Neumann condition in (6.6.1) we have:

$$\int_0^T \left(k_m(0)u_{m,x}(0,t)\right)^2 dt = \int_0^T \left(\int_0^l (k_m(x)u_{m,x}(x,t))_x dx\right)^2 dt.$$

Using now the equation $(k_m(x)u_{m,x}(x,t))_x = u_{m,t}(x,t) - F(x,t)$ we obtain:

$$\int_0^T \left(k_m(0)u_{m,x}(0,t)\right)^2 dt$$
$$\le 2l \left\{\int_0^T \int_0^l |u_{m,t}(x,t)|^2 dx dt + \int_0^T \int_0^l |F(x,t)|^2 dx dt\right\}.$$

By estimate (6.6.10) we conclude the sequence $\{k_m(0)u_{m,x}(0,t)\}$ is bounded in $L^2(0,T)$. We prove now that the sequence $\{k_m(0)u_{m,xt}(0,t)\}$ is also bounded in the norm of $L^2(0,T)$. Applying the similar technique we deduce:

$$\int_0^T \left(k_m(0) u_{m,xt}(0,t)\right)^2 dt$$
$$\leq 2l \left\{ \int_0^T \int_0^l |u_{m,tt}(x,t)|^2 dx dt + \int_0^T \int_0^l |F_t(x,t)|^2 dx dt \right\}.$$

The boundedness of the right hand side follows from estimate (6.6.15). This completes the proof of the lemma. □

Remark 6.6.1 In Sect. 5.5.2 we have proved that the input-output operator $\Phi[\cdot]$: $\mathcal{K} \subset H^1(0,l) \mapsto L^2(0,T)$ corresponding to the inverse coefficient problem with Dirichlet output is compact if $\mathcal{D}(\Phi) := \mathcal{K} \subset H^1(0,l)$ and $\mathcal{K}$ is defined by (6.6.5) as a subset of $H^1(0,l)$. But Lemma 6.6.3 asserts that *in the case of Neumann output the input-output operator* $\Psi[\cdot] : \mathcal{K} \subset H^1(0,l) \mapsto L^2(0,T)$ *can not be compact if* $\mathcal{D}(\Psi) \subset H^1(0,l)$. Namely, it is compact if only $\mathcal{D}(\Psi) := \mathcal{K}_c \subset H^2(0,l)$, where $\mathcal{K}_c$ is defined by (6.6.19) as a subset of $H^2(0,l)$. Taking into account the compactness of the embedding $H^2(0,l) \hookrightarrow H^1(0,l)$, we deduce that in the case of Neumann output the input-output operator Ψ is compact, if only it is defined on a compact set of the Sobolev space $H^1(0,l)$.

6.6.2 Lipschitz Continuity of the Input-Output Operator and Solvability of the Inverse Problem

To prove the Lipschitz continuity of the input-output operator $\Psi[\cdot] : \mathcal{K}_c \subset H^2(0,l) \mapsto L^2(0,T)$, first we need the following auxiliary result.

Lemma 6.6.4 *Let conditions (6.6.3) and (6.6.14) hold. Denote by $u(x,t;k_1)$ and $u(x,t;k_2)$ the regular weak solutions of the direct problem (6.6.1) corresponding to the admissible coefficients $k_1, k_2 \in \mathcal{K}$, respectively. Then the following estimate holds:*

$$\|u_t(\cdot,\cdot;k_1) - u_t(\cdot,\cdot;k_2)\|_{L^2(0,T;L^2(0,l))}^2$$
$$\leq \frac{T}{c_0^2} \left[\|F_t\|_{L^2(0,T;L^2(0,l))}^2 + 4c_1 C_2^2 \right] \|k_1 - k_2\|_{C[0,l]}^2$$
$$+ T \|((k_1 - k_2) u_0')'\|_{L^2(0,l)}^2. \tag{6.6.20}$$

where $c_0, c_1 > 0$ are the constants defined in (6.6.3) and $C_2 > 0$ is the constant defined by (6.6.17).

Proof For simplicity, we introduce the function by $v(x, t) := u(x, t; k_1) - u(x, t; k_2)$. Evidently this function solves the following initial boundary value problem:

$$\begin{cases} v_t = (k_1(x)v_x)_x + (\delta k(x)u_{2x})_x, & (x, t) \in \Omega_T, \\ v(x, 0) = 0, & 0 < x < l, \\ v(0, t) = 0, \ v_x(l, t) = 0, & 0 < t < T, \end{cases} \quad (6.6.21)$$

where $u_i(x, t) := u(x, t; k_i)$, $i = 1, 2$ and $\delta k(x) = k_1(x) - k_2(x)$. Differentiate equation (6.6.21) with respect to $t \in (0, T)$, multiply both sides by $v_t(x, t)$ and then integrate over the domain Ω_t. Applying then the integration by parts formula and using the homogeneous boundary conditions in (6.6.1) and the homogeneous initial condition in (6.6.21) we obtain

$$\int_0^l |v_t|^2 dx + 2 \iint_{\Omega_t} k_1(x) |v_{x\tau}|^2 dx d\tau$$
$$= -2 \iint_{\Omega_t} \delta k(x) u_{2x\tau} v_{x\tau} dx d\tau + \int_0^l |v_t(x, 0^+)|^2 dx, \ t \in [0, T].$$

Applying the ε-inequality $2ab \leq (1/\varepsilon)a^2 + \varepsilon b^2$ to the first right hand side integral and using the limit equation

$$\int_0^l |v_t(x, 0^+)|^2 dx = \int_0^l \left[(\delta k(x) u_0'(x))' \right]^2 dx$$

the second right hand side integral we conclude that

$$\|v_t\|_{L^2(0,l)}^2 + (2c_0 - \varepsilon) \|v_{x\tau}\|_{L^2(0,t;L^2(0,l))}^2$$
$$\leq \frac{1}{\varepsilon} \|\delta k\|_{C[0,l]}^2 \|u_{2x\tau}\|_{L^2(0,t;L^2(0,l))}^2 + \|(\delta k \, u_0')'\|_{L^2(0,l)}^2,$$

for all $t \in [0, T]$, $\varepsilon > 0$. Choosing here $\varepsilon = c_0 > 0$ and integrating then both sides over $(0, T)$ we deduce the estimate:

$$\|v_t\|_{L^2(0,T;L^2(0,l))}^2 \leq \frac{T}{c_0} \|\delta k\|_{C[0,l]}^2 \|u_{2xt}\|_{L^2(0,T;L^2(0,l))}^2 + T \|(\delta k \, u_0')'\|_{L^2(0,l)}^2.$$

With estimate (6.6.16) this yields the required estimate (6.6.20). □

Corollary 6.6.1 *Let conditions of Lemma 6.6.4 hold. Then the following estimate holds:*

$$\|u_t(\cdot, \cdot; k_1) - u_t(\cdot, \cdot; k_2)\|_{L^2(0,T;L^2(0,l))}^2 \leq M_L^2 \|k_1 - k_2\|_{H^1(0,l)}^2, \quad (6.6.22)$$

6.6 Identification of a Leading Coefficient in Heat Equation: Neumann ...

where

$$M_L^2 = \max\left\{\frac{M_C^2 T}{c_0^2}\left[\|F_t\|_{L^2(0,T;L^2(0,l))}^2 + 4c_1 C_2^2\right]; \right.$$
$$\left. T\left[\|u'\|_{C[0,l]}^2 + \|u''\|_{C[0,l]}^2\right]\right\}, \quad (6.6.23)$$

$M_C > 0$ *is the constant in the estimate* $\|k\|_{C[0,l]}^2 \le M_C \|k\|_{H^2(0,l)}$, c_0, $c_1 > 0$ *are the constants defined in* (6.6.3) *and* $C_2 > 0$ *is the constant defined by* (6.6.17).

Proof We use the inequality $(a_1b_1 + a_2b_2)^2 \le (a_1^2 + a_2^2)(b_1^2 + b_2^2)$ to estimate the last right hand side norm in (6.6.20) as follows:

$$\int_0^l \left[((k_1(x) - k_2(x))u_0'(x))'\right]^2 dx$$
$$\le \int_0^l \left[(u_0'(x))^2 + (u_0''(x))^2\right]\left[(k_1(x) - k_2(x))^2 + (k_1' - k_2')^2\right] dx$$
$$\le \left[\|u'\|_{C[0,l]}^2 + \|u''\|_{C[0,l]}^2\right] \|k_1 - k_2\|_{H^1(0,l)}^2.$$

Using this in estimate (6.6.20) we obtain estimate (6.6.22) with the constant $M_C > 0$ defined by (6.6.23). □

Having Corollary 6.6.1 we can prove the Lipschitz continuity of the input-output operator $\Psi[\cdot] : \mathcal{K}_c \subset H^2(0,l) \mapsto L^2(0,T)$ corresponding to the inverse coefficient problem with Neumann data.

Theorem 6.6.1 *Let conditions* (6.6.3) *and* (6.6.14) *hold. Then the input-output operator* $\Psi[\cdot] : \mathcal{K}_c \subset H^2(0,l) \mapsto L^2(0,T)$ *is Lipschitz continuous, that is,*

$$\|\Psi[k_1] - \Psi[k_2]\|_{L^2(0,T)} \le L_\Psi \|k_1 - k_2\|_{H^1(0,l)}, \quad (6.6.24)$$

where $L_\Psi = \sqrt{l}\, M_C$ *is the Lipschitz constant and* $M_C > 0$ *is defined by* (6.6.23).

Proof Let $u_m(x,t) := u(x,t;k_m)$, $m = 1,2$ be two solutions of the direct problem (6.6.1). Integrating these equations over $(0,l)$ and using the homogeneous Neumann condition we find:

$$-k_i(0)u_{m,x}(0,t) = \int_0^l u_{m,t}(x,t)dx - \int_0^l F(x,t)dx, \quad m = 1,2.$$

This yields:

$$[k_1(0)u_{1,x}(0,t) - k_2(0)u_{2,x}(0,t)]^2 = \left\{\int_0^l [u_{1,t}(x,t) - u_{2,t}(x,t)]dx\right\}^2$$
$$\le l \int_0^l [u_{1,t}(x,t) - u_{2,t}(x,t)]^2 dx. \quad (6.6.25)$$

Hence,

$$\|\Psi[k_1] - \Psi[k_2]\|_{L^2(0,T)}^2 := \int_0^T [k_1(0)u_{1,x}(0,t) - k_2(0)u_{2,x}(0,t)]^2 dt$$

$$\leq l \int_0^T \int_0^l \left(u_{1,t}(x,t) - u_{2,t}(x,t)\right)^2 dx dt.$$

With estimate (6.6.22) this yields the assertion of the theorem. □

Thus, Lemma 6.6.3 and Theorem 6.6.2 assert that if the set of admissible coefficients $\mathcal{K}_c$ is defined by (6.6.19) as a subset of the Sobolev space $H^2(0,l)$ and the input-output operator $\Phi[k] := (k(x)u_x(x,t;k))_{x=0}$ is defined from $\mathcal{K}_c \subset H^2(0,l)$ to $L^2(0,l)$, then this operator is compact and Lipschitz continuous. Hence, we can apply Theorem 6.5.2 from preceding section to the inverse coefficient problem with Neumann data.

Theorem 6.6.2 *Let conditions (6.6.3) and (6.6.14) hold. Then for any $\alpha > 0$, the functional (6.6.8) attains a minimizer $k_\alpha^\delta \in \mathcal{K}_c \subset H^2(0,l)$.*

Also, assuming injectivity of the input-output operator, we can apply the regularization Theorem 6.5.3 to the inverse coefficient problem (6.6.1) and (6.6.2) with noisy measured output.

6.6.3 Integral Relationship and Gradient Formula

Having above mathematical framework we only need to derive the Fréchet gradient $J'(k)$ of the Tikhonov functional (6.6.9) for subsequent application of the Conjugate Gradient Algorithm given in Sect. 6.5.4

Let $k_1, k_2 \in \mathcal{K}_c$ be admissible coefficients, $u_m(x,t) := u(x,t;k_m)$ the corresponding solutions of the direct problem (6.6.1) and $k_m(0)u_x(0,t;k_m)$, $m = 1, 2$, Neumann outputs. We derive first an important *integral relationship* relating the change $\delta k(x) := k_1(x) - k_2(x)$ in the coefficients to the change

$$k_1(0)u_{1x}(0,t) - k_2(0)u_{2x}(0,t) := k_1(0)u_x(0,t;k_1) - k_2(0)u_x(0,t;k_2)$$

in the Neumann outputs.

Lemma 6.6.5 *Let conditions (6.6.3) and (6.6.14) hold. Denote by $u_m(x,t) := u(x,t;k_m)$ and $k_m(0)u(0,t;k_m)$ the solutions of the direct problem (6.6.1) and the outputs, corresponding to the given admissible coefficients $k_m \in \mathcal{K}_c$, $m = 1, 2$. Then the following integral relationship holds:*

6.6 Identification of a Leading Coefficient in Heat Equation: Neumann ...

$$\int_0^T [k_1(0) u_{1x}(0, t) - k_2(0) u_{2x}(0, t)] q(t) dt$$
$$= - \iint_{\Omega_T} \delta k(x) u_{2x}(x, t) \psi_x(x, t) dx dt, \tag{6.6.26}$$

where $\delta k(x) = k_1(x) - k_2(x)$ and the function $\psi(x, t) = \psi(x, t; q)$ solves the adjoint problem

$$\begin{cases} \psi_t + (k_1(x)\psi_x)_x = 0, & (x, t) \in (0, l) \times [0, T), \\ \psi(x, T) = 0, & x \in (0, l), \\ \psi(0, t) = q(t), \ \psi_x(l, t) = 0, & t \in (0, T), \end{cases} \tag{6.6.27}$$

with an arbitrary input $q \in H^1(0, T)$ satisfying the consistency condition $q(T) = 0$.

Proof The function $\delta u(x, t) := u_1(x, t) - u_2(x, t)$ solves the initial boundary value problem (6.6.21). Multiply both sides of Eq. (6.6.21) by an arbitrary function $\psi \in L^\infty(0, T; H^2(0, l))$, with $\psi_t \in L^\infty(0, T; L^2(0, l)) \cap L^2(0, T; \mathcal{V})$, where $\mathcal{V}(0, l) := \{v \in H^1(0, l) : v(0) = 0\}$, integrate on Ω_T and use integration by parts formula multiple times. Then we get:

$$\int_0^l (\delta u(x, t) \psi(x, t))_{t=0}^{t=T} dx$$
$$- \int_0^T (k_1(x) \delta u_x(x, t) \psi(x, t) - k_1(x) \delta u(x, t) \psi_x(x, t))_{x=0}^{x=l} dt$$
$$- \iint_{\Omega_T} \delta u(x, t) [\psi_t(x, t) + (k_1(x) \psi_x(x, t))_x] dx dt \tag{6.6.28}$$
$$= - \iint_{\Omega_T} \delta k(x) u_{2x}(x, t) \psi_x(x, t) dx dt$$
$$+ \left(\delta k(x) \int_0^T u_{2x}(x, t) \psi(x, t) dt \right)_{x=0}^{x=l}.$$

The first left hand side integral in (6.6.28) is zero due to the initial and final conditions in (6.6.21) and (6.6.27). At $x = l$, the terms under the second left hand side and the last right hand side integrals drop out due to the homogeneous Neumann conditions. The third left hand side integral is also zero due to the adjoint equation $\psi_t(x, t) + (k_1(x) \psi_x(x, t))_x = 0$. Taking into account the nonhomogeneous Neumann conditions in (6.6.21) and (6.6.27) with the condition $\psi(0, t) = q(t)$ we conclude that

$$\int_0^T [k_1(0) \delta u_x(0, t) q(t)] dt$$
$$= - \iint_{\Omega_T} \delta k(x) u_{2x}(x, t) \psi_x(x, t) dx dt - \int_0^T [\delta k(0) u_{2x}(0, t) q(t)] dt.$$

Transforming the first and the last terms under the integrals we get:

$$k_1(0) \delta u_x(0, t) q(t) + \delta k(0) u_{2x}(0, t) q(t)$$
$$:= k_1(0) [u_{1x}(0, t) - u_{2x}(0, t)] + [k_1(0) - k_2(0)] u_{2x}(0, t)$$
$$= k_1(0) u_{1x}(0, t) - k_2(0) u_{2x}(0, t),$$

which is the change in Neumann outputs corresponding to the change $\delta k \in \mathcal{K}_c$ in coefficients. Substituting this in the above integral relation we arrive at the desired result. $\square$

To obtain gradient formula, it is convenient to rewrite the integral relationship (6.6.26) in terms of the increments $\delta k(x)$ and $\delta u(x, t; k) := u(x, t; k + \delta k) - u(x, t; k)$. To this end, we assume that $k_1(x) := k(x) + \delta k(x)$, $k_2(x) := k(x)$ and choose the arbitrary input $q(t)$ in the adjoint problem (6.6.27) as $q(t) = -[k(0)u_x(0, t; k) - f(t)]$. Then we obtain from the above lemma the following integral relationship which also plays a significant role in the numerical solution of inverse coefficient problems [37].

Corollary 6.6.2 *Let conditions of Lemma 6.6.5 hold. Assume that $k, k + \delta k \in \mathcal{K}_c$ are the admissible coefficients, $u(x, t; k)$, $u(x, t; k+\delta k)$ the corresponding solutions of the direct problem (6.6.1) and $k(0)u_x(0, t; k)$, $(k(0) + \delta k(0))u_x(0, t; k + \delta k)$ Neumann outputs. Then the following integral relationship, relating the change $\delta k \in \mathcal{K}_c$ in the coefficients to the change*

$$(k(0) + \delta k(0))u_x(0, t; k + \delta k) - k(0)u_x(0, t; k)$$

in the Neumann outputs, holds:

$$\int_0^T [(k(0) + \delta k(0))u_x(0, t; k + \delta k) - k(0)u_x(0, t; k)][k(0)u_x(0, t; k) - f(t)]dt$$
$$= \iint_{\Omega_T} \delta k(x) u_x(x, t; k) \psi_x(x, t; k + \delta k) dx dt, \quad (6.6.29)$$

where $\psi_x(x, t; k + \delta k)$ is the solution of the following adjoint problem:

$$\begin{cases} \psi_t + ((k(x) + \delta k(x))\psi_x)_x = 0, \ (x, t) \in (0, l) \times [0, T), \\ \psi(x, T) = 0, \ x \in (0, l), \\ \psi(0, t) = -[k(0)u_x(0, t; k) - f(t)], \ \psi_x(l, t) = 0, \quad t \in (0, T). \end{cases}$$

Now we use the integral relationship (6.6.29) to derive the Fréchet gradient $J'(k)$ of the Tikhonov functional (6.6.9).

Theorem 6.6.3 *Let conditions (6.6.3) and (6.6.14) hold. Then the Tikhonov functional (6.6.9) is Fréchet differentiable. Moreover, for the Fréchet gradient $J'(k)$ of the Tikhonov functional the following gradient formula holds:*

$$J'(k)(x) = \int_0^T u_x(x, t; k) \psi_x(x, t; k) dt, \ k \in \mathcal{K}_c, \quad (6.6.30)$$

6.6 Identification of a Leading Coefficient in Heat Equation: Neumann ...

where $\psi_x(x, t; k)$ is the solution of the adjoint problem

$$\begin{cases} \psi_t + (k(x)\psi_x)_x = 0, & (x,t) \in (0,l) \times [0,T), \\ \psi(x,T) = 0, & x \in (0,l), \\ \psi(0,t) = -[k(0)u_x(0,t;k) - f(t)], \ \psi_x(l,t) = 0, & t \in (0,T). \end{cases}$$

Proof Calculate the increment $\delta J(k) := J(k+\delta k) - J(k)$ of the Tikhonov functional (6.6.9). We have:

$$\delta J(k) = \\ \int_0^T [(k(0) + \delta k(0))u_x(0,t;k+\delta k) - k(0)u_x(0,t;k)][k(0)u_x(0,t;k) - f(t)]dt \\ + \frac{1}{2} \int_0^T [(k(0) + \delta k(0))u_x(0,t;k+\delta k) - k(0)u_x(0,t;k)]^2 dt.$$

Use here the integral relationship (6.6.29). Then we obtain:

$$\delta J(k) = \int_0^l \left(\int_0^T u_x(x,t;k)\psi_x(x,t;k+\delta k) dx\delta \right) k(x)dx \\ + \frac{1}{2} \int_0^T [(k(0) + \delta k(0))u_x(0,t;k+\delta k) - k(0)u_x(0,t;k)]^2 dt. \quad (6.6.31)$$

It follows from inequality (6.6.25) and estimate (6.6.20) that the second right hand side integral in (6.6.31) is of the order $\mathcal{O}\left(\|\delta k\|_{C[0,l]}^2\right)$. This completes the proof of the theorem. □

It is seen that the gradient formula (6.5.37) corresponding to the inverse coefficient problem with Dirichlet output, and the above gradient formula (6.6.30), corresponding to the inverse coefficient problem with Neumann output, have exactly the same form, although the functions $u_x(x, t; k)$ and $\psi_x(x, t; k)$ are the solutions of different direct and adjoint problems. This form of the gradients is convenient for computational experiments.

Chapter 7
Inverse Problems for Elliptic Equations

This chapter is an introduction to the basic inverse problems for elliptic equations. One class of these inverse problems arises when the Born approximation is used for scattering problem in quantum mechanics, acoustics or electrodynamics. In the first part of this chapter two inverse problems, the inverse scattering problem at a fixed energy and the inverse scattering problems at a fixed energy, are studied. The last problem is reduced to the tomography problem which is studied in the next chapter. In the second part of the chapter, the Dirichlet to Neumann operator is introduced. It is proved that this operator uniquely defines the potential $q(x)$ in $\Delta u(x) + q(x) = 0$.

7.1 The Inverse Scattering Problem at a Fixed Energy

Consider the stationary Schrödinger equation

$$-\Delta u + q(x)u = |k|^2 u, \quad x \in \mathbb{R}^3 \qquad (7.1.1)$$

of quantum mechanics at a fixed energy $E = |k|^2$. We look for a solution of this equation of the form:

$$u(x, k) = e^{ik \cdot x} + v(x, k). \qquad (7.1.2)$$

Here $q(x)$ is the potential and the vector $k = (k_1, k_2, k_3)$ defines the direction of the incident plane wave $e^{ik \cdot x}$. Function $v(x, k)$ satisfies the radiation conditions

$$v = \mathcal{O}\left(r^{-1}\right), \quad \frac{\partial v}{\partial r} - i|k|v = o\left(r^{-1}\right), \text{ as } r = |x| \to \infty. \qquad (7.1.3)$$

The boundary value problem (7.1.1)–(7.1.3) will be referred below as the *direct problem*.

Equations similar to (7.1.1) arise also in acoustics and electrodynamics. If $q(x) = 0$, then $v(x, k) = 0$. For $q(x) \neq 0$ the solution to the direct problem (7.1.1)–(7.1.3) defines a scattering wave on the potential. It is well known that the problem (7.1.1)–(7.1.3) is well posed, if $q(x) \in L^\infty(\mathbb{R}^3)$ and $q(x)$ decreases sufficiently rapidly, as $r \to \infty$. So, it has the unique solution. For simplicity, we will assume here that the potential $q(x)$ is a finite function with the compact domain $\Omega \subset \mathbb{R}^3$, having smooth boundary $\partial\Omega$. Moreover, we assume that $q(x) \in C(\Omega)$ and $q(x) = 0$ for $x \in \partial\Omega$.

It follows from (7.1.1) and (7.1.2) that the function $v(x, k)$ satisfies the equation

$$-\Delta v + q(x)(e^{\mathrm{i}k\cdot x} + v) = |k|^2 v, \quad x \in \mathbb{R}^3 \tag{7.1.4}$$

and the conditions (7.1.3). On the other hand, $v(x, k)$ is solution of the integral equation

$$v(x, k) = -\frac{1}{4\pi}\int_\Omega \frac{q(y)(e^{\mathrm{i}k\cdot y} + v(y, k))e^{\mathrm{i}|k||x-y|}}{|x - y|}\, dy. \tag{7.1.5}$$

Consider the asymptotic behavior of the function $v(x, k)$, as $|x| = r \to \infty$, so that $x/r = l/|k|$, where $l \in \mathbb{R}^3$ is an arbitrary vector with $|l| = |k|$. We have $|x - y| = r - (l \cdot y)/|k| + o(r^{-1})$ uniformly for all $y \in \Omega$. Then from (7.1.5) we get

$$v(x, k) = -\frac{e^{\mathrm{i}|k|r}}{4\pi r}\int_\Omega q(y)(e^{\mathrm{i}k\cdot y} + v(y, k))e^{-\mathrm{i}l\cdot y}\, dy$$
$$+ \mathcal{O}\left(\frac{1}{r^2}\right), \quad r \to \infty. \tag{7.1.6}$$

From (7.1.6) we see that $v(x, k)$ can be represented in the form

$$v(x, k) = -\frac{e^{\mathrm{i}|k|r}}{4\pi r} f(k, l) + \mathcal{O}\left(\frac{1}{r^2}\right), \quad r \to \infty. \tag{7.1.7}$$

The function $f(k, l)$ is called the *scattering amplitude*.

The *inverse scattering problem* here is to determine the unknown coefficient $q(x)$ in (7.1.4) from given $f(k, l)$ for all k and l, such that $|k| = |l| = \sqrt{E}$. This version of the problem is called the *inverse scattering problem in the Born approximation*.

This problem has been studied in the various papers and books (see, for example, the books [17, 75, 84] and papers [25, 26, 41, 76, 77]).

Following [78], consider here the problem of a reconstruction of the potential $q(x)$ in the Born approximation using scattering amplitude at a fixed energy E. We assume that $q(x)$ is sufficiently small and we can neglect by the second order terms. So we can use the linear approximation of the problem. It is seen form Eqs. (7.1.6) and (7.1.7) that in this case the scattering amplitude is defined the following formula:

7.1 The Inverse Scattering Problem at a Fixed Energy

$$f(k,l) = \int_\Omega q(y) e^{i(k-l)\cdot y}\, dy = \hat{q}(k-l), \tag{7.1.8}$$

where $\hat{q}(\lambda)$ means the Fourier transform of the function $q(x)$:

$$\hat{q}(\lambda) = \int_\Omega q(y) e^{i\lambda\cdot y}\, dy, \quad \lambda \in \mathbb{R}^3.$$

Note that the function $f(k,l)$ is given for all k and l so that $|k|^2 = |l|^2 = E$ define $\hat{q}(\lambda)$ for all $|\lambda| \leq \rho_0 = 2\sqrt{E}$. Indeed, let $\lambda = \rho\nu(\theta,\varphi)$, where $\nu(\theta,\varphi) = (\sin\theta\cos\varphi, \sin\theta\sin\varphi, \cos\theta)$. Define k, l by the formulae

$$\begin{aligned} k &= \tfrac{1}{2}\left(\rho\nu(\theta,\varphi) + \sqrt{4E-\rho^2}\,\nu_\theta(\theta,\varphi)\right), \\ l &= \tfrac{1}{2}\left(-\rho\nu(\theta,\varphi) + \sqrt{4E-\rho^2}\,\nu_\theta(\theta,\varphi)\right), \end{aligned} \tag{7.1.9}$$

where $\nu_\theta(\theta,\varphi) = (\cos\theta\cos\varphi, \cos\theta\sin\varphi, -\sin\theta)$. Then for any λ, $|\lambda| \geq \rho_0$ we have: $k - l = \lambda$ and $|k|^2 = |l|^2 = E$. Hence, if the energy E is sufficiently large, the Fourier image of $q(x)$ is defined by the scattering amplitude for the large ball $B_{\rho_0} = \{\lambda \in \mathbb{R}^3 : |\lambda| \leq \rho_0\}$. Since the Fourier transform of finite continuous function is an analytical function, values of $\hat{q}(\lambda)$ inside the ball B_{ρ_0} define uniquely $\hat{q}(\lambda)$, for all $\lambda \in \mathbb{R}^3$.

This conclusion leads to the following uniqueness theorem.

Theorem 7.1.1 *Let the scattering amplitude $f(k,l)$ in (7.1.7) is given at a fixed energy for all k and l. Then the inverse scattering problem in the Born approximation has at most an one solution.*

Moreover, we can construct the *approximate solution* $q_{appr}(x)$ of the inverse problem by following formula

$$q_{appr}(x) = \frac{1}{(2\pi)^3}\int_{B_{\rho_0}} f(k,l) e^{-i\rho\nu(\theta,\varphi)\cdot x} \rho^2\, d\rho \sin\theta\, d\theta\, d\varphi, \tag{7.1.10}$$

where k, l given by (7.1.9). The error of the approximation $q_{err}(x)$ can be estimated under some a-priory assumption on the function $q(x)$. Let us assume, for example, that the function $q(x)$ with the compact support in Ω, contained in the ball centered at the origin of radius R, belongs to $C^m(\mathbb{R}^3)$, $m > 3$. Suppose that its norm satisfies the condition: $\|q\|_{C^m(\mathbb{R}^3)} \leq q_0$. Then the Fourier image $\hat{q}(\lambda)$ of $q(x)$ satisfies the estimate:

$$|\hat{q}(\lambda)| \leq Cq_0 |\lambda|^{-m}, \quad |\lambda| \geq \rho_0, \tag{7.1.11}$$

where $C = C(R,m) > 0$. Indeed,

$$\hat{q}(\lambda) = \int_{\mathbb{R}^3} q(x) e^{i\rho\nu\cdot x}\, dx = \int_{-R}^{R} Q(s,\nu) e^{i\rho s}\, ds. \qquad (7.1.12)$$

Here $\nu = \nu(\theta, \varphi)$ and $Q(s, \nu)$ is the *Radon transform* of $q(x)$ given by formula

$$Q(s, \nu) = \int_{\nu\cdot x = s} q(x)\, d\sigma,$$

where $d\sigma$ is the square element. The variable x under the last integral can be derived by the formula

$$x = s\nu + r\left(\nu_\theta \cos\phi + \frac{\nu_\varphi}{\sin\theta} \sin\phi\right), \quad \nu_\theta = \frac{\partial \nu}{\partial \theta},\ \nu_\varphi = \frac{\partial \nu}{\partial \varphi}.$$

Substituting this with $d\sigma = r\,dr\,d\phi$ into the integral (7.1.12) we conclude that

$$\hat{q}(\lambda) = \frac{(-1)^m}{(i\rho)^m} \int_{-R}^{R} \frac{\partial^m Q(s,\nu)}{\partial \rho^m} e^{i\rho s}\, ds. \qquad (7.1.13)$$

Evidently,

$$\left|\frac{\partial^m Q(s,\nu)}{\partial \rho^m}\right| \le Cq_0,$$

with the above defined constant $C = C(R, m)$, depending only on R and m. Using this in (7.1.13) we arrive at the estimate (7.1.11).

Then the error of the approximation $q_{err}(x)$ can be estimate as

$$|q_{err}(x)| \le \frac{1}{(2\pi)^3} \int_{\mathbb{R}^3 \setminus B_{\rho_0}} |\hat{q}(\lambda)|\, \rho^2\, d\rho\, \sin\theta\, d\theta\, d\varphi \le \frac{Cq_0}{2\pi^2} \int_{\rho_0}^{\infty} \frac{d\rho}{\rho^{m-2}}$$

$$\le \frac{Cq_0}{2\pi^2(m-1)\rho_0^{m-1}} = \frac{Cq_0}{2\pi^2(m-1)E^{(m-1)/2}}.$$

Therefore, the error of the approximate formula (7.1.10) tends to zero, as $E \to \infty$.

7.2 The Inverse Scattering Problem with Point Sources

In the previous section we have studied the inverse scattering problem with incident plane wave. Here we discuss the case when the incident wave is produced by *point sources*.

Let $q(x)$ is compactly supported in the ball $\Omega = \{x \in \mathbb{R}^3 : |x| \le R\}$ function and $q(x) \in C(\mathbb{R}^3)$. Consider the equation

7.2 The Inverse Scattering Problem with Point Sources

$$-\Delta u + q(x)u = k^2 u + \delta(x - y), \quad x \in \mathbb{R}^3, \tag{7.2.1}$$

where $\delta(z)$ is a Dirac delta function. In the contrast to the previous section, here k is a positive number, $k^2 = E$, and the point $y \in \mathbb{R}^3$ is the variable parameter. Let $u(x, k, y)$ be a solution of this equation of the form

$$u(x, k, y) = \frac{e^{ik|x-y|}}{4\pi |x - y|} + v(x, k, y), \tag{7.2.2}$$

where the function $v(x, k, y)$ satisfies the Sommerfeld conditions

$$v = \mathcal{O}(r^{-1}), \quad \frac{\partial v}{\partial r} - ikv = o(r^{-1}), \text{ as } r = |x| \to \infty. \tag{7.2.3}$$

Then, using Eq. (7.2.1), we deduce that $v(x, k, y)$ is the solution of the following equation

$$-\Delta v + q(x)\left(\frac{e^{ik|x-y|}}{4\pi |x - y|} + v(x, k, y)\right) = k^2 v, \quad x \in \mathbb{R}^3. \tag{7.2.4}$$

Inverting the Helmholtz operator $-(\Delta + k^2)$, we obtain that the function $v(x, k, y)$ is the solution of the integral equation

$$v(x, k, y) = v_0(x, k, y) - \frac{1}{4\pi} \int_\Omega \frac{e^{ik|\xi-x|}}{|\xi - x|} q(\xi) v(\xi, k, y) \, d\xi, \tag{7.2.5}$$

where

$$v_0(x, k, y) = -\frac{1}{(4\pi)^2} \int_\Omega \frac{e^{ik(|\xi-x|+|\xi-y|)}}{|\xi - x||\xi - y|} q(\xi) \, d\xi. \tag{7.2.6}$$

We again will restrict our analysis to the case of the Born approximation for this equation assuming that $q(x)$ is enough small, which allows to make the linearization of the problem. Then $v(x, k, y) \approx v_0(x, k, y)$.

Now we state the following theorem.

Theorem 7.2.1 *Let $q(x)$ be a function with a compact support in the ball $\Omega = \{x \in \mathbb{R}^3 : |x| \le R\}$ and $q(x) \in C^2(\mathbb{R}^3)$. Then for any fixed y and k, $v_0(x, k, y) \in C(\mathbb{R}^3)$ and*

$$v_0(x, k, y) = \frac{e^{ik|x-y|}}{8ik\pi} \int_{L(x,y)} q(\xi) \, ds + o\left(\frac{1}{k}\right), \text{ as } k \to \infty. \tag{7.2.7}$$

where $L(x, y) = \{\xi \in \mathbb{R}^3 : \xi = y(1 - s) + sx, s \in [0, 1]\}$ is the segment of the straight line passing through points x and y.

Proof Denote by $E(x, y, t)$ the ellipsoid

$$E(x, y, t) = \{\xi \in \mathbb{R}^3 : |\xi - x| + |\xi - y| = t\},$$

Let us fix points x and y. Represent x as $x = y + \rho\nu(\theta, \varphi)$, where $\nu(\theta, \varphi)$ is the unite vector and θ and φ are its spherical coordinates. Here ρ, θ and φ are fixed. Then the ellipsoid $E(x, y, t)$ is defined only for $t \geq |x - y| = \rho$. If $t \to |x - y|$, then the ellipsoid degenerates into the segment $L(x, y)$. If $t > \rho$, then the main axis of this ellipsoid passes through the points x and y and the intersection points of this axis with the ellipsoid are $\xi^1 = y - \nu(\theta, \varphi)(t - \rho)/2$ and $\xi^2 = x + \nu(\theta, \varphi)(t - \rho)/2$. Using these observations, we represent the variable point $\xi \in E(x, y, t)$ as follows:

$$\xi = y + \frac{1}{2}(\rho + tz)\nu(\theta, \varphi) + r\left(\nu_\theta(\theta, \varphi) \cos\varphi + \frac{\nu_\varphi(\theta, \varphi)}{\sin\theta}\sin\psi\right), \quad (7.2.8)$$

where $r = r(z; \rho, t) = r(z, |x - y|, t)$ in (7.2.8) is given by the formula

$$r(z; \rho, t) = \frac{1}{2}\sqrt{(t^2 - \rho^2)(1 - z^2)} \quad (7.2.9)$$

and

$$\nu_\theta(\theta, \varphi) = \frac{\partial \nu(\theta, \varphi)}{\partial \theta}, \quad \nu_\varphi(\theta, \varphi) = \frac{\partial \nu(\theta, \varphi)}{\partial \varphi}.$$

In formulae (7.2.8) and (7.2.9) z and ψ are variable parameters: $z \in [-1, 1]$, $\psi \in [0, 2\pi)$. Hence, $\xi = \xi(z, \psi; x, y, t)$. Note that the unite vectors ν, ν_θ, $\nu_\varphi/\sin\theta$ are mutually orthogonal. Then each point $\xi \in E(x, y, t)$ is uniquely defined by $z \in [-1, 1]$ and $\psi \in [0, 2\pi)$. Indeed, in this case for arbitrary $z \in [-1, 1]$ and $\psi \in [0, 2\pi)$ the following equalities hold

$$|\xi - y| = \sqrt{r^2 + \frac{1}{4}(tz + \rho)^2} = \frac{1}{2}(t + z\rho),$$
$$|\xi - x| = \sqrt{r^2 + \frac{1}{4}(tz - \rho)^2} = \frac{1}{2}(t - z\rho).$$

Hence, $|\xi - y| + |\xi - x| = t$, i.-e. the point $\xi \in E(x, y, t)$. On the other hand, for any ξ given by the formula (7.2.8) we have $(\xi - y) \cdot \nu = (\rho + tz)/2$. It means that the coordinate z characterize the cross-section of the ellipsoid $E(x, y, t)$ by the plane orthogonal to ν. For $z = -1$ this cross-sections degenerates into the point ξ^1, for $z = 1$ into ξ^2. Since the vectors ν_θ, $\nu_\varphi/\sin\theta$ are orthogonal to ν, in any cross-section defined by $z \in (-1, 1)$ the coordinate ψ in (7.2.8) is uniquely determines the position of ξ in this cross-section.

For the sake of brevity, we denote $e_1 = \nu, e_2 = \nu_\theta, e_3 = \nu_\varphi/\sin\theta$. Then, it follows from (7.2.8) that each $\xi \in E(x, y, t)$ can be presented as

7.2 The Inverse Scattering Problem with Point Sources

$$\xi = y + \frac{1}{2}(\rho + tz)e_1 + \frac{1}{2}\sqrt{(t^2 - \rho^2)(1 - z^2)}(e_2 \cos\psi + e_3 \sin\psi). \quad (7.2.10)$$

For each fixed x and y family of the ellipsoids $E(x, y, t), t \geq \rho = |x - y|$, covers all space $\mathbb{R}^3$. So, every point $\xi \in \mathbb{R}^3$ can be uniquely defined by z, t, ψ. The Jacobian $J = \frac{\partial(\xi_1, \xi_2, \xi_3)}{\partial(z, t, \psi)}$ can be easily calculated:

$$J = \frac{\partial(\xi_1, \xi_2, \xi_3)}{\partial(z, t, \psi)} := \begin{vmatrix} \partial\xi_1/\partial z & \partial\xi_2/\partial z & \partial\xi_3/\partial z \\ \partial\xi_1/\partial t & \partial\xi_2/\partial t & \partial\xi_3/\partial t \\ \partial\xi_1/\partial\psi & \partial\xi_2/\partial\psi & \partial\xi_3/\partial\psi \end{vmatrix}$$

$$= \begin{vmatrix} te_1/2 + r_z(e_2\cos\psi + e_3\sin\psi) \\ ze_1/2 + r_t(e_2\cos\psi + e_3\sin\psi) \\ r(-e_2\sin\psi + e_3\cos\psi) \end{vmatrix}$$

$$= \frac{1}{2}r(tr_t - zr_z) = \frac{1}{8}(t^2 - z^2\rho^2).$$

At the same time

$$|\xi - x||\xi - y| = \frac{1}{4}(t^2 - z^2\rho^2).$$

Therefore,

$$\frac{d\xi}{|\xi - x||\xi - y|} = \frac{1}{2}dzd\psi dt.$$

Using the above transformation in the right hand side integral of (7.2.6) we arrive at the formula

$$v_0(x, k, y) = -\frac{1}{32\pi^2}\int_\rho^\infty \int_0^{2\pi}\int_{-1}^1 e^{ikt}q(\xi)\,dzd\psi dt, \quad (7.2.11)$$

where ξ is defined by formula (7.2.10). Since the function $q(x)$ has a compact support in the ball Ω, the integral in (7.2.11) with respect to t is taken along a bounded interval depended on x and y. Further, the function $v_0(x, k, y)$ in (7.2.11) is continuous, due to the continuity of the integrand. Using in (7.2.11) the integration by parts formula respect to t, we get:

$$v_0(x, k, y) = \frac{e^{ik\rho}}{32ik\pi^2}\left(\int_0^{2\pi}\int_{-1}^1 q(\xi)\,dzd\psi\right)_{t=\rho+0}$$

$$+ \frac{1}{32ik\pi^2}\int_\rho^\infty \int_0^{2\pi}\int_{-1}^1 e^{ikt}\nabla q(\xi) \cdot \xi_t\, dzd\psi dt$$

$$= \frac{e^{ik\rho}}{32ik\pi^2}I_1 + \frac{1}{32ik\pi^2}I_2. \quad (7.2.12)$$

Since $\xi \to y + \nu\rho(1+z)/2$ as $t \to \rho$, we have

$$I_1 = \left(\int_0^{2\pi} \int_{-1}^1 q(\xi)\, dz\, d\psi\right)_{t=\rho+0}$$

$$= 2\pi \int_{-1}^1 q(y + \nu\rho(1+z)/2)\, dz$$

$$= 4\pi \int_0^1 q(y + s(x-y))\, ds = 4\pi \int_{L(x,y)} q(\xi)\, ds. \qquad (7.2.13)$$

For the second integral I_2 we use formula (7.2.10) for ξ to deduce

$$\xi_t = \frac{z}{2} e_1 + \frac{t}{2}\sqrt{\frac{1-z^2}{t^2-\rho^2}}(e_2 \cos\psi + e_3 \sin\psi).$$

Then we can transform the integral I_2 as follows:

$$I_2 = \int_\rho^\infty \int_0^{2\pi} \int_{-1}^1 e^{ikt} \nabla q(\xi) \cdot \xi_t\, dz\, d\psi\, dt$$

$$= \int_\rho^\infty \int_0^{2\pi} \int_{-1}^1 e^{ikt} \left(\frac{t}{2}\nabla_\xi q(\xi)\cdot e_1\right.$$

$$\left. + \frac{t}{2}\sqrt{\frac{1-z^2}{t^2-\rho^2}}\Big[\nabla_\xi q(\xi)\cdot e_2 \cos\psi + \nabla_\xi q(\xi)\cdot e_3 \sin\psi\Big]\right) dz\, d\psi\, dt.$$

Using here the integration by parts formula with respect to ψ, we get

$$I_2 = \int_\rho^\infty \int_0^{2\pi} \int_{-1}^1 e^{ikt} \left(\frac{t}{2}\nabla_\xi q(\xi)\cdot e_1\right.$$

$$- \frac{t}{2}\sqrt{\frac{1-z^2}{t^2-\rho^2}}\Big[\nabla_\xi [\nabla_\xi q(\xi)\cdot e_2]\cdot \xi_\psi \sin\psi$$

$$\left. + \nabla_\xi [\nabla_\xi q(\xi)\cdot e_3]\cdot \xi_\psi \cos\psi\Big]\right) dz\, d\psi\, dt.$$

It follows from (4.2.10) that

$$\sqrt{\frac{1-z^2}{t^2-\rho^2}}\, \xi_\psi = \frac{1}{2}(1-z^2)(-e_2 \sin\psi + e_3 \cos\psi)$$

is a continuous function of z and ψ. As a consequence, the integrand I_2 in (7.2.12) is also continuous function of the variables t, z and ψ. Therefore the double integral with respect to z, ψ is continuous function of t. Then integral with respect to t is the

7.2 The Inverse Scattering Problem with Point Sources

Fourier transform of continuous finite function. Therefore $I_2 \to 0$ as $k \to \infty$. This conclusion with formulae (7.2.12) and (7.2.13) imply the proof of the theorem. $\square$

Let us now assume that the scattering field is given for every $(x, y) \in (S \times S)$ and all $k \geq k_0$, where k_0 a positive number:

$$v(x, k, y) = f(x, k, y), \quad (x, y) \in (S \times S), \; k \geq k_0 \qquad (7.2.14)$$

Then, in the Born approximation, we obtain the following asymptotic formula:

$$f(x, k, y) = \frac{e^{ik|x-y|}}{8ik\pi} \int_{L(x,y)} q(\xi)\,ds + o\left(\frac{1}{k}\right), \text{ as } k \to \infty. \qquad (7.2.15)$$

where $L(x, y)$ is the segment defined in Theorem 7.2.1. Calculating the integral of $q(x)$ over this segment we get:

$$\int_{L(x,y)} q(\xi)\,ds = 8i\pi \lim_{k\to\infty} \left[e^{-ik|x-y|} k f(x, k, y) \right]$$
$$= g(x, y), \; (x, y) \in (S \times S). \qquad (7.2.16)$$

Therefore we arrive at the *tomography problem*: find $q(x)$ in Ω from given integrals along arbitrary strait lines jointing the boundary of Ω. Note that this problem can be solved separately for each cross-section of the ball Ω by any plane. We will study this problem in the next chapter. It follows from this consideration that the tomography problem has, at most, an unique solution, and this solution can be derived in an explicit form.

7.3 Dirichlet to Neumann Map

Let $\Omega \subset \mathbb{R}^n$ be a bounded domain with the smooth boundary $\partial\Omega$. Consider the Dirichlet problem

$$\Delta u + q(x)u = 0, \; x \in \Omega, \quad u|_{\partial\Omega} = f(x). \qquad (7.3.1)$$

Denote by $u(x; q)$ the unique weak solution $u(\cdot; q) \in H^1(\Omega)$ of this problem, corresponding to a given coefficient $q(x)$ from some class of admissible coefficients in $L^2(\Omega)$. It is well known that if $\Omega \subset \mathbb{R}^n$ is a bounded domain with the C^1-smooth boundary $\partial\Omega$, then there exists a bounded linear operator $T_r : H^1(\Omega) \mapsto H^{1/2}(\partial\Omega)$, called the *trace* of u on $\partial\Omega$, such that

$$T_r u := u(\cdot; q)|_{\partial\Omega}, \text{ and } \|T_r u\|_{L^2(\partial\Omega)} \leq C_0 \|u\|_{H^1(\Omega)}, \; C_0 = C_0(\Omega) > 0.$$

Now, taking (formally) the derivative of the solution $u(\cdot; q) \in H^1(\Omega)$ of problem (7.3.1) along the outward normal n to the boundary $\partial\Omega$ at $x \in \partial\Omega$, we can find the trace:

$$\left.\frac{\partial u(\cdot; q)}{\partial n}\right|_{\partial\Omega}, \quad q \in L^2(\Omega). \tag{7.3.2}$$

According to [62], this trace is uniquely determined as an element of $H^{-1/2}(\partial\Omega)$. Thus, we have constructed a mappings which transforms each element (coefficient) $q \in \mathcal{Q}$ to the unique element (trace of the derivative of the solution) of $H^{1/2}(\partial\Omega)$ given by (7.3.2). In this transformation, under the assumption that $q \in \mathcal{Q}$ is given, we first used the *Dirichlet data* given in (7.3.1), which uniquely defines the solution, then we used the normal derivative of the solution in (7.3.2), which is usually associated with the *Neumann data*. Subsequently, this gave rise to the terminology "Dirichlet-Neumann operator".

Definition 7.3.1 Let $u(x; q)$ be the unique weak solution $u(\cdot; q) \in H^1(\Omega)$ of problem (7.3.1), corresponding to a given coefficient $q \in L^2(\Omega)$. Then the mapping

$$\Lambda_q : H^{1/2}(\partial\Omega) \mapsto H^{-1/2}(\partial\Omega) \tag{7.3.3}$$

defined by (7.3.2) is called the Dirichlet-Neumann operator, that is,

$$\Lambda_q u := \left.\frac{\partial u(\cdot; q)}{\partial n}\right|_{\partial\Omega}, \quad q \in L^2(\Omega). \tag{7.3.4}$$

Let now $f \in L^2(\partial\Omega)$ be a given *input* in (7.3.1). It follows from the above conclusions that for a given coefficient $q \in L^2(\Omega)$ the Dirichlet-Neumann operator is uniquely defined by the equality

$$\Lambda_q f(x) := \left.\frac{\partial u(\cdot; q)}{\partial n}\right|_{\partial\Omega} = g(x), \quad x \in \partial\Omega. \tag{7.3.5}$$

We assume now that $g \in L^2(\Omega)$ be a *given output data*. Then we can naturally raise the question: *does the operator Λ_q define uniquely the function $q(x)$?* The answer on this question is positive [91].

Theorem 7.3.1 *Let $\Omega \subset \mathbb{R}^n$, $n \geq 3$, be a bounded domain with smooth boundary and $q_k(x) \in L^2(\Omega)$ for $k = 1, 2$. Then $\Lambda_{q_1} = \Lambda_{q_2}$ implies $q_1(x) = q_2(x)$ in Ω.*

Proof of this theorem is based on a construction of a special solution of the equation $\Delta u + q(x)u = 0$, which depends on a complex parameter $\zeta = \xi + i\eta$ satisfying the condition $\zeta \cdot \zeta = 0$. This solution has the form $u_\zeta(x) = (1 + v_\zeta(x))\exp(x \cdot \zeta)$, where $v_\zeta(x)$ satisfies the estimate $\|v_\zeta\|_{L^2(\Omega)} \leq C/|\zeta|$ for large $|\zeta|$. Note that this solution is unbounded for $x \cdot \eta < 0$ and $|\eta| \to \infty$. We refer the reader to [91] for details of the proof.

7.3 Dirichlet to Neumann Map

Now we consider a linearized inverse problem. Let $\Omega \in \mathbb{R}^2$ be the unite disk $|x| < 1$. Denote by (r, φ) polar coordinates of point $x \in \Omega$. We assume that the input data $f(x)$ in (7.3.1) is given as the Fourier harmonics, that is, $f(x) := \exp(im\varphi)$, for $m = 0, \pm 1, \pm 2, \ldots$. Denote the solution of the problem (7.3.1) corresponding the such Dirichlet data by $u^m := u^m(r, \varphi; q)$. Then given operator Λ_q is equivalent to the following information

$$\frac{\partial u^m(r, \varphi; q)}{\partial r}\bigg|_{r=1} = g^m(\varphi), \quad \varphi \in [0, 2\pi], \quad m = 0, \pm 1, \pm 2, \ldots. \quad (7.3.6)$$

Consider the inverse problem of recovering $q(r, \varphi)$ from the given by (7.3.6) output data $g^m(\varphi)$, $m = 0, \pm 1, \pm 2, \ldots$. Assume that $q(r, \varphi)$ and its derivative $q_\varphi(r, \varphi)$ are continues functions in Ω. Assume also that function $q(r, \varphi)$ is small, i.e. $|q(r, \varphi)| << 1$. Then, in the linear approximation, we can represent solution to the problem

$$\Delta u^m + q(x)u^m = 0, \ x \in \Omega, \ u^m|_{r=1} = e^{im\varphi}, \quad (7.3.7)$$

in the form

$$u^m(r, \varphi) = \bar{u}^m(r, \varphi) + v^m(r, \varphi), \quad (7.3.8)$$

where $\bar{u}^m(r, \varphi)$ is the solution to the problem

$$\frac{1}{r}\frac{\partial}{\partial r}\left(r\frac{\partial \bar{u}^m}{\partial r}\right) + \frac{1}{r^2}\frac{\partial^2 \bar{u}^m}{\partial \varphi^2} = 0, \ \bar{u}^m|_{r=1} = e^{im\varphi}, \quad (7.3.9)$$

and $v^m(r, \varphi)$ solves the problem

$$\frac{1}{r}\frac{\partial}{\partial r}\left(r\frac{\partial v^m}{\partial r}\right) + \frac{1}{r^2}\frac{\partial^2 v^m}{\partial \varphi^2} + q(x)\bar{u}^m = 0, \ v^m|_{r=1} = 0. \quad (7.3.10)$$

The solution of the problem (7.3.9) is given by the formula

$$\bar{u}^m = r^{|m|}e^{im\varphi}, \quad (7.3.11)$$

that can be checked directly. Represent the functions $v^m(r, \varphi)$ and $q(r, \varphi)$ as the Fourier series with respect to φ:

$$v^m(r, \varphi) = \sum_{n=-\infty}^{\infty} v_n^m(r)e^{in\varphi}, \quad q(r, \varphi) = \sum_{n=-\infty}^{\infty} q_n(r)e^{in\varphi}. \quad (7.3.12)$$

Then the Fourier coefficients $v_n^m(r)$ satisfy the following equations

$$\frac{1}{r}\frac{d}{dr}\left(r\frac{dv_n^m}{dr}\right) - \frac{n^2}{r^2}v_n^m + q_{n-m}r^{|m|} = 0, \quad v_n^m|_{r=1} = 0. \tag{7.3.13}$$

We can represent the solution of problem (7.3.13) via the Green function $G_n(r, \rho)$ as

$$v_n^m(r, \varphi) = -\int_0^1 G_n(r, \rho) q_{n-m}(\rho) \rho^{|m|} d\rho. \tag{7.3.14}$$

Here the $G_n(r, \rho)$ a is bounded function for all $r \in [0, 1]$ and satisfies for $r \in [0, \rho]$ and $r \in [\rho, 1]$ to the equation

$$\frac{1}{r}\frac{d}{dr}\left(r\frac{dG_n}{dr}\right) - \frac{n^2}{r^2}G_n = 0 \tag{7.3.15}$$

and the following conditions

$$G_n(1, \rho) = 0, \quad G_n(\rho - 0, \rho) = G_n(\rho + 0, \rho),$$
$$\left.\frac{dG_n(r, \rho)}{dr}\right|_{r=\rho+0} - \left.\frac{dG_n(r, \rho)}{dr}\right|_{r=\rho-0} = 1. \tag{7.3.16}$$

The such function can be easily constructed. Noting that the Eq. (7.3.15) has for $n \neq 0$ two linearly independent solutions $r^{|n|}$ and $r^{-|n|}$, we can find the bounded function G_n for $n \neq 0$ in the form

$$G_n(r, \rho) = \begin{cases} C_{1n}r^{|n|} + C_{2n}r^{-|n|}, & \rho \leq r \leq 1, \\ C_{3n}r^{|n|}, & 0 \leq r \leq \rho. \end{cases} \tag{7.3.17}$$

Choosing $C_{kn} = C_{kn}(\rho), k = 1, 2, 3$, from conditions (7.3.16) we come to the linear system

$$\begin{aligned} C_{1n} + C_{2n} &= 0, \\ C_{1n}\rho^{|n|} + C_{2n}\rho^{-|n|} - C_{3n}\rho^{|n|} &= 0, \\ C_{1n}\rho^{|n|-1} - C_{2n}\rho^{-|n|-1} - C_{3n}\rho^{|n|-1} &= 1/|n|. \end{aligned} \tag{7.3.18}$$

The solution of this system is

$$C_{1n} = \frac{\rho^{1+|n|}}{2|n|}, \quad C_{2n} = -\frac{\rho^{1+|n|}}{2|n|}, \quad C_{3n} = \frac{\rho^{1+|n|} - \rho^{1-|n|}}{2|n|}. \tag{7.3.19}$$

Hence,

$$G_n(r, \rho) = \frac{1}{2|n|} \begin{cases} (r^{|n|} - r^{-|n|})\rho^{1+|n|}, & \rho \leq r \leq 1, \\ r^{|n|}(\rho^{1+|n|} - \rho^{1-|n|}), & 0 \leq r \leq \rho, \end{cases} \tag{7.3.20}$$

7.3 Dirichlet to Neumann Map

if $n \neq 0$.

If $n = 0$ then (7.3.15) has two linearly independent solutions 1 and $\ln r$. Then we can find the bounded function $G_0(r, \rho)$ in the form

$$G_0(r, \rho) = \begin{cases} C_{10} + C_{20} \ln r, & \rho \leq r \leq 1, \\ C_{30}, & 0 \leq r \leq \rho. \end{cases} \quad (7.3.21)$$

Easily to check that in this case convenient C_{k0}, for which conditions (7.3.16) hold, are found as:

$$C_{10} = 0, \quad C_{20} = \rho, \quad C_{30} = \rho \ln \rho. \quad (7.3.22)$$

Therefore,

$$G_0(r, \rho) = \begin{cases} \rho \ln r & \rho \leq r \leq 1, \\ \rho \ln \rho, & 0 \leq r \leq \rho. \end{cases} \quad (7.3.23)$$

Note that data of the inverse problem allows to find dv_n^m/dr at $r = 1$. Indeed, using equalities (7.3.6), (7.3.8) and (7.3.11), we obtain

$$\left. \frac{dv_n^m}{dr} \right|_{r=1} = g_n^m - \delta_{mn} = \bar{g}_n^m, \quad (7.3.24)$$

where g_n^m are Fourier coefficients of given functions $g^m(\varphi)$ and δ_{mn} is the Krönecker delta. Then differentiating equality (7.3.14) and putting $r = 1$, we get

$$\bar{g}_n^m = -\int_0^1 q_{n-m}(\rho) \rho^{|m|+|n|+1} d\rho, \quad n, m = 0, \pm 1, \pm 2, \ldots. \quad (7.3.25)$$

Put here $n - m = k$ and fix k. The above equalities can be rewritten as

$$\bar{g}_{m+k}^m = -\int_0^1 q_k(\rho) \rho^{|m|+|m+k|+1} d\rho, \quad m = 0 \pm 1, \pm 2, \ldots. \quad (7.3.26)$$

Since for real-valued potential $q(r, \varphi)$ the equalities $q_{-k} = \overline{q_k}$ hold, we can consider only nonnegative values of k. Assuming $k \geq 0$, consider (7.3.26) for $m \geq 0$. Then

$$\bar{g}_{m+k}^m = -\int_0^1 q_k(\rho) \rho^{k+1+2m} d\rho, \quad m = 0, 1, 2, \ldots, \quad (7.3.27)$$

or

$$\bar{g}_{m+k}^m = -\frac{1}{2} \int_0^1 \eta_k(z) z^m dz, \quad m = 0, 1, 2, \ldots, \quad (7.3.28)$$

where $\eta_k(z) = q_k(\sqrt{z})\sqrt{z^k}$.

Thus, we arrive at the well known the moment's problem: given the moments (7.3.28) for $m = 0, 1, 2, \ldots$ find $\eta_k(z)$. This problem has a unique solution. Furthermore, all $q_k(r)$ are uniquely defined from (7.3.28). Then the solution $q(r, \varphi)$ of the inverse problem can be derived via the Fourier series.

Chapter 8
Inverse Problems for the Stationary Transport Equations

Inverse problems related to the transport equations arise in many areas of applied sciences and have various applications in medical imaging and tomography. At least the three techniques—X ray tomography, single particle emission tomography and positron emission tomography—are based on the transport equations. On the other hand, mathematical models governed by the transport equations have some advantages over the traditional integral geometric approach, at least by accurate modeling and inversion formulae, which we derive in this chapter.

8.1 The Transport Equation Without Scattering

Let $\Omega \in \mathbb{R}^3$ be a compact domain with smooth boundary $\partial \Omega$, $S^2 = \{\nu \in \mathbb{R}^3 | |\nu| = 1\}$, $x \in \mathbb{R}^3$ and $\nu \in S^2$. For function $u(x, \nu)$ consider the equation

$$\nu \cdot \nabla_x u + \sigma(x)u + \int_{S^2} K(x, \nu, \nu')u(x, \nu')d\nu' = F(x, \nu), \ x \in \Omega, \ \nu \in S^2. \quad (8.1.1)$$

Equation (8.1.1) is called the *transport equation*. The function $u(x, \nu)$ is the density of particles $u(x, \nu)$ in the space of positions $x \in \Omega$ travelling in direction $\nu \in S^2$, $\sigma(x)$ is the attenuation coefficient, $K(x, \nu, \nu')$ is the dispersion index and function $F(x, \nu)$ is the source term. Equation (8.1.1) describes a transport of particles in a medium which absorbs and scatterers the particles.

Since the domain Ω in the Eq. (8.1.1) is compact, we need to add a boundary condition in order to define a unique solution to (8.1.1). For this goal we introduce the set $\partial_- \Omega(\nu) = \{x \in \partial \Omega | n(x) \cdot \nu \leq 0\}$, for a fixed ν, where $n(x)$ is an outward normal to $\partial \Omega$ at the point $x \in \partial \Omega$. Then the boundary condition has the form

$$u(x,\nu) = h(x,\nu) \text{ for all } x \in \partial_-\Omega(\nu) \text{ and for all } \nu \in S^2. \qquad (8.1.2)$$

The relations (8.1.1) and (8.1.2) form the *direct problem* for the transport equation: given $\sigma(x)$, $K(x,\nu,\nu')$ and $F(x,\nu)$ find $u(x,\nu)$ for $x \in \Omega$ and $\nu \in S^2$. This is a well posed problem in appropriate functional spaces of functions and for the desired function $u(x,\nu)$.

For Eq. (8.1.1) there exist different formulations of the inverse problems of determining an unknown attenuation coefficient $\sigma(x)$ and dispersion index $K(x,\nu,\nu')$, from a given information on the solutions to direct problems (see, for instance the book [86] and references therein). In this chapter, we concentrate only on the inverse problem related to *the transport equation without scattering*. Namely, instead of Eq. (8.1.1) we will consider the equation

$$\nu \cdot \nabla_x u + \sigma(x)u = 0, \ x \in \Omega, \ \nu \in S^2. \qquad (8.1.3)$$

For this equation we consider the following *inverse coefficient problem*: find $\sigma(x)$ in Ω from the information

$$g(x,\nu) := u(x,\nu) \text{ for all } x \in \partial_+\Omega(\nu) \text{ and for all } \nu \in S^2, \qquad (8.1.4)$$

about the solution of the problem (8.1.3) and (8.1.2) on the boundary $\partial_+\Omega(\nu) = \{x \in \partial\Omega | n(x) \cdot \nu > 0\}$. This is a well known *tomography problem* which occurs in tomography when a physical body is exposed to X rays and the radiation is measured outside the body in a tomographic fashion. Note that the scalar product $\nu \cdot \nabla_x u$ represents the derivative of function $u(x,\nu)$ in the direction ν. Assuming that $h(x,\nu) > 0$ one can invert Eq. (8.1.3) using the boundary data (8.1.2). Then we obtain the following formula for the function $u(x,\nu)$:

$$u(x,\nu) = h(\xi^*(x,\nu),\nu)e^{-\int_{L(x,\nu)} \sigma(\xi)\,ds}, \qquad (8.1.5)$$

where $L(x,\nu) = \{\xi \in \Omega | \xi = x - s\nu\}$ is the segment of the direct line going from point x in the direction $-\nu$ and belonging to Ω, $\xi^*(x,\nu)$ is the intersection point of $L(x,\nu)$ with $\partial_-\Omega(\nu)$ and ds is the arc length element. Then from (8.1.5) we obtain the relation

$$\int_{L(x,\nu)} \sigma(\xi)\,ds = f(x,\nu), \quad x \in \partial_+\Omega(\nu), \ \nu \in S^2, \qquad (8.1.6)$$

where $f(x,\nu) = -\ln(g(x,\nu)/h(\xi^*(x,\nu),\nu))$.

Thus, the tomography problem is transformed to the problem of recovering a coefficient $\sigma(x)$ from the given integrals along the segments $L(x,\nu)$ of arbitrary direct lines crossing domain Ω. Note that we can consider the problem as a two-dimensional one for an arbitrary cross-section domain Ω by a plane and choosing $L(x,\nu)$ belonging to this plane only. For example, one can consider the one-parametric family of cross-sections of Ω by the planes $x_3 = const$ and solve the tomography problem for each of this cross-sections in order to find $\sigma(x)$ for all $x \in \Omega$.

8.1 The Transport Equation Without Scattering

Another class of inverse problems, close to the tomography problem, is related to the determination of an unknown source term $F(x)$ in the transport equation

$$\nu \cdot \nabla_x u + a(x)u = F(x), \quad x \in \Omega, \; \nu \in S^2, \tag{8.1.7}$$

when the function $a(x) \geq 0$ is assumed to be known. Adding to this equation the homogeneous boundary condition

$$u(x, \nu) = 0, \quad x \in \partial_- \Omega(\nu), \; \nu \in S^2, \tag{8.1.8}$$

we formulate the following *the inverse source problem* of recovering an unknown source in the transport equation: Find $F(x)$ in (8.1.7) and (8.1.8) from the measured data $g(x, \nu)$ given by (8.1.4).

Remark that the inverse source problem is linear, while the inverse coefficient problem defined by (8.1.3) and (8.1.4) is non-linear.

Let us derive an integral representation for the solution of the inverse source problem. Along the half-line $L(x, \nu) = \{\xi \in \Omega \subset \mathbb{R}^3 : \xi = x - s\nu, \; s \in (0, \infty)\}$ the Eq. (8.1.7) can be rewritten in the form

$$-\frac{d}{ds}\left[u(x - s\nu, \nu)e^{-\int_0^s a(x-s'\nu)ds'}\right] = F(x - s\nu)e^{-\int_0^s a(x-s'\nu)ds'}. \tag{8.1.9}$$

Let $s(x, \nu)$ be the arc length of the segment of $L(x, \nu)$ that belong Ω. Then integrating (8.1.9) along $L(x, \nu)$ for $s \in [0, s(x, \nu)]$ and using the condition (8.1.8) we get

$$u(x, \nu) = \int_0^{s(x,\nu)} F(x - s\nu)e^{-\int_0^s a(x-s'\nu)ds'} ds, \quad (x, \nu) \in \Omega \times S^2.$$

Substituting here $x \in \partial_+ \Omega(\nu)$ and using the additional condition (8.1.4), we obtain we integral equation for unknown function $F(x)$:

$$\int_0^{s(x,\nu)} F(x - s\nu)e^{-\int_0^s a(x-s'\nu)ds'} ds = g(x, \nu), \quad x \in \partial_+ \Omega(\nu), \; \nu \in S^2.$$

Assuming that $F(x) = 0$ outside Ω, the latter equation can also be rewritten in the form

$$g(x, \nu) = \int_{L(x,\nu)} F(x - s\nu)e^{-\int_0^s a(x-s'\nu)ds'} ds, \quad x \in \partial_+ \Omega(\nu), \; \nu \in S^2.$$

Hence, the inverse source problem is reduced to *the attenuated Radon transform*: find $F(x)$ from given integrals along arbitrary half-lines $L(x, \nu)$, $x \in \partial_+ \Omega(\nu)$, $\nu \in S^2$, with the attenuation defined by the coefficient $a(x)$.

The last problem can also be reduced to the two-dimension one for any cross-section of Ω by a plane and choosing $L(x, \nu)$, $x \in \partial_+ \Omega(\nu)$, $\nu \in S^2$, belonging to this cross-section only. An inversion formula for this problem was obtained by R. Novikov (see [79]).

8.2 Uniqueness and a Stability Estimate in the Tomography Problem

Consider the tomography problem (8.1.6) on the plane $(x_1, x_2) \in \mathbb{R}^2$. Let $\Omega = \{x \in \mathbb{R}^2 | \, |x| \leq 1\}$ be the circle and $\partial\Omega = \{x \in \mathbb{R}^2 | \, |x| = 1\}$ be its boundary. Assume here that $\sigma \in C^1(\overline{\Omega})$. Then substituting $x = x(\varphi) = (\cos\varphi, \sin\varphi)$ for $x \in \partial_+\Omega(\nu)$ and $\nu = \nu(\theta) = (\cos\theta, \sin\theta)$, we can rewrite the Eq. (8.1.6) in the following form:

$$\int_{L(x,\nu)} \sigma(\xi)\,ds = \hat{f}(\varphi, \theta), \quad (\varphi, \theta) \in [0, 2\pi] \times [0, 2\pi], \tag{8.2.1}$$

where $\hat{f}(\varphi, \theta) = f(x(\varphi), \nu(\theta))$. For $x \in \Omega$ and $\nu = (\cos\theta, \sin\theta)$ introduce the function

$$v(x, \theta) = \int_{L(x,\nu)} \sigma(x - s\nu)\,ds. \tag{8.2.2}$$

Then this function satisfy the equation

$$\nu(\theta) \cdot \nabla_x v(x, \theta) = \sigma(x), \quad x \in \Omega, \ \theta \in [0, 2\pi] \tag{8.2.3}$$

and the condition

$$v(x(\varphi), \theta) = \hat{f}(\varphi, \theta), \quad (\varphi, \theta) \in [0, 2\pi] \times [0, 2\pi]. \tag{8.2.4}$$

Taking derivative with respect to θ, we exclude the function $\sigma(x)$ from the Eq. (8.2.3) and obtain

$$\frac{\partial}{\partial \theta}(\nu \cdot \nabla_x v(x, \theta)) = 0, \quad x \in \Omega, \ \theta \in [0, 2\pi]. \tag{8.2.5}$$

Use Mukhometov's identity [70], which can be checked directly:

$$2(\nu_\theta \cdot \nabla_x v(x, \theta)) \frac{\partial}{\partial \theta}(\nu \cdot \nabla_x v(x, \theta)) = \frac{\partial}{\partial x_1}(v_\theta v_{x_2}) - \frac{\partial}{\partial x_2}(v_\theta v_{x_1})$$
$$+ |\nabla_x v(x, \theta)|^2 + \frac{\partial}{\partial \theta}[(\nu(\theta) \cdot \nabla_x v(x, \theta))(\nu_\theta(\theta) \cdot \nabla_x v(x, \theta))], \tag{8.2.6}$$

where $\nu_\theta = (-\sin\theta, \cos\theta)$. Then it follows from (8.2.5) and (8.2.6) that

8.2 Uniqueness and a Stability Estimate in the Tomography Problem

$$\frac{\partial}{\partial x_1}(v_\theta v_{x_2}) - \frac{\partial}{\partial x_2}(v_\theta v_{x_1}) + |\nabla_x v(x,\theta)|^2$$
$$+ \frac{\partial}{\partial \theta}[(\nu \cdot \nabla_x v(x,\theta))(\nu_\theta \cdot \nabla_x v(x,\theta))] = 0. \tag{8.2.7}$$

Integrating the equality (8.2.7) over Ω and $[0, 2\pi]$, and then taking into account that $v(x,\theta))$ is a periodic function with respect to θ we get:

$$\int_0^{2\pi} \int_\Omega \left[\frac{\partial}{\partial x_1}(v_\theta v_{x_2}) - \frac{\partial}{\partial x_2}(v_\theta v_{x_1}) + |\nabla_x v(x,\theta)|^2 \right] dx d\theta = 0.$$

Applying Gauss's formula, we obtain

$$\int_0^{2\pi} \int_\Omega |\nabla_x v(x,\theta)|^2 dx d\theta = - \int_0^{2\pi} \int_0^{2\pi} v_\theta(x(\varphi),\theta) v_\varphi(x(\varphi),\theta) d\theta d\varphi.$$

Using (8.2.4) and the relations $\sigma^2(x) = |\nu \cdot \nabla_x v(x,\theta)|^2 \leq |\nabla_x v(x,\theta)|^2$, we find the stability estimate for the tomography problem as follows:

$$\int_\Omega \sigma^2(x) dx \leq -\frac{1}{2\pi} \int_0^{2\pi} \int_0^{2\pi} \hat{f}_\varphi(\varphi,\theta) \hat{f}_\theta(\varphi,\theta) d\theta d\varphi.$$

Thus,

$$\|\sigma\|_{L^2(\Omega)} \leq \frac{1}{2\sqrt{\pi}} \|\hat{f}\|_{H^1([0,2\pi]\times[0,2\pi])}. \tag{8.2.8}$$

The *uniqueness of the solution of the tomography problem* follows from estimate (8.2.8). That is, if data $\hat{f}_k(x,\nu)$ corresponds to $\sigma_k(x)$ for $k = 1, 2$, and $\hat{f}_1(x,\nu) = \hat{f}_2(x,\nu)$, then $\sigma_1(x) = \sigma_2(x)$, $x \in \overline{\Omega}$.

8.3 Inversion Formula

Let $\sigma(x) \in C(\mathbb{R}^2)$ and $\sigma(x) = 0$ for $x \in \mathbb{R}^2 \setminus \Omega$. Assume $\nu = (\cos\theta, \sin\theta)$ and $\nu_\theta = (-\sin\theta, \cos\theta)$, $p \in [0, \infty)$ and define function $f(\theta, p)$ by the formula

$$f(\theta, p) = \int_{-\infty}^\infty \sigma(s\nu_\theta + p\nu) ds, \quad \theta \in [0, 2\pi], \ p \in [0, \infty). \tag{8.3.1}$$

The mapping $\mathfrak{R} : \sigma \to f$ is called the *Radon transform*. It is clear that the Radon transform for fixed (θ, p) is the integral from $\sigma(x)$ along the direct line $\nu \cdot \xi = p$. So, the tomography problem consists in the inverting of the Radon transform. We derive now the inversion formula following [28].

For a fixed $x \in \Omega$ and $p \in [0, \infty)$ consider the equality

$$\frac{1}{2\pi} \int_0^{2\pi} f(\theta, p + x \cdot \nu) d\theta = \frac{1}{2\pi} \int_{-\infty}^{\infty} \int_0^{2\pi} \sigma(s\nu_\theta + p\nu + (x \cdot \nu)\nu) d\theta \, ds$$

$$= \frac{1}{2\pi} \int_{-\infty}^{\infty} \int_0^{2\pi} \sigma((s - x \cdot \nu_\theta)\nu_\theta + p\nu + x) d\theta \, ds$$

$$= \frac{1}{2\pi} \int_{-\infty}^{\infty} \int_0^{2\pi} \sigma(t\nu_\theta + p\nu + x) d\theta \, dt$$

$$= \int_{-\infty}^{\infty} F\left(x, \sqrt{p^2 + t^2}\right) dt, \quad (8.3.2)$$

where

$$F\left(x, \sqrt{p^2 + t^2}\right) = \frac{1}{2\pi} \int_0^{2\pi} \sigma(t\nu_\theta + p\nu + x) d\theta. \quad (8.3.3)$$

Note that for a fixed x the function $F(x, r)$ represents an average value of $\sigma(\xi)$ over the circumference with the center at x and the radius $r = \sqrt{p^2 + t^2}$ since $|\xi - x| = |t\nu_\theta + p\nu| = r = \sqrt{p^2 + t^2}$, and therefore depends on x and r only. Evidently, this is a continuous function of x and r, for all $x \in \Omega$ and $r \in [0, \infty)$. Furthermore, $F(x, r)$ is a finite-valued function of r and $F(x, 0) = \sigma(x)$. Thus, to obtain an inversion formula, we need to calculate $F(x, 0)$. For this goal, we use the following identity, obtained by the change of variable:

$$\int_{-\infty}^{\infty} F\left(x, \sqrt{p^2 + t^2}\right) dt = \int_p^{\infty} F(x, r) \frac{2r \, dr}{\sqrt{r^2 - p^2}}. \quad (8.3.4)$$

Hence,

$$\frac{1}{2\pi} \int_0^{2\pi} f(\theta, p + x \cdot \nu) d\theta = \int_p^{\infty} F(x, r) \frac{2r \, dr}{\sqrt{r^2 - p^2}}. \quad (8.3.5)$$

Applying to the right hand side of the latter equality the operator

$$Lg(s) = -\frac{1}{\pi s} \frac{\partial}{\partial s} \int_s^{\infty} \frac{g(p) p \, dp}{\sqrt{p^2 - s^2}},$$

we find

8.3 Inversion Formula

$$-\frac{1}{\pi s}\frac{\partial}{\partial s}\int_s^\infty \int_p^\infty F(x,r)\frac{2r\,dr}{\sqrt{(r^2-p^2)(p^2-s^2)}} p\,dp$$

$$= -\frac{1}{\pi s}\frac{\partial}{\partial s}\int_s^\infty F(x,r)\int_s^r \frac{2p\,dp}{\sqrt{(r^2-p^2)(p^2-s^2)}} r\,dr$$

$$= -\frac{1}{s}\frac{\partial}{\partial s}\int_s^\infty F(x,r)r\,dr = F(x,s). \qquad (8.3.6)$$

Then from (8.3.5) we obtain the inversion formula as follows

$$\sigma(x) = -\lim_{s\to 0^+}\frac{1}{2\pi^2 s}\frac{\partial}{\partial s}\int_s^\infty \int_0^{2\pi} f(\theta, p+x\cdot\nu)d\theta\frac{p\,dp}{\sqrt{p^2-s^2}}. \qquad (8.3.7)$$

This formula valid and for not a finite-valued function $\sigma(x)$. It is still true if we assume that $\sigma(x)$ tends to zero as $|x|\to\infty$ together with $|x|^k\sigma(x)$ for any positive k.

Various aspects of inverse problems for transport equations can be found in [4, 5, 18].

Chapter 9
The Inverse Kinematic Problem

The problem of recovering the sound speed (or index of refraction) from travel time measurements is an important issue in determining the substructure of the Earth. The same inverse problem can also be defined as the problem of reconstructing of a Riemannian metric inside a compact domain Ω from given distances of geodesics joining arbitrary couples of points belonging to the boundary of Ω. This chapter studies some basic inverse kinematic problems. Specifically, in this chapter we give an analysis of the inverse kinematic problem for one- and two-dimensional cases.

9.1 The Problem Formulation

Let a compact domain $\Omega \in \mathbb{R}^n$ be filled with an isotropic inhomogeneous substance in which waves propagate with the speed $c(x)$. We shall assume that $c(x)$ is an uniformly bounded positive smooth function in the closed domain $\overline{\Omega}$. Introduce the Riemannian metric $d\tau = |dx|/c(x), |dx| = (\sum_{i=1}^{n} dx_i^2)^{1/2}$ and denote by $\tau(x, y)$ the length of the geodesic line $\Gamma(x, y)$ joining points x and y. Physically, the function $\tau(x, y)$ means travel time between points x and y. It is well known that $\tau(x, y)$ is a symmetric function of their arguments, i.-e. $\tau(x, y) = \tau(y, x)$ and satisfies to the eikonal equation

$$|\nabla_x \tau(x, y)|^2 = n^2(x), \tag{9.1.1}$$

where $n(x) = 1/c(x)$ is the refractive index. Moreover, $\tau(x, y)$ satisfy the additional condition

$$\tau(x, y) \sim n(y)|x - y|, \text{ as } x \to y. \tag{9.1.2}$$

The latter condition means that the travel time $\tau(x, y)$ can be calculated as in homogeneous medium with speed equal to $c(y)$ if x close to y. In the next subsection we consider how to construct the function $\tau(x, y)$ and the geodesic lines $\Gamma(x, y)$ for a given function $c(x)$.

Now we formulate the inverse problem: find $c(x)$ in Ω from $\tau(x, y)$ given for all $(x, y) \in (\partial\Omega \times \partial\Omega)$. Here $\partial\Omega$ is a smooth boundary of Ω.

The above problem is called *the inverse kinematic problem*. It has various applications in the geophysics and other sciences. It was noted for a long time ago that the inverse kinematics problem has no unique solution if the medium has inner wave guides. Therefore this problem is studied usually under assumption that field of geodesic lines is regular inside Ω. The latter means that each couple of points x, y can be joined in Ω by only one geodesic line and boundary $\partial\Omega$ of Ω is convex with respect to these geodesics.

9.2 Rays and Fronts

Here we construct a solution to problem (9.1.1) and (9.1.2) following the well known method of solving the first order partial differential equations. We assume that $\tau(x, y)$ is a twice continuously differentiable function for all $x \neq y$. Introduce the function $p = p(x, y) := \nabla_x \tau(x, y)$. Then Eq. (9.1.1) can be rewritten in the form

$$\sum_{j=1}^{n} p_j^2(x, y) = n^2(x), \qquad (9.2.1)$$

where p_j denote j-th component of the vector p. Taking the derivative of both sides of (9.2.1) with respect to x_k, one obtains

$$2\sum_{j=1}^{n} \frac{\partial p_j}{\partial x_k} p_j = 2n(x) n_{x_k}(x), \quad k = 1, 2, \ldots, n. \qquad (9.2.2)$$

Since

$$\frac{\partial p_j}{\partial x_k} = \frac{\partial^2 \tau}{\partial x_j \partial x_k} = \frac{\partial p_k}{\partial x_j}$$

Equation (9.2.2) can be transformed to the form

$$\sum_{j=1}^{n} \frac{\partial p_k}{\partial x_j} p_j = n(x) n_{x_k}(x), \quad k = 1, 2, \ldots, n. \qquad (9.2.3)$$

Introduce the new variable parameter s such that value $s = 0$ corresponds to point y and consider in $\mathbb{R}^n$ the curves determined by the equations

9.2 Rays and Fronts

$$\frac{dx_j}{ds} = \frac{p_j}{n^2(x)}, \quad i = 1, 2, \ldots, n. \tag{9.2.4}$$

Along these curves one has

$$\sum_{j=1}^{n} \frac{\partial p_k}{\partial x_j} p_j = n^2(x) \sum_{j=1}^{n} \frac{\partial p_k}{\partial x_j} \frac{dx_j}{ds} = n^2(x) \frac{dp_k}{ds}.$$

Hence, Eq. (9.2.3) are the ordinary differential equations

$$\frac{dp_k}{ds} = \frac{\partial \ln n(x)}{\partial x_k}, \quad k = 1, 2, \ldots, n. \tag{9.2.5}$$

Equations (9.2.4) and (9.2.5) form the system of differential equations for x and p that determines x, p as functions of s. Let us explain the meaning of the variable s. To this end, calculate the derivative of τ with respect to s along the curves determined by Eqs. (9.2.4) and (9.2.5). We find:

$$\frac{d\tau}{ds} = \sum_{j=1}^{n} \tau_{x_j} \frac{dx_j}{ds} = \sum_{j=1}^{n} \frac{p_j^2}{n^2(x)} = 1.$$

Hence, $\tau = s$ because s and τ equal zero at point y. Therefore Eqs. (9.2.4) and (9.2.5) can be written in the form

$$\frac{dx}{d\tau} = \frac{p}{n^2(x)}, \quad \frac{dp}{d\tau} = \nabla \ln n(x). \tag{9.2.6}$$

Equation (9.2.6) are called *Euler's equations*. The curves determined by Euler's equations are called the *characteristic lines*. In space $\mathbb{R}^n$ they determine the geodesic lines $\Gamma(x, y)$. In geophysical applications these lines are defined as *rays*. In order to find these lines consider the solution to Eq. (9.2.6) with the following initial data

$$x|_{\tau=0} = y, \quad p|_{\tau=0} = p^0, \tag{9.2.7}$$

where $p^0 = (p_1^0, \ldots, p_n^0)$ is an arbitrary vector satisfying to the condition

$$|p^0| = n(y). \tag{9.2.8}$$

Then the solution to the Cauchy problem (9.2.6) and (9.2.8) gives the geodesic $\Gamma(x, y)$ which goes from point y in the direction determined by the vector

$$\zeta^0 = \left.\frac{dx}{d\tau}\right|_{\tau=0} = \frac{p^0}{n^2(y)}. \tag{9.2.9}$$

Hence, $\zeta^0 = (\zeta_1^0, \ldots, \zeta_n^0)$ satisfies to the relation

$$n(y)|\zeta^0| = 1. \tag{9.2.10}$$

Problem (9.2.6) and (9.2.7) has a unique solution at least in the vicinity of the point y if $n(x)$ is a smooth function. Namely, if $n(x) \in \mathbf{C}^k(\mathbb{R}^n)$ with $k \geq 2$ then the solution can be represented in the form

$$x = f_1(\tau, y, \zeta^0), \quad p = f_2(\tau, y, \zeta^0),$$

where f_1, f_2 are $\mathbf{C}^{k-1}$-smooth functions of τ, y, ζ^0. Moreover, these functions can be represented in the following form

$$x = f(\tau\zeta^0, y), \quad p = \hat{f}(\tau\zeta^0, y)/\tau,$$

where $f(\tau\zeta^0, y) = f_1(1, y, \tau\zeta^0)$, $\hat{f}(\tau\zeta^0, y) = f_2(1, y, \tau\zeta^0)$ The latter is a consequence of the following observation: if we introduce $\hat{p}, \hat{\tau}$ and $\hat{p}^0, \hat{\zeta}^0$ by the relations $p = \lambda\hat{p}, \tau = \hat{\tau}/\lambda, p^0 = \lambda\hat{p}^0, \zeta^0 = \lambda\hat{\zeta}^0$, where $\lambda > 0$, and simultaneously replace p, τ, p^0 on $\hat{p}, \hat{\tau}, \hat{p}^0$ in (9.2.6) and (9.2.7), these equations are not changed. Hence, $x, \hat{p}$ can be given in the form

$$x = f_1(\hat{\tau}, y, \hat{\zeta}^0) = f_1(\lambda\tau, y, \zeta^0\lambda^{-1}), \quad \hat{p} = f_2(\hat{\tau}, y, \hat{\zeta}^0) = f_2(\lambda\tau, y, \zeta^0\lambda^{-1}).$$

Taking here $\lambda = 1/\tau$, one gets the above representation $x = f(\zeta, y), p = \hat{f}(\zeta, y)/\tau$, where $\zeta = \tau\zeta^0$. The components $\zeta_1, \ldots, \zeta_n$ of the variable ζ are called *the Riemannian coordinates* of the point x with respect to fixed point y. From relations (9.2.6) and the definition (9.2.9) of ζ^0 follows that in a vicinity of y one has

$$x = f(\zeta, y) = y + \left.\frac{\partial x}{\partial \tau}\right|_{\tau=0} \tau + \mathcal{O}(\tau^2) = y + \zeta + \mathcal{O}(|\zeta|^2). \tag{9.2.11}$$

Hence,

$$\det\left(\frac{\partial x}{\partial \zeta}\right)\bigg|_{\zeta=0} = 1. \tag{9.2.12}$$

It means that function $x = f(\zeta, y)$ has an inverse $\zeta = g(x, y)$ determined, at least, in the vicinity of point the y. Furthermore, the inverse function is smooth, that is, if $n(x) \in \mathbf{C}^k(\mathbb{R}^n)$, then it belongs to $\mathbf{C}^{k-1}$. It follows from (9.2.10) that the following representation holds for the function $\tau^2(x, y)$:

$$\tau^2(x, y) = n^2(x)|g(x, y)|^2. \tag{9.2.13}$$

Since $p = \nabla\tau$, one also has

9.2 Rays and Fronts

$$\nabla \tau^2(x, y) = 2\hat{f}(g(x, y), y). \tag{9.2.14}$$

Now it is obvious that $\tau^2(x, y)$ is C^k-smooth function of x and y, at least, for x close to y. For the case, when section curvatures of the Riemannian metric are non-positive, the geodesic lines $\Gamma(x, y)$ passing through the point y have no point of intersection, for any x, except the starting point y, and in addition, $\tau^2(x, y)$ is a smooth function anywhere. The condition $\tau(x, y) = \mathcal{O}(|x - y|)$ as $x \to y$ is a consequence of (9.2.11) which is equivalent to the relation

$$\zeta = x - y + \mathcal{O}(|x - y|^2) \quad \text{as } x \to y. \tag{9.2.15}$$

Thus, in order to find the geodesic line which passes through the point y in direction ζ^0 satisfying condition (9.2.10), one should solve the Cauchy problem (9.2.6) and eq6.9), where $p^0 = \zeta^0 n^2(y)$. Each of these geodesic lines with the common point y can be represented by the equation $x = f(\zeta, y)$, where $\zeta = \zeta^0 \tau$, where $f(\zeta, y)$ is a smooth function and has the inverse one $\zeta = g(x, y)$. Then formula (9.2.13) determines $\tau(x, y)$ satisfying relations (9.1.1) and 9.1.2). Physically, the Riemannian sphere $\tau(x, y) = constant$, presents the front of a wave propagating from a point source placed at y. Since $p = \nabla_x \tau(x, y)$ is the normal to this sphere, it follows from (7.2.6) that geodesic lines $\Gamma(x, y)$ are orthogonal to the front. Briefly speaking, the rays and fronts are orthogonal one to other.

Note that p^0 can be represented in the form

$$p^0 = -\nabla_y \tau(x, y). \tag{9.2.16}$$

Indeed, the vector $\nabla_y \tau(x, y)$ is directed at the point y in the direction of outward normal to the Riemannian sphere $\tau(x, y) = constant$ centered at x. On the other hand, this vector is in the tangent direction to $\Gamma(x, y)$ at y, but in opposite to p^0 direction. Hence, we need to put sign $(-)$ in (9.2.16) to obtain p^0.

The formulae (7.2.9) and (7.2.16) imply the following relation:

$$\zeta = -\frac{1}{2n^2(y)} \nabla_y \tau^2(x, y), \tag{9.2.17}$$

which expresses the Riemannian variable ζ through the geodesic distance between the points x and y.

9.3 The One-Dimensional Problem

Let $\mathbb{R}^2_+ = \{(x, y) \in \mathbb{R}^2 | y \geq 0\}$ and a speed c in this half-space be a positive function of y, i.e., $c = c(y)$. Suppose that $c \in \mathbf{C}^2[0, \infty)$ and its derivative $c'(y)$ is positive on $[0, \infty)$. Then a perturbation produced at the origin $(0, 0)$ reaches to a point $(\xi, 0)$, $\xi > 0$, along a smooth curve, which belongs (except of its ends) to $\mathbb{R}^2_+$ and it is symmetric with respect to the line $x = \xi/2$. In geophysics this curve is called the *ray*. From a mathematical point of view, it is a geodesic line for the Riemannian metric with an element of a length $d\tau$ determined by the formula $d\tau = \sqrt{dx^2 + dy^2}/c(y)$. Denote by $t(\xi)$ the corresponding travel time along the ray. For ξ small enough, $t(\xi)$ is a single-valued monotonic increasing function. It may be false if ξ is not small. Assume here that $t(\xi)$ is a single-valued function on the interval $(0, \xi_0)$.

Consider the following problem: given $t(\xi)$ for $\xi \in (0, \xi_0)$ find $c(y)$ in $\mathbb{R}_+^2$, where it is possible. The problem is called *the one-dimensional inverse kinematic problem*. This problem has been solved by G. Herglotz [42] in the beginning of the last century. He obtains an explicit formula for the solution to the problem. Below we explain the main ideas, which solve the problem.

Consider the function $\tau(x, y)$, which means here the travel time from the origin to point $(x, y) \in \mathbb{R}_+^2$. Then eikonal equation has the form

$$\left(\frac{\partial \tau(x,y)}{\partial x}\right)^2 + \left(\frac{\partial \tau(x,y)}{\partial y}\right)^2 = n^2(y), \qquad (9.3.1)$$

where $n(y) = 1/c(y)$. Use now Euler's equations (9.2.6). Then the equations can be rewritten as follows

$$\frac{dx}{d\tau} = \frac{p}{n^2(y)}, \quad \frac{dy}{d\tau} = \frac{q}{n^2(y)},$$
$$\frac{dp}{d\tau} = 0, \quad \frac{dq}{d\tau} = (\ln n(y))'. \qquad (9.3.2)$$

Here $p = \tau_x(x, y)$ and $q = \tau_y(x, y)$. It follows from these equations that p is a constant along the geodesic line $\Gamma(x, y)$ joined the origin and the point (x, y). Particularly, for $\Gamma(\xi, 0)$ the parameter p is determined by ξ, i.e., $p = p(\xi)$.

From equation (9.3.1) we deduce that $q = \pm\sqrt{n^2(y) - p^2}$. The sign $(+)$ should be taken at those points of the geodesic line where $\tau_y(x, y) > 0$ and sign $(-)$ for the points in which $\tau_y(x, y) < 0$. From (9.3.2) we find that

$$\frac{dx}{dy} = \pm\frac{p}{\sqrt{n^2(y) - p^2}}, \quad \frac{d\tau}{dy} = \pm\frac{n^2(y)}{\sqrt{n^2(y) - p^2}}. \qquad (9.3.3)$$

Suppose that the equation $n(y) = p$ has the solution $y = \eta > 0$, $\eta = \eta(p)$. Then the ray $\Gamma(\xi, 0)$ tangents to the straight line $y = \eta$ and represents the symmetric *arc* passing through the origin and the point $(\xi, 0)$. This *arc* has its vertex at the line $y = \eta$ and belongs to the strip $\{(x, y) | 0 \leq y \leq \eta\}$. The coordinates of the vertex are $(\xi/2, \eta)$. In the Eq. (9.3.3) the sign $(+)$ corresponds to the part of the ray from the origin to point $(\xi/2, \eta)$ and $(-)$ to the remain part. Integrating (9.3.3), one finds the relations for ξ and $t(\xi)$ in the form

9.3 The One-Dimensional Problem

$$\xi = 2\int_0^\eta \frac{p\,dy}{\sqrt{n^2(y)-p^2}} \quad t(\xi) = 2\int_0^\eta \frac{n^2(y)\,dy}{\sqrt{n^2(y)-p^2}}. \qquad (9.3.4)$$

Recall that here $n(\eta) = p$.

The Eq. (9.3.4) form the complete system of relations for the inverse problems. To give an analysis of this system, we make a change of variables in the integrals by introducing the new integration variable $z = n(y)$. Since $n(y)$ is a monotone decreasing function, there exists an inverse monotone function $y = f(z)$ such that $n(f(z)) \equiv z$. Note that the derivative $f'(z)$ is negative and $f(p_0) = 0$, where $p_0 = n(0)$. Then the relations (9.3.4) can be rewritten in the following form

$$\xi = 2\int_{p_0}^p \frac{p\,f'(z)\,dz}{\sqrt{z^2-p^2}}, \quad t(\xi) = 2\int_{p_0}^p \frac{z^2\,f'(z)\,dz}{\sqrt{z^2-p^2}}, \quad p \le p_0. \qquad (9.3.5)$$

The first relation in (9.3.5) presents ξ as a function of p, i.e., $\xi = \xi(p)$, while the second relation determines the other function $t = \hat{t}(p) := t(\xi(p))$. The pair $\langle \xi(p), t(\xi(p)) \rangle$ gives the parametric representation of the function $t = t(\xi)$. Note that the function $\hat{t}(p)$ can be presented in the form

$$\hat{t}(p) = p\,\xi(p) + 2\int_{p_0}^p \sqrt{z^2-p^2}\,f'(z)\,dz, \quad p \le p_0. \qquad (9.3.6)$$

From (9.3.6) we deduce:

$$t'(\xi) = \frac{\hat{t}'(p)}{\xi'(p)} = p, \quad p \le p_0. \qquad (9.3.7)$$

Hence, having the function $t(\xi)$ we can find the correspondence between ξ and p, i.e. the function $\xi = \xi(p)$. This function is defined for $p \in [p_1, p_0]$, where $p_1 = t'(\xi_0)$. Then we can use the first equation (9.3.5) in order to find $f(z)$ for $z \in [p_1, p_0]$. Namely,

$$f(s) = \frac{1}{\pi}\int_s^{p_0} \frac{\xi(p)\,dp}{\sqrt{p^2-s^2}}, \quad s \in [p_1, p_0]. \qquad (9.3.8)$$

Indeed,

$$\frac{1}{\pi}\int_s^{p_0} \frac{\xi(p)\,dp}{\sqrt{p^2-s^2}} = \frac{2}{\pi}\int_s^{p_0}\int_{p_0}^p \frac{f'(z)\,dz}{\sqrt{(p^2-s^2)(z^2-p^2)}} p\,dp$$

$$= \frac{1}{\pi}\int_{p_0}^s\int_s^z \frac{2p\,dp}{\sqrt{(p^2-s^2)(z^2-p^2)}}f'(z)\,dz = \int_{p_0}^s f'(z)\,dz = f(s).$$

This means that $n(y) = f^{-1}(y)$ is determined for $y \in [0, f(p_1)]$. Thus, given $t(\xi)$ for $\xi \in [0, \xi_0]$ one can uniquely find function $f(z)$ for $z \in [t'(\xi_0), p_0]$ and then $n(y)$ inside the layer $y \in [0, f(t'(\xi_0))]$.

If $c'(y) \le 0$ for all y, then the ray $\Gamma(\xi, 0)$ reaches the point $(\xi, 0)$ along axis x. Hence, $t(\xi) = |\xi|/c(0)$ and only $c(0)$ can be found in the inverse problem. A more interesting case is when the function $c'(y)$ is positive for $[0, y_0)$ and changes sign at $y = y_0 > 0$. As was noted, in this case $c(y)$ can be determined uniquely for $y \in [0, y_0]$. Remark that it is impossible uniquely find $c(y)$ also for $y > y_0$ using the data of the inverse problem. A characterization of a set of all possible solutions to the inverse problem in this case was given by Gerver and Markushevich [29].

9.4 The Two-Dimensional Problem

Let $\Omega \subset \mathbb{R}^2$ be a compact domain with C^1-smooth boundary S. Assume, furthermore, that the positive function $n(x) \subset C^2(\overline{\Omega})$ is defined in $\overline{\Omega} = \Omega \cup S$. Suppose that the family of geodesic lines $\Gamma(x, y)$, $x \in \overline{\Omega}$, $y \in S$, of the Riemannian metric $d\tau = n(x)|dx|$ is regular in $\overline{\Omega}$ and S is convex with respect to geodesics $\Gamma(x, y)$, $x \in S$, $y \in S$. Such functions $n(x)$ we shall call *admissible*. We define the function $\tau(x, y)$ for $x \in \overline{\Omega}$, $y \in S$, as the *Riemannian distance* (travel time, from a physical point of view) between x and y.

Consider the problem: find $n(x)$ in $\overline{\Omega}$ from $\tau(x, y)$ given for $x \in S$ and $y \in S$.

We will give now a stability estimate and uniqueness theorem for the solution of this inverse problem, following to [70]. Let S be given in the parametric form $S = \{y \in \mathbb{R}^2 | y = \chi(t), t \in [0, T]\}$, where $\chi(t)$ is a C^1-smooth periodic function with the period T. We assume that an increase of t corresponds to mowing point $y = \chi(t)$ in counterclockwise. Then the output data $\tau(x, y)$, $x \in S$ and $y \in S$, of the inverse problem can be written as follows:

$$\tau(\chi(s), \chi(t)) = g(s, t), \quad (s, t) \in [0, T] \times [0, T], \tag{9.4.1}$$

where $g(s, t)$ is a given function.

Theorem 9.4.1 *Let $n_1(x)$ and $n_2(x)$ be admissible functions and $g_1(s, t)$ and $g_2(s, t)$ be the output data (9.4.1) for the Riemannian metrics $d\tau_1 = n_1(x)|dx|$ and $d\tau_2 = n_2(x)|dx|$. Then the following stability estimate holds*

$$\|n_1 - n_2\|_{L^2(\Omega)} \le \frac{1}{2\sqrt{\pi}} \|g_1 - g_2\|_{H^1([0,T] \times [0,T])}. \tag{9.4.2}$$

Proof. Let $n_j(x)$ for $j = 1, 2$ be admissible functions and $\tau_j(x, y)$ corresponding them Riemannian distances between x and y. Introduce the functions $\hat{\tau}_j(x, t) = \tau_j(x, \chi(t))$. Then

9.4 The Two-Dimensional Problem

$$\hat{\tau}_j(\chi(s),t) = g_j(s,t), \quad (s,t) \in [0,T] \times [0,T], \quad j=1,2, \quad (9.4.3)$$

and

$$\nabla_x \hat{\tau}_j(x,t) = n_j(x)\nu(\theta_j), \quad \nu(\theta_j) = (\cos\theta_j, \sin\theta_j), \quad j=1,2, \quad (9.4.4)$$

where $\theta_j = \theta_j(x,t)$ is the angle between the tangent line to the geodesic $\Gamma_j(x,\chi(t))$ and axis x_1. Denote

$$\tilde{\tau}(x,t) = \hat{\tau}_1(x,t) - \hat{\tau}_2(x,t), \quad \tilde{n}(x) = n_1(x) - n_2(x),$$
$$\tilde{\theta}(x,t) = \theta_1(x,t) - \theta_2(x,t), \quad \tilde{g}(s,t) = g_1(s,t) - g_2(s,t).$$

From Eq. (9.4.4) we get

$$\nabla_x \tilde{\tau}(x,t) = \tilde{n}(x)\nu(\theta_1) + n_2(x)(\nu(\theta_1) - \nu(\theta_2)). \quad (9.4.5)$$

Multiplying both sides on $\nu(\theta_1)$ we obtain

$$\nu(\theta_1) \cdot \nabla_x \tilde{\tau}(x,t) = \tilde{n}(x) + n_2(x)(1 - \cos\tilde{\theta}). \quad (9.4.6)$$

Then taking derivative with respect to t, we exclude $\tilde{n}(x)$ and obtain the equation

$$\frac{\partial}{\partial t}[\nu(\theta_1) \cdot \nabla_x \tilde{\tau}(x,t)] = n_2(x)\sin\tilde{\theta}\,\frac{\partial\tilde{\theta}}{\partial t}, \quad x \in \Omega,\ t \in [0,T]. \quad (9.4.7)$$

Let $\nu^\perp(\theta_1) = (-\sin\theta_1, \cos\theta_1)$. Multiply both sides of (9.4.7) on $2\nu^\perp(\theta_1) \cdot \nabla_x \tilde{\tau}(x,t)$, we get

$$2(\nu^\perp(\theta_1) \cdot \nabla_x \tilde{\tau}(x,t))\frac{\partial}{\partial t}[\nu(\theta_1) \cdot \nabla_x \tilde{\tau}(x,t)]$$
$$= 2n_2(x)(\nu^\perp(\theta_1) \cdot \nabla_x \tilde{\tau}(x,t))\sin\tilde{\theta}\,\frac{\partial\tilde{\theta}}{\partial t}. \quad (9.4.8)$$

Transform both sides of the latter equality separately. For the left hand side use the identity

$$2(\nu^\perp(\theta_1) \cdot \nabla_x \tilde{\tau}(x,t))\frac{\partial}{\partial t}[\nu(\theta_1) \cdot \nabla_x \tilde{\tau}(x,t)]$$
$$= \frac{\partial}{\partial t}\{[\nu(\theta_1) \cdot \nabla_x \tilde{\tau}(x,t)](\nu^\perp(\theta_1) \cdot \nabla_x \tilde{\tau}(x,t))\}$$
$$+ \frac{\partial}{\partial x_1}[\tilde{\tau}_t(x,t)\tilde{\tau}_{x_2}(x,t)] - \frac{\partial}{\partial x_2}[\tilde{\tau}_t(x,t)\tilde{\tau}_{x_1}(x,t)] + |\nabla_x \tilde{\tau}(x,t)|^2 \frac{\partial\theta_1}{\partial t}, \quad (9.4.9)$$

which can be checked directly. On the other hand, we have

$$2n_2(x)(\nu^\perp(\theta_1) \cdot \nabla_x \tilde{\tau}(x,t)) \sin \tilde{\theta} \frac{\partial \tilde{\theta}}{\partial t}$$

$$= 2n_2(x)[\nu^\perp(\theta_1) \cdot (n_1(x)\nu(\theta_1) - n_2(x)\nu(\theta_2))] \sin \tilde{\theta} \frac{\partial \tilde{\theta}}{\partial t}$$

$$= -2n_2^2(x) \sin^2 \tilde{\theta} \frac{\partial \tilde{\theta}}{\partial t} = -\frac{\partial}{\partial t}\left\{n_2^2(x)\left[\tilde{\theta} - \frac{1}{2}\sin(2\tilde{\theta})\right]\right\}. \qquad (9.4.10)$$

As a result of the equalities (9.4.8) and (9.4.10) we arrive at the relation

$$\frac{\partial}{\partial t}\left\{[\nu(\theta_1) \cdot \nabla_x \tilde{\tau}(x,t)](\nu^\perp(\theta_1) \cdot \nabla_x \tilde{\tau}(x,t)) + n_2^2(x)\left[\tilde{\theta} - \frac{1}{2}\sin(2\tilde{\theta})\right]\right\}$$
$$+ \frac{\partial}{\partial x_1}[\tilde{\tau}_t(x,t)\tilde{\tau}_{x_2}(x,t)] - \frac{\partial}{\partial x_2}[\tilde{\tau}_t(x,t)\tilde{\tau}_{x_1}(x,t)] + |\nabla_x \tilde{\tau}(x,t)|^2 \frac{\partial \theta_1}{\partial t} = 0. \qquad (9.4.11)$$

Note that in the curly brackets of the latter equality stands the T periodic function of t. Therefore, if one takes integral with respect to t over $[0, T]$ of the first therm, it vanishes. Integrating (9.4.11) over Ω with respect to x and over $[0, T]$ with respect to t, we get

$$\int_0^T \int_\Omega \left\{ \frac{\partial}{\partial x_1}[\tilde{\tau}_t(x,t)\tilde{\tau}_{x_2}(x,t)] - \frac{\partial}{\partial x_2}[\tilde{\tau}_t(x,t)\tilde{\tau}_{x_1}(x,t)] \right.$$
$$\left. + |\nabla_x \tilde{\tau}(x,t)|^2 \frac{\partial \theta_1}{\partial t} \right\} dx dt = 0. \qquad (9.4.12)$$

Using Gauss's formula and the condition (9.4.3), we obtain

$$\int_0^T \int_0^T \tilde{g}_t(s,t)\tilde{g}_s(s,t) ds dt + \int_\Omega \int_0^T |\nabla_x \tilde{\tau}(x,t)|^2 \frac{\partial \theta_1}{\partial t} dt dx = 0. \qquad (9.4.13)$$

In the latter equality $\partial \theta_1/\partial t \geq 0$, due the regularity condition of geodesics and

$$|\nabla_x \tilde{\tau}(x,t)|^2 = |n_1(x)\nu(\theta_1) - n_2(x)\nu(\theta_2)|^2$$
$$= \tilde{n}^2(x) + 2n_1(x)n_2(x)(1 - \cos \tilde{\theta}) \geq \tilde{n}^2(x).$$

Therefore

$$2\pi \int_\Omega \tilde{n}^2(x) dx \leq -\int_0^T \int_0^T \tilde{g}_t(s,t)\tilde{g}_s(s,t) ds dt.$$

Here $H^1([0, T] \times [0, T])$ is the Sobolev space of square integrable functions $g(s, t)$ in the domain $\{(s, t) : s \in [0, T], t \in [0, T]\}$ together with first partial derivatives with respect to s and t. The required estimate (9.4.2) follows from this inequality. □

As a consequence of Theorem 9.4.1 we obtain the following uniqueness theorem.

9.4 The Two-Dimensional Problem

Theorem 9.4.2 *Let $n_1(x)$, $n_2(x)$ and $g_1(s,t)$, $g_2(s,t)$ be the functions defined in Theorem 9.4.1 and $g_1(s,t) = g_2(s,t)$. Then $n_1(x) = n_2(x)$ for all $x \in \overline{\Omega}$.*

We refer to the book [86] and papers [29, 42, 70, 82] for some other aspects of the inverse kinematic problems. Recently, an effective iterative method for reconstruction of the refractive index of a medium from time-off-flight measurements has been proposed in [89].

Appendix A
Invertibility of Linear Operators

A.1 Invertibility of Bounded Below Linear Operators

The analysis given in Introduction allows us to conclude that most inverse problems are governed by compact operators. More specifically, all inverse problems related to PDEs are governed by corresponding input-output operators which are compact operator. The last property mainly resulted from the Sobolev embedding theorems. Therefore *compactness of an operator, being an integral part of inverse problems, is a main source of ill-posedness of inverse problems.*

We begin some well-known fundamental results related to invertibility of linear operators in Banach and, in particular, in Hilbert spaces, since compact operators on a Banach space are always completely continuous. Details of these results can be found in textbooks on functional analysis (see, for example, [3, 20, 103, 104]).

Let $A : \mathcal{D} \subset B \mapsto \tilde{B}$ be a linear continuous operator on the Banach space B *into* the Banach space $\tilde{B}$. Consider the operator equation $Au = F$, $F \in \mathcal{R}(A) \subset \tilde{B}$, where $\mathcal{R}(A)$ is the range of the operator A. We would like to ask whether the linear continuous operator A has a bounded inverse, because this is a first step for solving the operator equation $Au = F$. By definition of an inverse operator, if for any $F \in \mathcal{R}(A)$ there is at most one $u \in \mathcal{D}(A)$ such that $Au = F$, i.e. if the operator A is injective, then the correspondence from $\tilde{B}$ to B defines an operator, which is called an *inverse operator*. This definition directly implies that A^{-1} exists if and only if the equation $Au = 0$ has only the unique solution $u = 0$, i.e. the null space $\mathcal{N}(A) := \{u \in B : Au = 0\}$ of the operator A consists of only zero element: $\mathcal{N}(A) = \{0\}$. Evidently, if A^{-1} exists, it is also a linear operator. Note that if H is a Hilbert space, then by the *orthogonal decomposition theorem*, $H = \mathcal{N}(A) \bigoplus \mathcal{N}(A)^\perp$, where $\mathcal{N}(A) \bigoplus \mathcal{N}(A)^\perp := \{z \in H : z = u + v, \ u \in \mathcal{N}(A), v \in \mathcal{N}(A)^\perp\}$ and $\mathcal{N}(A)^\perp$ is the orthogonal complement of $\mathcal{N}(A)$, i.e. $(u, v)_H = 0$ for all $u \in \mathcal{N}(A)$ and for all $v \in \mathcal{N}(A)^\perp$.

The first group of results in this direction is related to invertibility of bounded below linear operators defined on Banach space.

Definition A.1.1 A linear operator A defined on the Banach space B into the Banach space $\tilde{B}$ is called bounded below if there exists such a positive constant $C > 0$ that

$$\|Au\|_{\tilde{B}} \geq C\|u\|_B, \quad \forall u \in \mathcal{D}(A). \tag{A.1.1}$$

The theorem below shows an important feature of bounded below linear operators. Specifically, a bounded below linear operator A defined between the Banach spaces B and $\tilde{B}$ has always a continuous inverse defined on the range $\mathcal{R}(A)$ of A, if even A is not continuous.

Theorem A.1.1 *Let B and $\tilde{B}$ be Banach spaces and A be a linear continuous operator defined on B into $\tilde{B}$. Then the inverse operator A^{-1}, defined on $\mathcal{R}(A)$, exists and bounded if and only if it is bounded below.*

Proof Assume, first, that the inverse operator A^{-1} exists and bounded on $\mathcal{R}(A)$. Then there exists a constant $M > 0$ such that $\|A^{-1}F\|_B \leq M\|F\|_{\tilde{B}}$, for all $F \in \mathcal{R}(A)$. Substituting here $u := A^{-1}F$ we get: $\|u\|_B \leq M\|Au\|_{\tilde{B}}$. This implies (A.1.1) with $C = 1/M$, i.e. A is bounded below.

Assume now condition (A.1.1) holds. This, first of all, implies injectivity of the operator A. Indeed, let $u_1, u_2 \in B$ be arbitrary elements and $Au_1 = Au_2 = v \in \mathcal{R}(A)$. Then, by condition (A.1.1),

$$0 = \|Au_1 - Au_2\|_{\tilde{B}} = \|A(u_1 - u_2)\|_{\tilde{B}} \geq C\|u_1 - u_2\|_B.$$

This $\|u_1 - u_2\|_B = 0$, i.e. implies $u_1 = u_2$. Hence for any $F \in \mathcal{R}(A)$ there is at most one $u \in \mathcal{D}(A)$ such that $Au = F$, i.e. the inverse operator A^{-1} exists. To prove the boundedness of this operator, we substitute $u = A^{-1}F$ in (A.1.1). Then we have: $\|F\|_{\tilde{B}} \geq C\|A^{-1}F\|_B$. This means the boundedness of the inverse operator. □

Thus, this theorem gives a necessary and sufficient condition for solvability of the operator equation $Au = F$ with bounded below linear operator A. A simple illustration of this theorem can be given by the following example.

Example A.1.1 An existence of a bounded inverse.

Consider the following boundary value problem.

$$\begin{cases} (Au)(x) := -u''(x) = F(x), \ x \in (0, 1), \\ u(0) = u'(1) = 0. \end{cases} \tag{A.1.2}$$

We rewrite problem (A.1.2) in the operator equation form

$$Au = F, \quad A : \mathcal{D}(A) \xrightarrow{\text{onto}} C[0, 1],$$

which $\mathcal{D}(A) := C^2(0, 1) \cap C_0[0, 1]$, $C_0[0, 1] := \{u \in C[0, 1] : u(0) = u'(1) = 0\}$ and $\mathcal{R}(A) := C[0, 1]$. We can show easily that the operator A is bounded below. Indeed, the identities

$$u(x) \equiv \int_0^x u'(\xi)d\xi, \ u'(x) \equiv -\int_x^1 u''(\xi)d\xi, \ \forall x \in [0, 1]$$

imply

$$|u(x)| \leq \sup_{[0,1]} |u'(x)|, \ |u'(x)| \leq \sup_{[0,1]} |u''(x)|.$$

Hence,

$$\|Au\|_{C[0,1]} := \sup_{[0,1]} |u''(x)| \geq \|u\|_{C[0,1]},$$

which implies that the operator A is bounded below.

Integration equation (A.1.2) and using boundary conditions, we find

$$u(x) = \int_0^x \int_\xi^1 F(\eta) d\eta d\xi =: (A^{-1}F)(x), \ x \in [0, 1].$$

By definition of an inverse operator, the inverse operator $A^{-1} : \mathcal{R}(A) \mapsto \mathcal{D}(A)$, defined by (A.1.3), exists. Evidently, A^{-1} is bounded, although the operator $A := d^2/dx^2$ is unbounded. □

As the second part of the proof of the above theorem shows, condition (A.1.1) contains within itself two conditions: injectivity of the operator A and continuity of the inverse operator A^{-1}. So, injectivity is a part of conditions for invertibility of linear operators. Further, if, in addition to injectivity, $\mathcal{R}(A) \equiv \tilde{B}$, i.e. the operator A is defined on B onto $\tilde{B}$ or, equivalently, operator A is surjective, then the linear continuous operator $A : \mathcal{D} \subset B \mapsto \tilde{B}$ is continuously invertible *for all $F \in \tilde{B}$* and $\mathcal{D}(A^{-1}) = \tilde{B}$. In this case the operator equation $Au = F$ has a unique solution *for all $F \in \tilde{B}$*, while Theorem A.1.1 asserts that the operator equation $Au = F$ solvable *only for $F \in \mathcal{R}(A)$*. In other words, under the conditions of only Theorem A.1.1, the operator equation $Au = F$ *may not have a solution for all $F \in \tilde{B}$*. As a result of these considerations, we conclude that the best case for unique solvability of the operator equation $Au = F$ is the case when the linear continuous operator A is bijective, i.e. injective and surjective. The following result, which follows from Open Mapping Theorem, is exactly in this direction.

Theorem A.1.2 *Let $A : B \mapsto \tilde{B}$ be a linear continuous and bijective operator from a Banach space B to a Banach space $\tilde{B}$. Then the inverse operator $A^{-1} : \tilde{B} \mapsto B$ is also continuous (and hence bounded).*

A.2 Invertibility of Linear Compact Operators

Let us return again to the surjectivity property of the linear continuous operator A, i.e. to the condition $\mathcal{R}(A) \equiv \tilde{B}$. This condition means, first of all, *the range $\mathcal{R}(A)$ needs to be a closed set*, since $\tilde{B}$ being a Banach space is a closed set. In this case, i.e. if $\mathcal{R}(A)$ is a closed set, we can replace $\tilde{B}$ by $\mathcal{R}(A)$, and define the operator as $A : \mathcal{D} \subset B \mapsto \mathcal{R}(A)$. However, as we will see below, the condition $\mathcal{R}(A) \equiv \tilde{B}$ is a very strong restriction, and is *more difficult to satisfy for compact operators*. Even for non-compact continuous operators, this condition may not be fulfilled, as the following simple example shows.

Example A.2.1 A linear bounded operator with non closed range

Consider the bounded linear operator $A : L^2(0, 1) \mapsto L^2(0, 1)$, defined as

$$(Au)(x) := \int_0^x u(y)dy, \ u \in H := L^2(0, 1).$$

The range of this operator is the subspace $\overset{\circ}{H}{}^1(0, 1) := \{v \in H^1(0, 1) : u(0) = 0\}$ of the Sobolev space $H^1(0, 1)$. Since the space $\overset{\circ}{H}{}^1(0, 1) \subset L^2(0, 1)$ is not closed in $L^2(0, 1)$, the range $R(A) = \overset{\circ}{H}{}^1(0, 1)$ of the above bounded linear operator is not closed, i.e. $\mathcal{R}(A) \neq \overline{\mathcal{R}(A)}$. □

Thus, this example shows that not only the condition $\mathcal{R}(A) \equiv \tilde{B}$, but even more weaker condition $\mathcal{R}(A) \equiv \overline{\mathcal{R}(A)}$ may not be satisfied. If the condition $\mathcal{R}(A) \equiv \overline{\mathcal{R}(A)}$ holds, then the linear bounded operator A is defined as a *closed range operator*. Otherwise, i.e. if $\mathcal{R}(A) \neq \overline{\mathcal{R}(A)}$, A is defined as a *non-closed range operator*. Note that the operator equation $Au = F$ with non-closed range operator A is defined *ill-posed*, according to the concept of M.Z. Nashed [73].

The following fundamental theorem shows what means the condition $\mathcal{R}(A) \equiv \tilde{B}$ in terms of invertibility of linear compact operators. In particular, this theorem implies that *a compact linear operator $A : B \mapsto \tilde{B}$ defined in infinite-dimensional Banach space B is not invertible*.

Theorem A.2.1 *Let $A : B \mapsto \tilde{B}$ be a linear compact operator defined between the Banach spaces B and $\tilde{B}$. Assume that $\mathcal{R}(A) \equiv \tilde{B}$, i.e. A is surjective. If A has a bounded inverse A^{-1} on $\tilde{B}$ onto B, then B is finite dimensional.*

Note that a compact operator $A : H \mapsto \tilde{H}$ is called *finite dimensional (or finite-rank) operator*, if its range $\mathcal{R}(A)$ is a finite dimensional set in $\tilde{H}$. As we noted above, if a bounded linear operator is of finite dimensional, then it is a compact operator.

From this theorem we can deduce some important properties related to invertibility of linear compact operators.

Corollary A.2.1 *A surjective linear compact operator $A : B \mapsto \tilde{B}$ defined on a infinite-dimensional Banach space B can not have a bounded inverse.*

Appendix A: Invertibility of Linear Operators

Let us compare now Theorem A.1.2 with Theorem A.2.1, where in both the linear continuous operator A is surjective, i.e. is *onto* $\tilde{B}$. Theorem A.1.2 asserts that the inverse operator $A^{-1} : \tilde{B} \mapsto B$ is continuous, while Theorem A.2.1 asserts that if the inverse operator $A^{-1} : \tilde{B} \mapsto B$ is continuous, then $\tilde{B}$ must be finite dimensional. This leads to the following conclusion.

Corollary A.2.2 *If $\tilde{B}$ is infinite-dimensional, there is no injective compact linear operator $A : B \mapsto \tilde{B}$, defined from B onto $\tilde{B}$.*

Let us analyze now what does mean surjectivity condition for linear compact operators, imposed in Theorem A.2.1. As it was noted above, this condition ($\mathcal{R}(A) \equiv \tilde{B}$) is a very strong restriction and may not be satisfied for some classes of compact operators. What happens if the this condition does not hold, i.e. A is not surjective, and is only a closed range compact operator, i.e. $\mathcal{R}(A) \equiv \overline{\mathcal{R}(A)}$? Next theorem shows that, at least in Hilbert spaces, the closedness and finite dimensionality of the range of a linear compact operators are equivalent properties.

Theorem A.2.2 *Let $A : H \mapsto \tilde{H}$ be a linear compact operator defined between two infinite-dimensional Hilbert spaces H and $\tilde{H}$. Then the range $\mathcal{R}(A)$ is closed if and only if it is finite dimensional.*

Proof The proof of this theorem is based on the well-known elementary result that the identity operator $I : \mathcal{R}(A) \mapsto \mathcal{R}(A)$ is compact if and only if its range $\mathcal{R}(A)$ is finite dimensional. Indeed, if $\mathcal{R}(A)$ is closed, then it is complete in $\tilde{H}$, i.e. is also a Hilbert space. Then we can use decomposition theorem, $H = \mathcal{N}(A) \bigoplus \mathcal{N}(A)^\perp$, and define the restriction $A|_{\mathcal{N}(A)^\perp}$ of the compact operator A on $\mathcal{N}(A)^\perp$, according the construction given in Sect. 1.3. Evidently, this restriction is continuous and bijective onto the Hilbert space $\mathcal{R}(A)$. Hence we may apply Theorem A.1.2 to conclude that the restricted operator $A|_{\mathcal{N}(A)^\perp}$ has a bounded inverse. Then the identity operator

$$I := A \left(A|_{\mathcal{N}(A)^\perp}\right)^{-1} : \mathcal{R}(A) \mapsto \mathcal{R}(A),$$

defined as a superposition of compact and continuous operators, is also compact. This completes the proof. □.

Besides the above mentioned importance, the above theorems are directly related to ill-posed problems. Indeed, if $A : H \mapsto \tilde{H}$ be a linear compact operator between the infinite-dimensional Hilbert spaces H and $\tilde{H}$, and *its range $\mathcal{R}(A)$ is also infinite-dimensional*, then $\mathcal{R}(A)$ is not closed, by the assertion of the theorem. This means that *the equation $Au = F$ is ill-posed* in the sense of the condition (**p1**) of Hadamard's ill-posedness, defined in Introduction. Note that the range of almost all integral (compact) operators related to integral equations with non-degenerate kernel, is infinite-dimensional. Below we will illustrate how an infinite-dimensional range compact operator, defined by the integral equation with Hilbert-Schmidt kernel function, can uniformly be approximated by a sequence of finite rank compact operators. In inverse problems this approach, i.e. the restriction of infinite-dimensional range compact operator A to a finite dimensional subspace of $\tilde{H}$, is defined as *regularization by discretization*. Although this approach yields a well-posed problem, since linear operators on finite dimensional spaces are always bounded (i.e. continuous), the condition number of the finite dimensional problem may be very large unless the finite dimensional subspace is chosen properly.

Thus, a finite dimensionality of a range of a linear compact operator A may "remove" the ill-posedness of the equation $Au = F$. On the other hand, many compact operators, associated with function spaces that occur in differential and integral equations, can be characterized as being the limit of a sequence of bounded finite dimensional operators. The following theorem gives a construction of a sequence of bounded finite dimensional operators in separable Hilbert space [61].

Theorem A.2.3 *Let $A : H \mapsto \tilde{H}$ be an infinite-dimensional linear compact operator defined on a separable Hilbert space H. Then there exists a sequence of finite dimensional linear operators $\{A_N\}$ such that A is a norm limit of this sequence, i.e. $\|A - A_N\|_{\tilde{H}} \to 0$, as $n \to \infty$.*

Proof As a separable Hilbert space, H has an orthonormal basis, defined as $\{\psi_n \in H : n \in \mathcal{N}\}$, and for any $u \in H$ we can write

$$u = \sum_{n=1}^{\infty} \langle u, \psi_n \rangle_H \psi_n; \quad Au = \sum_{n=1}^{\infty} \langle u, \psi_n \rangle_H A\psi_n.$$

We use this orthonormal basis to define the finite dimensional linear $A_N : H \mapsto \tilde{H}$ operator as follows:

$$A_N u := \sum_{n=1}^{N} \langle u, \psi_n \rangle_H A\psi_n. \tag{A.2.2}$$

Since both operators A and A_N are compact, the operator $R_N := A - A_N$ is also compact. To prove the theorem, we need to show that $\|R_N\| \to 0$ or equivalently,

$$r_N := \sup_{\|u\|_H \leq 1} \|R_N u\|_{\tilde{H}} \to 0, \quad \text{as } N \to \infty. \tag{A.2.3}$$

By definition of the supremum, there exists a sequence $\{u^{(n)}\} \subset H$, such that $\|u^{(n)}\|_H \leq 1$, for all $n \in \mathcal{N}$, and $\|R_N u^{(n)}\|_{\tilde{H}} \to r_N$, as $n \to \infty$. Further, by Banach-Alaoglu theorem, there is a weakly convergent subsequence $\{u^{(n_m)}\} \subset \{u^{(n)}\}$ such that $u^{(n_m)} \rightharpoonup u_N^*$, and $\|u_N^*\|_H \leq 1$. Then the compact operator R_N transforms this weakly convergent subsequence to the strongly convergent one: $\|R_N u^{(n_m)} - R_N u_N^*\|_{\tilde{H}} \to 0$, as $n_m \to \infty$. In particular, $\|R_N u^{(n_m)}\|_{\tilde{H}} \to \|R_N u_N^*\|_{\tilde{H}}$. On the other hand, $\|R_N u^{(n)}\|_{\tilde{H}} \to r_N$, as $n \to \infty$. This implies that $r_N = \|R_N u_N^*\|_{\tilde{H}}$.

Let us define now the difference $R_N u_N^* := (A - A_N) u_N^*$ using (A.2.2) and (A.2.3):

$$R_N u_N^* = A \sum_{n=N+1}^{\infty} \langle u_N^*, \psi_n \rangle_H A \psi_n =: A S_N^*, \quad S_N^* := \sum_{i=N+1}^{\infty} \langle u_N^*, \psi_n \rangle_H A \psi_n.$$

This show that in order to prove $r_N \to 0$, as $N \to \infty$, we need to prove the weak convergence $S_N \rightharpoonup 0$, as $N \to \infty$, due to the compactness of the operator R_N. Using the Hölder inequality and the Parseval's equality, we have:

$$|\langle S_N, u \rangle_H| = \left\langle \sum_{n=N+1}^{\infty} \langle u_N^*, \psi_n \rangle_H \psi_n, \sum_{n=1}^{\infty} \langle u, \psi_n \rangle_H \psi_n \right\rangle_H$$

$$= \left\langle \sum_{n=N+1}^{\infty} \langle u_N^*, \psi_n \rangle_H \psi_n, \sum_{n=N+1}^{\infty} \langle u, \psi_n \rangle_H \psi_n \right\rangle_H$$

$$\leq \left(\sum_{n=N+1}^{\infty} |\langle u_N^*, \psi_n \rangle_H| \right)^{1/2} \left(\sum_{n=N+1}^{\infty} |\langle u, \psi_n \rangle_H| \right)^{1/2}$$

$$\leq \|u_N^*\|_H \left(\sum_{n=N+1}^{\infty} |\langle u, \psi_n \rangle_H| \right)^{1/2}.$$

The $\|u_N^*\|_H \leq 1$ and the second multiplier in the last inequality tends to zero, as $N \to 0$. Hence, for any $u \in H$, $|\langle S_N, u \rangle_H| \to 0$, as $N \to 0$. This completes the proof. $\square$

This theorem implies that the set of finite dimensional linear operators is dense in the space of compact operators defined on a separable Hilbert space.

Example A.2.2 An approximation of a compact operator associated with an integral equation

Consider the linear bounded operator defined by (1.3.2):

$$(Au)(x) := \int_a^b K(x, y) u(y) dy, \quad x \in [a, b],$$

assuming $H = \tilde{H} = L^2(a, b)$, $K \in L^2((a, b) \times (a, b))$ and $\|K\|_{L^2((a,b) \times (a,b))} \leq M$. Evidently, under these assumptions the integral operator $A : H \mapsto \tilde{H}$, defined as a *Hilbert-Schmidt operator with non-degenerate (or non-separable) kernel function*, is a compact operator. Let $\{\phi_n \in L^2(a, b) : n \in \mathcal{N}\}$ be an orthonormal basis for $L^2(a, b)$. Then

$$\{\psi_{m,n} \in L^2((a, b) \times (a, b)) : \psi_{m,n}(x, y) := \phi_n(x) \phi_n(y), n, m \in \mathcal{R}\}$$

forms an orthonormal basis for $L^2((a, b) \times (a, b))$. Hence

$$K_N(x, y) := \sum_{n,m=1}^{N} \langle K, \psi_{m,n} \rangle_{L^2((a,b)\times(a,b))} \psi_{m,n}(x, y), \ x, y \in [a, b].$$

Now we define the operator

$$(A_N u)(x) := \int_a^b K_N(x, y) u(y) dy, \ x \in [a, b].$$

Evidently, $A_N : H \mapsto \tilde{H}_N \subset \tilde{H}$ is a finite dimensional compact operator with the range $\mathcal{R}_N(A_N) \subset \text{span}\{\psi_{m,n}(x, y)\}_{m,n=1}^{N}$ and $\|A - A_N\|_{\tilde{H}} \to 0$, as $n \to \infty$, as follows from Theorem A.2.3 in Introduction. □

As was noted above, the range of a compact operator is "almost finite dimensional". Specifically, not only a compact operator itself can be approximated by finite dimensional linear (i.e. compact) operators, but also the range of a linear compact operator can be approximated by a finite dimensional subspace. This gives a further justification for the presentation of the following theorem.

Theorem A.2.4 *Let* $A : B \mapsto \tilde{B}$ *be a linear operator between two infinite-dimensional Banach spaces* B *and* $\tilde{B}$. *Then for any* $\varepsilon > 0$ *there exists a finite dimensional subspace* $\mathcal{R}_N(A) \subset \mathcal{R}(A)$ *such that, for any* $u \in B$

$$\inf_{v \in \mathcal{R}_N(A)} \|Au - v\|_B \leq \varepsilon \|u\|_B.$$

This well-known result can be found in the above mentioned books on functional analysis.

Remark finally that the approaches proposed in Theorems A.2.3 and A.2.4 are widely used within the framework of regularization methods based on discretization.

Appendix B
Some Estimates For One-Dimensional Parabolic Equation

We prove here some basic estimates for the weak and regular weak solutions of the direct problem

$$\begin{cases} u_t(x,t) = (k(x)u_x(x,t))_x, & (x,t) \in \Omega_T, \\ u(x,0) = 0, & 0 < x < l, \\ -k(0)u_x(0,t) = g(t), \ u_x(l,t) = 0, & 0 < t < T, \end{cases} \quad (B.0.1)$$

discussed in Sect. 5.5.

Note first of all, that energy estimates for the second order parabolic equations are given in general form in the books [24, 57]. A priori estimates given below for the weak and regular weak solutions of problem (B.0.1) cannot be obtained directly from those given in general form. Our aim here to derive all necessary a priori estimates for the weak and regular weak solutions and for their derivatives via the Neumann datum $g(t)$. To derive these energy estimates as well as to prove an existence of weak solutions by the Galerkin approximation one needs indeed to proceed three steps:

1. Construction of the approximate solution,

$$u_n(x,t) = \sum_{i=1}^{n} d_{n,i}(t)\varphi_i(x), \quad (x,t) \in \Omega_T.$$

where $\{\varphi_i\}_{i=1}^{\infty}$ are orthonormal eigenvectors of the differential operator $\ell : H^2(0,l) \mapsto L^2(0,l)$ defined by

$$\begin{cases} (\ell\varphi)(x) := -(k(x)\varphi'(x))' = \lambda\varphi(x), & x \in (0,l); \\ \varphi'(0) = 0, \ \varphi'(l) = 0, \end{cases}$$

corresponding to eigenvalues $\{\lambda_i\}_{i=1}^{\infty}$.

2. Derivation of energy estimates for the approximate solution $u_n(x,t)$;
3. Convergence of the approximate solution to a weak solution in appropriate sense.

However, we will derive here necessary for our goals energy estimates directly for a weak solution, omitting these steps. Readers can find details of these steps in the books ([57], Sect. 3) and ([24], Sect. 7.1).

We assume here, as in Sect. 5.5, that the functions $k(x)$ and $g(t)$ satisfy the following conditions:

$$\begin{cases} k \in H^1(0,l), \ 0 < c_0 \leq k(x) \leq c_1 < \infty; \\ g \in H^1(0,T), \ g(t) > 0, \ \text{for all } t \in (0,T), \ \text{and } g(0) = 0. \end{cases} \quad (B.0.2)$$

Under these conditions the regular weak solution of problem (B.0.1) with improved regularity defined as

$$\begin{cases} u \in L^\infty(0,T; H^2(0,l)), \\ u_t \in L^\infty(0,T; L^2(0,l)) \cap L^2(0,T; H^1(0,l)), \\ u_{tt} \in L^2(0,T; H^{-1}(0,l)), \end{cases} \quad (B.0.3)$$

exists and unique, according to [24, 57].

B.1 Estimates For the Weak Solution

First we derive here some necessary estimates for the weak solution $u \in L^2(0,T; H^1(0,l))$ with $u_t \in L^2(0,T; H^{-1}(0,l))$ of problem (B.0.1) which are used in Sect. 5.5. To obtain these estimates we need the following Gronwall-Bellman inequality [13].

Lemma B.1.1 *Let $v(t)$ and $f(t)$ be nonnegative continuous functions defined on $[\alpha, \beta]$, and $r(t)$ be positive continuous and nondecreasing function defined on $[\alpha, \beta]$. Then the inequality*

$$v(t) \leq \int_\alpha^t f(\tau)v(\tau)d\tau + r(t), \ t \in [\alpha, \beta]$$

implies that

$$v(t) \leq r(t) \exp\left(\int_\alpha^t f(\tau)d\tau\right), \ t \in [\alpha, \beta]. \quad (B.1.1)$$

Lemma B.1.2 *Let the conditions hold:*

$$\begin{cases} k \in L^\infty(0,l), \ 0 < c_0 \leq k(x) \leq c_1, \\ g \in L^2(0,T). \end{cases} \quad (B.1.2)$$

Then for the weak solution of the parabolic problem (B.0.1) the following estimate holds:

Appendix B: Some Estimates For One-Dimensional Parabolic Equation

$$\max_{[0,T]} \|u\|_{L^2(0,l)} + \|u_x\|_{L^2(0,T;L^2(0,l))} \leq C_0 \|g\|_{L^2(0,T)}, \quad \text{(B.1.3)}$$

where $C_0 = \max\{C_1; C_2\}$,

$$\begin{cases} C_1 = (l/c_0)^{1/2} \exp(Tc_0/l^2), \\ C_2 = \left(2TC_1^2/l^2 + l/c_0^2\right)^{1/2}. \end{cases} \quad \text{(B.1.4)}$$

and the constant $c_0 > 0$ is defined by (B.0.2).

Proof Multiply both sides of Eq. (B.0.1) by $u(x, t)$, integrate on $[0, l]$ and then use the integration by parts formula. Taking into account the boundary conditions in (B.0.1) we get:

$$\frac{1}{2}\frac{d}{dt}\int_0^l u^2(x,t)dx + \int_0^l k(x)u_x^2(x,t)dx = g(t)u(0,t).$$

Integrating both sides on $[0, t]$, $t \in [0, T]$ and using the initial condition $u(x, 0) = 0$ obtain the following energy identity:

$$\int_0^l u^2(x,t)dx + 2\int_0^t \int_0^l k(x)u_x^2(x,\tau)dxd\tau = 2\int_0^t g(\tau)u(0,\tau)d\tau,$$

for a.e. $t \in [0, T]$. Using now the ε-inequality $2ab \leq (1/\varepsilon)a^2 + \varepsilon b^2$ in the last right hand side integral to get:

$$\int_0^l u^2(x,t)dx + 2c_0 \int_0^t \int_0^l u_x^2(x,\tau)dxd\tau$$
$$\leq \frac{1}{\varepsilon}\int_0^t g^2(\tau)d\tau + \varepsilon \int_0^t u^2(0,\tau)d\tau, \quad \text{(B.1.5)}$$

for a.e. $t \in [0, T]$, where $c_0 > 0$ is the constant in condition (B.1.2). To estimate the last integral in (B.1.5) we use the identity

$$u^2(0,t) = \left(u(x,t) - \int_0^x u_\xi(\xi,t)d\xi\right)^2,$$

apply to the right hand side the inequality $(a - b)^2 \leq 2(a^2 + b^2)$ and then use the Hölder inequality. Integrating then on $[0, l]$ we obtain:

$$\int_0^t u^2(0,\tau)d\tau \leq 2\int_0^t u^2(x,\tau)d\tau + 2x\int_0^t \int_0^l u_x^2(x,\tau)dxd\tau.$$

Integrating again on $[0, l]$ and then dividing by $l > 0$ both sides we arrive at the inequality:

$$\int_0^t u^2(0,\tau)d\tau \le \frac{2}{l}\int_0^t\int_0^l u^2(x,\tau)d\tau dx + l\int_0^t\int_0^l u_x^2(x,\tau)dxd\tau. \quad (B.1.6)$$

Using this we estimate in (B.1.5) we deduce:

$$\int_0^l u^2(x,t)dx + (2c_0 - l\varepsilon)\int_0^t\int_0^l u_x^2(x,\tau)dxd\tau$$
$$\le \frac{2\varepsilon}{l}\int_0^l\int_0^t u^2(x,\tau)d\tau dx + \frac{1}{\varepsilon}\int_0^t g^2(\tau)d\tau, \quad (B.1.7)$$

for a.e. $t \in [0, T]$. We require that the arbitrary parameter $\varepsilon > 0$ satisfies the condition $0 < \varepsilon < 2c_0/l$. Taking for convenience this parameter as $\varepsilon = c_0/l$ we obtain:

$$\int_0^l u^2(x,t)dx + c_0\int_0^t\int_0^l u_x^2(x,\tau)dxd\tau$$
$$\le \frac{2c_0}{l^2}\int_0^l\int_0^t u^2(x,\tau)d\tau dx + \frac{l}{c_0}\int_0^t g^2(\tau)d\tau. \quad (B.1.8)$$

The first consequence of (B.1.8) is the inequality

$$\int_0^l u^2(x,t)dx \le \frac{2c_0}{l^2}\int_0^t\int_0^l u^2(x,\tau)dxd\tau dx + \frac{l}{c_0}\int_0^t g^2(\tau)d\tau,$$

for a.e. $t \in [0, T]$. Applying Lemma B.1.1 to this inequality we deduce that

$$\int_0^l u^2(x,t)dx \le C_1^2\int_0^t g^2(\tau)d\tau, \ t \in [0, T], \quad (B.1.9)$$

where $C_1 > 0$ is the constant defined by (B.1.4). Then the norm $\|u\|_{L^2(0,l)}$ is estimated as follows:

$$\|u\|_{L^2(0,l)} \le C_1\|g\|_{L^2(0,T)}, \ t \in [0, T]. \quad (B.1.10)$$

Remember that according to Theorem 3.1.1 in Sect. 2.1, the weak solution is in $C(0, T; L^2(0, l))$. Hence (B.1.10) implies the following estimate for the weak solution of the direct problem (B.1.5):

$$\max_{[0,T]}\|u\|_{L^2(0,l)} \le C_1\|g\|_{L^2(0,T)}. \quad (B.1.11)$$

The second consequence of (B.1.8) is the inequality

$$c_0\int_0^T\int_0^l u_x^2(x,\tau)dxdt \le \frac{2c_0}{l^2}\int_0^T\int_0^l u^2(x,t)dxdtdx + \frac{l}{c_0}\int_0^T g^2(t)dt.$$

Using estimate (B.1.9) we deduce from this inequality the following estimate:

$$\iint_{\Omega_T} u_x^2(x,t)dxdt \leq C_2^2 \int_0^T g^2(t)dt, \tag{B.1.12}$$

where $C_2 = C_2(l, T, c_0) > 0$ is the constant defined by (B.1.4).

Estimates (B.1.11) and (B.1.12) imply the required estimate (B.1.3). □

B.2 Estimates for the Regular Weak Solution

In addition to the above estimates for the weak solution, we use in Sect. 5.5 some estimates for the regular weak solution with improved regularity defined by (B.0.3).

Lemma B.2.1 *Let conditions (B.0.2) hold. Then for the regular weak solution of the parabolic problem (B.0.1) defined by (B.0.3) the following estimates hold:*

$$\begin{cases} \operatorname*{ess\,sup}_{[0,T]} \|u_t\|_{L^2(0,l)} \leq C_1 \|g\|_{\mathcal{V}(0,T)}, \\ \|u_{xt}\|_{L^2(0,T;L^2(0,l))} \leq C_2 \|g\|_{\mathcal{V}(0,T)}, \end{cases} \tag{B.2.1}$$

where $\mathcal{V}(0,T) := \{g \in H^1(0,T) : g(0) = 0\}$ *and the constants* $C_1 > 0, C_2 > 0$ *are defined by (B.1.4).*

Proof Differentiate formally Eq. (B.0.1) with respect to $t \in (0, T)$, multiply both sides by $u_t(x, t)$, integrate over $[0, l]$ and use the integration by parts formula. Taking then into account the initial and boundary conditions we obtain the following integral identity:

$$\frac{1}{2}\frac{d}{dt}\int_0^l u_t^2(x,t)dx + \int_0^l k(x)u_{xt}^2(x,t)dx = g'(t)u_t(0,t).$$

Integrating this identity on $[0, t]$, $t \in [0, T]$ and using the limit equation

$$\int_0^l u_t^2(x, 0^+)dx = \lim_{t \to 0^+} \int_0^l \left((k(x)u_x(x, 0^+))_x\right)^2 dx = 0,$$

to deduce that

$$\int_0^l u_t^2(x,t)dx + 2\int_0^t \int_0^l k(x)u_{x\tau}^2(x,\tau)dxd\tau = 2\int_0^t g'(\tau)u_\tau(0,\tau)d\tau,$$

for all $t \in [0, T]$. Applying now the ε-inequality $2ab \leq (1/\varepsilon) b^2 + \varepsilon a^2$ to the last right hand side integral we obtain the inequality:

$$\int_0^l u_t^2(x,t)dx + 2c_0 \int_0^t \int_0^l k(x) u_{x\tau}^2(x,\tau) dx d\tau$$
$$\leq \frac{1}{\varepsilon} \int_0^t (g'(\tau))^2 d\tau + \varepsilon \int_0^t u_\tau^2(0,\tau) d\tau, \ t \in [0,T].$$

Using the same argument as in the proof of inequality (B.1.6), we estimate the last right hand side integral as follows:

$$\int_0^t u_\tau^2(0,\tau) d\tau \leq \frac{2}{l} \int_0^t \int_0^l u_\tau^2(x,\tau) d\tau dx + l \int_0^t \int_0^l u_{x\tau}^2(x,\tau) dx d\tau, \quad (B.2.2)$$

for all $t \in [0,T]$. Substituting this in above inequality we conclude:

$$\int_0^l u_t^2(x,t) dx + (2c_0 - l\varepsilon) \int_0^t \int_0^l u_{x\tau}^2(x,\tau) dx d\tau$$
$$\leq \frac{2\varepsilon}{l} \int_0^t \int_0^l u_\tau^2(x,\tau) dx d\tau + \frac{1}{\varepsilon} \int_0^t (g'(\tau))^2 d\tau, \ t \in [0,T] \quad (B.2.3)$$

This is the same inequality (B.1.7) with $u(x,t)$ replaced by $u_t(x,t)$ and $g(t)$ replaced by $g'(t)$. Repeating the same procedure as in the proof of the previous lemma we deduce that

$$\int_0^l u_t^2(x,t) dx \leq C_1^2 \int_0^t (g'(\tau))^2 d\tau, \ t \in [0,T], \quad (B.2.4)$$

and then

$$\iint_{\Omega_T} u_{xt}^2(x,t) dx dt \leq C_2^2 \int_0^T (g'(t))^2 dt, \quad (B.2.5)$$

where $C_1, C_2 > 0$ are the constants defined by (B.1.4).

With the inequality $\|g'\|_{L^2(0,T)} \leq \|g\|_{\mathcal{V}(0,T)}$ estimates (B.2.4) and (B.2.5) imply the required estimate (B.2.1). □

References

1. R.A. Adams, J.J.F. Fournier, *Sobolev Spaces*, 2nd edn. (Academic Press, New York, 2003)
2. R.C. Aster, B. Borchers, C.H. Thurber, *Parameter Estimation and Inverse Problems*, 2nd edn. (Academic Press, Amsterdam, 2013)
3. K. Atkinson, W. Han, *Theoretical Numerical Analysis: A Functional Analysis Framework, Texts in Applied Mathematics* (Springer, New York, 2009)
4. G. Bal, Inverse transport theory and applications. Inverse Problems **25**, 053001 (2009)
5. G. Bal, A. Jollivet, Stability estimates in stationary inverse transport. Inverse Problems Imaging **2**(4), 427–454 (2008)
6. G. Bal, *Introduction to Inverse Problems*. Lecture Notes, Columbia University, New York, (2012), http://www.columbia.edu/~gb2030/PAPERS/
7. H. Bateman, *Tables of Integral Transformations*, vol. I (McGraw-Hill Book Company, New York, 1954)
8. J. Baumeister, *Stable Solution of Inverse Problems* (Vieweg, Braunschweig, 1987)
9. L. Beilina, M.V. Klibanov, *Approximate Global Convergence and Adaptivity for Coefficient Inverse Problems* (Springer, New York, 2012)
10. M.I. Belishev, Wave bases in multidimensional inverse problems. Sbornic Math. **67**(1), 23–41 (1990)
11. M.I. Belishev, Recent progress in the boundary control method. Inverse Problems **23**(5), R1–R67 (2007)
12. M.I. Belishev, A.S. Blagovestchenskii, *Dynamic Inverse Problems of Wave Theory*. St.-Petersburg University (1999) (in Russian)
13. R. Bellman, Asymptotic series for the solutions of linear differential-difference equations. Rendicondi del circoloMatematica Di Palermo **7**, 19 (1958)
14. A.L. Buhgeim, M.V. Klibanov, Global uniqueness of a class of multidimensional inverse problems. Siviet Math. Dokl. **24**(2), 244–247 (1981)
15. A.P. Calderón, On an inverse boundary problem, *Seminar on Numerical Analysis and its Applications to Continuum Physics*, Soc (Brasiliera de Matematica, Rio de Janeiro, 1980), pp. 61–73
16. J.R. Cannon, P. DuChateau, An inverse problem for a nonlinear diffusion equation. SIAM J. Appl. Math. **39**, 272–289 (1980)
17. K. Chadan, P.C. Sabatier, *Inverse Problems in Quantum Scattering Theory, Texts and Monographs in Physics* (Springer, New York, 1977)
18. M. Choulli, P. Stefanov, Reconstruction of the coefficients of the stationary transport equation from boundary measurements. Inverse Problems **12**, L19–L23 (1996)

19. D. Colton, R. Kress, *Inverse Acoustic and Electromagnetic Scattering Theory*, 2nd edn. (Springer, Heidelberg, 1998)
20. J.B. Conway, *A Course in Functional Analysis*, 2nd edn. (Springer, New York, 1997)
21. R. Courant, D. Hilbert, *Methods of Mathematical Physics*, vol. 1 (Julius Springer, Heidelberg, 1937)
22. P. DuChateau, Monotonicity and invertibility of coefficient-to-data mappings for parabolic inverse problems. SIAM J. Math. Anal. **26**(6), 1473–1487 (1995)
23. H.W. Engl, M. Hanke, A. Neubauer, *Regularization of Inverse Problems* (Kluwer Academic Publishers, Dordrecht, 1996)
24. L.C. Evans, *Partial Differential Equations*, vol. 19, 2nd edn., Graduate Studies in Mathematics (American Mathematical Society, Providence, 2010)
25. L.D. Faddeev, Growing solutions of the Schrödinger equation. Sov. Phys. Dokl. **10**, 1033–1035 (1966)
26. L.D. Faddeev, The inverse problem in the quantum theory of scattering II. J. Sov. Math. **5**, 334–396 (1976)
27. I.M. Gelfand, Some problems of functional analysis and algebra. in *Proceedings of the International Congress of Mathematicians* (Amsterdam, 1954), pp. 253–276
28. I.M. Gel'fand, S.G. Gindikin, M.I. Graev, *Selected Problems of Integral Geometry* (Dobrosvet, Moscow, 2012). (in Russian)
29. M.L. Gerver, V.M. Markushevich, An investigation of a speed of seismic wave ambiguity by a godograph. *Dokl. Akad. Nauk SSSR* **163**(6), 1377–1380 (1965) (in Russian)
30. E.A. González-Velasco, *Fourier Analysis and Boundary Value Problems* (Academic Press, San Diego, 1995)
31. C.W. Groetsch, *Inverse Problems in the Mathematical Sciences* (Vieweg Verlag, Wiesbaden-Braunschweig, Germany, 1993)
32. J. Hadamard, *Le probléme de Cauchy et les équations aux dérivées partielles linéaires hyperboliques* (Hermann, Paris, 1932)
33. J. Hadamard, *La théorie des équations aux dérivées partielles* (Editions Scientifiques, Peking, 1964)
34. M. Hanke, *Conjugate Gradient Type Methods for Ill-Posed Problems* (Longman House, Harlow, 1995)
35. P.C. Hansen, The truncated SVD as a method for regularization. BIT **27**, 534–553 (1987)
36. D.N. Háo, *Methods for Inverse Heat Conduction Problems* (Peter Lang, Frankfurt, 1998)
37. A. Hasanov, P. DuChateau, B. Pektas, An adjoint problem approach and coarse-fine mesh method for identification of the diffusion coefficient in a linear parabolic equation. J. Inverse Ill-Posed Problems **14**, 435–463 (2006)
38. A. Hasanov, Simultaneous determination of source terms in a linear parabolic problem from the final overdetermination: weak solution approach. J. Math. Anal. Appl. **330**, 766–779 (2007)
39. A. Hasanov, Simultaneous determination of source terms in a linear hyperbolic problem from the final overdetermination: weak solution approach. IMA J. Appl. Math. **74**, 1–19 (2009)
40. A. Hasanov, B. Mukanova, Relationship between representation formulas for unique regularized solutions of inverse source problems with final overdetermination and singular value decomposition of inputoutput operators. IMA J. Appl. Math. **80**(3), 676–696 (2015)
41. G.M. Henkin, R.G. Novikov, The $\bar{\partial}$-equation in the multidimensional inverse scattering problem. Russ. Math. Surv. **42**(3), 109–180 (1987)
42. G. Herglotz, Uber die Elastizität der Erde bei Borücksichtigung ihrer variablen Dichte. Zeitschr. für Math. Phys. **52**(3), 275–299 (1905)
43. B. Hofmann, *Regularization for Applied Inverse and Ill-Posed Problems* (B. G. Teubner, Leipzig, 1986)
44. B. Hofmann, Approximate source conditions in Tikhonov-Phillips regularization and consequences for inverse problems with multiplication operators. Math. Methods Appl. Sci. **29**(3), 351–371 (2006)

45. V. Isakov, *Inverse Problems for Partial Differential Equations*. Applied Mathematical Sciences, vol. 127 (Springer, New York 2006)
46. V. Isakov, *Inverse Source Problems* (American Mathematical Society, Providence, 1990)
47. V. Isakov, Inverse parabolic problems with the final over determination. Commun. Pure Appl. Math. **44**(2), 185–209 (1991)
48. K. Ito, B. Jin, *Inverse Problems: Tikhonov Theory and Algorithms*. Series on Applied Mathematics, vol. 22 (World Scientific Publishing, Hackensack, 2015)
49. V.K. Ivanov, On ill-posed problems. Mat. USSR Sb. **61**, 211–223 (1963)
50. S.I. Kabanikhin, *Inverse and Ill-Posed Problems: Theory and Applications*. Walter de Gruyter (2011)
51. J. Kaipio, E. Somersalo, *Statistical and Computational Inverse Problems* (Springer, New York, 2005)
52. B. Kaltenbacher, A. Neubauer, O. Scherzer, *Iterative Regularization Methods for Nonlinear Ill-Posed Problems*. Radon Series on Computational and Applied Mathematics, vol. 6 (Walter de Gruyter, Boston, 2008)
53. J.B. Keller, Inverse Problem. Amer. Math. Monthly **83**, 107–118 (1976)
54. A. Kirsch, *Introduction to Mathematical Theory of Inverse Problems*, 2nd edn. (Springer, New York, 2011)
55. M.V. Klibanov, Carlemates estimates for global uniqueness, stability and numerical methods for coefficients inverse problems. J. inverse and Ill-Posed Problems **21**, 477–560 (2013)
56. M.V. Klibanov, V.G. Kamburg, Globally strictly convex cost functional for an inverse parabolic problem. Math. Methods Appl. Sci. **39**, 930–940 (2016)
57. O.A. Ladyzhenskaya, *The Boundary Value Problems of Mathematical Physics* (Springer, New York, 1985)
58. M.M. Lavrentiev, *Some Improperly Posed Problems of Mathematical Physics*, vol. 11 (Springer Tracts in Natural Philosophy (Springer, Heidelberg, 1967)
59. M. M. Lavrentiev, V. G. Romanov, S. P. Shishatski, *Ill-Posed Problems of Mathematical Physics and Analysis*. Translations of Mathematical Monographs, vol. 64, (American Mathematical Society, Providence, 1986)
60. M.M. Lavrentiev, V.G. Romanov, V.G. Vasiliev, *Multidimensional Inverse Problems for Differential Equations*, Lecture Notes in Mathematics (Springer, Heidelberg, 1970)
61. L.P. Lebedev, I.I. Vorovich, G.M.L. Gladwell, *Functional Analysis, Applications in Mechanics and Inverse Problems* (Kluwer Academic Publishers, Dordrecht, 2002)
62. J.-L. Lions, *Some Methods of Solving Non-Linear Boundary Value Problems* (Dunod-Gauthier-Villars, Paris, 1969)
63. A.K. Louis, *Inverse und Schlecht Gestellte Probleme* (Teubner, Stuttgart, 1989)
64. L.A. Lusternik, V.J. Sobolev, *Elements of Functional Analysis*. International Monographs on Advanced Mathematics & Physics. (Wiley, New York, 1975)
65. V.A. Marchenko, Some questions in the theory of one-dimensional linear differential operators of the second order, I. *Trudy Moskov. Mat. Obšč*, **1**, 327-420, (1952) (Russian); English transl. in *Amer. Math. Soc. Transl.*, **101**(2), 1–104 (1973)
66. V.A. Marchenko, *Sturm-Liouville Operators and Applications* (Birkhäuser, Basel, 1986)
67. V.A. Morozov, On the solution of functional equations by the method of regularization. Soviet Math. Dokl. **7**, 414–417 (1966)
68. V.A. Morozov, *Methods for Solving Incorrectly Posed Problems* (Springer, New York, 1984)
69. E.H. Moore, On the reciprocal of the general algebraic matrix. Bull. Am. Math. Soc. **26**, 394–395 (1920)
70. R.G. Mukhometov, The reconstruction problem of a two-dimensional Riemannian metric and integral geometry. Soviet Math. Dokl. **18**(1), 32–35 (1977)
71. R.G. Mukhometov, V.G. Romanov, On the problem of finding an isotropic Riemannian metric in an n-dimensional space (Russian). Dokl. Akad. Nauk SSSR **243**(1), 41–44 (1978)
72. G. Nakamura, R. Potthast, *Inverse Modeling - An introduction to the Theory and Methods of Inverse Problems and Data Assimilation* (IOP Ebook Series, London, 2015)

73. M. Z. Nashed, A new approach to classification and regularization of ill-posed operator equations. In: *Inverse and Ill-posed Problems*, (Sankt Wolfgang, 1986), volume 4 of Notes Rep. Math. Sci. Engrg., pages 53-75. Academic Press, Boston, MA, 1987
74. F. Natterer, *The Mathematics of Computerized Tomography* (Wiley, New York, 1986)
75. R.G. Newton, *Inverse Schrödinger Scattering in Three Dimensions, Texts and Monographs in Physics* (Springer, Heidelberg, 1989)
76. R.G. Novikov, Multidimensional inverse spectral problem for the equation $-\Delta\psi + (v(x) - Eu(x))\psi = 0$. Funct. Anal. Appl. **22**, 263–272 (1988)
77. R.G. Novikov, The inverse scattering problem on a fixed energy level for the two-dimensional Schrödinger operator. J. Funct. Anal. **103**, 409–463 (1992)
78. R.G. Novikov, The $\bar{\partial}$-approach to monochromatic inverse scattering in three dimensions. J. Geom. Anal. **18**(2), 612–631 (2008)
79. R.G. Novikov, An inversion formula for the altenuated X-ray transform. Ark. Mat. **40**, 145–167 (2002)
80. R. Penrose, A generalized inverse for matrices. Proc. Camb. Philosophical Soc. **51**, 406–413 (1955)
81. R. Penrose, On best approximate solution of linear matrix equations. Proc. Camb. Philosophical Soc. **52**, 17–19 (1956)
82. L. Pestov, G. Uhlmann, H. Zhou, An inverse kinematic problem with internal sources. Inverse Problems **31**, 055006 (2015)
83. D.L. Phillips, A technique for the numerical solution of certain integral equation of the first kind. J. Assoc. Comput. Mach. **9**, 84–97 (1962)
84. A.G. Ramm, *Multidimensional Inverse Scattering Problems* (Longman Scientific/Wiley, New York, 1992)
85. V.G. Romanov, *Inverse Problems of Mathematical Physics* (VNU Science Press, Utrecht, 1987)
86. V.G. Romanov, *Investigation Methods for Inverse Problems* (VSP, Utrecht, 2002)
87. V.G. Romanov, S.I. Kabanikhin, *Inverse Problems for Maxwell's Equations* (VSP, Utrecht, 1994)
88. V.G. Romanov, An example of an absence of a global solution in an inverse problem for a hyperbolic equation. Siberian Math. J. **44**(3), 867–868 (2003)
89. U. Schröder, T. Schuster, *Regularization Methods in Banach Spaces*. Radon Series on Computational and Applied Mathematics, vol. 10, (Walter de Gruyter, Boston, 2012)
90. T. Schuster, B. Kaltenbacher, B. Hofmann, K. S. Kazimierski, *Regularization methods in Banach spaces*. Radon Series on Computational and Applied Mathematics, vol. 10, (Walter de Gruyter, Boston, 2012)
91. J. Sylvester, G. Uhlmann, A global uniqueness theorem for an inverse boundary value problem. Ann. Math. **125**, 153–169 (1987)
92. A. Tarantola, *Inverse Problem Theory and Methods for Model Parameter Estimation* (SIAM, Philadelphia, 2004)
93. A.N. Tikhonov, Theoréms d'unicite pour léquation de la chaleur. Mat. Sb. **42**, 199–216 (1935)
94. A.N. Tikhonov, On the stability of inverse problems. *Doklady Akademii Nauk SSSR*, **39**(5), 195–198 (1943) (in Russian)
95. A. Tikhonov, Regularization of incorrectly posed problems. Soviet Math. Dokl. **4**, 1624–1627 (1963)
96. A.N. Tikhonov, Solution of incorrectly formulated problems and the regularization method. Soviet Math. Dokl. **4**, 1035–1038 (1963)
97. A.N. Tikhonov, V.Y. Arsenin, *Solutions of Ill-posed Problems* (Winston and Sons, Washington, 1977)
98. A.N. Tikhonov, A.V. Concharsky, V.V. Stepanov, A.G. Yagola, *Numerical Methods for the Solution of Ill-Posed Problems* (Springer Science & Business Media, New York, 1995)
99. A.N. Tikhonov, A.A. Samarskii, *Equations of Mathematical Physics*, 2nd edn. (Pergamon Press, Oxford, 1963)
100. F.G. Tricomi, *Integral Equations* (Interscience, New York, 1957)

101. V. Thomée, *Galerkin Finite Element Methods for Parabolic Problems*. Springer Series in Computational Mathematics, 2nd Ed., (Springer, New York, 2010)
102. C.R. Vogel, *Computational Methods for Inverse Problems*. SIAM Frontiers in Applied Mathematics (Philadelphia, 2002)
103. E. Zeidler, *Applied Functional Analysis, Applications to Mathematical Physics* (Springer, New York, 1995)
104. E. Zeidler, *Applied Functional Analysis, Main Principles and Their Applications* (Springer, New York, 1995)

Index

A
Accuracy error, 44, 88, 107, 188
Adjoint operator, 34
A-posteriori parameter choice rule, 50
Approximate solution, 44, 87, 207, 247
A-priori parameter choice rule, 50
Attenuated Radon transform, 221

B
Backward parabolic problem, 96
Banach contraction mapping principle, 138, 162
Best approximate solution, 46, 51, 182
Best approximation, 25
Born approximation, 207

C
Characteristic lines, 229
Compactness
    dual role in ill-posedness, 6, 38
    input-output operator, 67, 74, 92, 100, 177
Computational noise level, 104, 116, 188
Conjugate Gradient Algorithm, 107, 186
Consistency conditions for output data, 72, 125, 188
Convergence error, 107, 187
Convergent regularization strategy, 49, 55, 60, 80

D
Data error, 45
Descent direction parameter, 108

Direct problem, 65, 91, 124, 129, 141, 166, 173, 177, 220
Duhamels Principle, 102

E
Eigen system
    input-output operator, 73, 75, 93, 99
    self-adjoint operator, 38
Error estimate, 87
Eulers equations, 229

F
Filter function, 49, 76
Fredholm integral equation, 13
Fredholm operator, 13

G
Galerkin Finite Element Method, 105
Global stability, 138
Gradient formula, 47, 68, 71, 97, 186, 202

I
Ill-posedness
    degree, 41, 43
    Hadamard's definition, 8
Ill-posed problem
    mildly ill-posed, 44
    moderately ill-posed, 43
    severely ill-posed, 44, 100
Input, 1, 66
Input-output mapping/operator, 71, 181, 186
Input-output relationship, 15

Inverse coefficient problem, 2, 3, 92, 129, 141, 168, 175, 192, 220
Inverse conductivity problem, 161
Inverse crime, 104
Inverse kinematic problem, 227, 232
Inverse Laplace transform, 166
Inverse scattering problem, 206
Inverse source problem
   with Dirichlet boundary data, 124
   with final data, 15, 70
   with Neumann data, 162
Inverse spectral problem, 163
Inversion formula, 207, 209

L
Landweber iteration algorithm, 112
Laplace transform, 163
Lavrentiev regularization, 53

M
Measured output data
   Dirichlet type, 173
   final time, 64, 90
Moore-Penrose inverse, 36
Morozov's Discrepancy Principle, 57

N
Noisy data, 55
Normal equation, 34, 68

O
Orthogonal complement, 24
Orthogonal decomposition, 25
Orthogonal projection, 26
Orthogonal Projection Theorem, 26
Output, 67

P
Parameter choice rule, 49
Parameter identification problem, 2
Parameter-to-solution map, 2
Picard criterion, 79
Picards Theorem, 42
Point source, 208
Projection operator, 27

Q
Quasi-solution, 46

R
Radon transform, 208
Recovering the potential, 166
Reflection method, 124
Regular weak solution, 65, 175, 251
Regularization error, 45
Regularization strategy, 45, 49
Regularized Tikhonov functional, 46
Regularizing filter, 84
Residual, 33
Riemannian coordinates, 25
Riemannian distance, 234
Riemannian invariants, 148
Riemannian metric, 227

S
Scattering amplitude, 206
Schrödinger equation, 205
Set of admissible coefficients, 177, 192
Singular system, 38
   non-self-adjoint operator, 41
   self-adjoint operator, 40, 74, 100
Singular value, 39
Singular value decomposition, 44
Singular value expansion, 75
Source condition, 60
Sourcewise element, 62
Spectral function, 116
Spectral representation, 38
Stability estimate, 3, 123, 140, 234
Stopping condition, 108, 187
Stopping rule, 62

T
Termination index, 111
Tikhonov functional, 68, 97, 181, 193
Tikhonov-Phillips regularization, 46
Tikhonov regularization, 46
Tikhonov regularization with a Sobolev norm, 182
Tomography problem, 213, 220
Transformed inverse problem, 141
Transport equation, 219
Truncated singular value decomposition, 44, 87

V
Volterra equation, 125, 126

W

Weak solution
    heat equation, 64, 173
    Sturm-Liouville problem, 9
    wave equation, 92

Well-posedness, 2

Well-posed problem, 16, 124